CAPILLARY
ELECTROPHORESIS

CAPILLARY ELECTROPHORESIS

THEORY & PRACTICE

Edited by

Paul D. Grossman
Applied Biosystems, Inc.
Foster City, California

Joel C. Colburn
Applied Biosystems, Inc.
Foster City, California

Academic Press, Inc.
Harcourt Brace Jovanovich, Publishers

San Diego   New York   Boston   London   Sydney   Tokyo   Toronto

Copyright © 1992 by ACADEMIC PRESS, INC.

All Rights Reserved.
No part of this publication may be reproduced or transmitted in any form or by any means, electronic or mechanical, including photocopy, recording, or any information storage and retrieval system, without permission in writing from the publisher.

Academic Press, Inc.
1250 Sixth Avenue, San Diego, California 92101-4311

United Kingdom Edition published by
Academic Press Limited
24–28 Oval Road, London NW1 7DX

Library of Congress Cataloging-in-Publication Data

Capillary electrophoresis : theory & practice / edited by Paul D.
Grossman and Joel C. Colburn.
     p. cm.
    Includes bibliographical references and index.
    ISBN 0-12-304250-X
    1. Capillary electrophoresis.   I. Grossman, Paul D.   II. Colburn,
Joel C.
  QP519.9.C36C37   1992
  574.19'285--dc20   92-6481
                                                                CIP

PRINTED IN THE UNITED STATES OF AMERICA
92 93 94 95 96 97   BC   9 8 7 6 5 4 3 2 1

To Bonnie, Carol, Danielle, Laura,
and to Henk Lauer for introducing us
to this exciting field.

Contents

5 Capillary Gel Electrophoresis

Robert S. Dubrow

6 Fundamentals of Micellar Electrokinetic Capillary Chromatography

Michael J. Sepaniak, A. Craig Powell, David F. Swaile, and Roderic O. Cole

PART **III** APPLICATIONS OF CAPILLARY ELECTROPHORESIS

9 Capillary Electrophoresis Separations of Peptides: Practical Aspects and Applications
Joel C. Colburn

10 Protein Analysis by Capillary Electrophoresis
John E. Wiktorowicz and Joel C. Colburn

11 Separation of Small Molecules by High-Performance Capillary Electrophoresis

Charles W. Demarest, Elizabeth A. Monnot-Chase, James Jiu, and Robert Weinberger

APPENDIX: Troubleshooting Guide to Capillary Electrophoresis Operations   343

Joel C. Colburn and Paul D. Grossman

Contributors

Numbers in parentheses indicate the pages on which the authors' contributions begin.

Joel C. Colburn (237, 273, 343), Applied Biosystems, Inc., Foster City, California 94404

Roderic O. Cole (159), Applied Analytical Industries, Inc., Wilmington, North Carolina 28405

Charles W. Demarest (301), Searle R&D, Skokie, Illinois 60077

Robert S. Dubrow (133), Applied Biosystems, Inc., Foster City, California 94404

Andrew G. Ewing (45), Davey Laboratory, Pennsylvania State University, University Park, Pennsylvania 16802

Paul D. Grossman (3, 111, 215, 343), Applied Biosystems, Inc., Foster City, California 94404

Stellan Hjertén (191), Department of Biochemistry, University of Uppsala, Biomedical Center, S-751 23 Uppsala, Sweden

James Jiu (301), Searle R&D, Skokie, Illinois 60077

Elizabeth A. Monnot-Chase (301), Searle R&D, Skokie, Illinois 60077

Stephen E. Moring (89), Applied Biosystems, Inc., Foster City, California 94404

Teresa M. Olefirowicz (45), Davey Laboratory, Pennsylvania State University, University Park, Pennsylvania 16802

A. Craig Powell (159), Presbyterian College, Clinton, South Carolina 29325

Michael J. Sepaniak (159), Department of Chemistry, The University of Tennessee, Knoxville, Tennessee 37996

David F. Swaile (159), The Proctor & Gamble Company, Sharon Woods Technical Center, Cincinnati, Ohio 45421

Robert Weinberger (301), CE Technologies, Chappaqua, New York 10514

John E. Wiktorowicz (273), Applied Biosystems, Inc., Foster City, California 94404

Preface

Capillary electrophoresis (CE) is recognized as a powerful new analytical separation technique that brings speed, quantitation, reproducibility, and automation to the inherently highly resolving but labor intensive methods of electrophoresis. CE has established itself as an important and widely utilized technique for routine analytical separations. It is our commitment to the future of CE that has driven us to this project.

Our purpose in creating this volume is to draw together a rapidly evolving, diverse, and multidisciplinary subject. The explosive growth of CE research has led to some 600 articles and numerous instrument company application notes, as well as many local, national, and international meetings concerned with CE. In order to make this vast amount of information more accessible, we have undertaken a project in which the key aspects of this technology could be presented in a concise and logical fashion.

This book is designed to be a practical guide, used by a wide audience, including those new to CE, those more experienced, routine users, those interested in technology development, and those involved with applications research. References have been emphasized to allow the reader to explore the detailed specifics and theoretical foundations.

With these goals in mind, we have covered the field with relatively few chapters written by experts in their respective areas. We asked these contributors for practical approaches to their discussions where possible, including opinions and observations that often do not appear in the literature. The aim of each chapter is to present the capabilities, limitations, potentials, and future challenges facing each area of CE.

This volume is composed of eleven chapters organized into three major sections—Background Concepts, Modes of CE, and Applications of CE—and an Appendix. The section on Background Concepts includes discussions of concepts common to all CE experiments, including rigorous treatments of electroendosmosis, separation efficiency, and Joule heating, which are meant to be comprehensive and meaningful to the interested practitioner (Chapter 1). This section also presents a review of detection methods (Chapter 2), a discussion of analytical figures of merit, and data handling considerations (Chapter 3). The next section treats the five major modes of CE. The respective separation principles and advantages of each mode are presented. Traditional slab gel methods such as sieving-gel electrophoresis (Chapter 5) and iso-electric focusing (Chapter 7), are discussed in terms of their CE analogs. The modes unique to the capillary format, free solution (Chapter 4), micellar electrokinetic capillary chromatography (Chapter 6), and entangled polymer matrix-based separations (Chapter 8), are included. The last section is concerned with applications of CE to peptide (Chapter 9), protein (Chapter 10), and small molecule analyses (Chapter 11); these chapters serve as reviews with extensive illustrative examples for "hands-on" value. DNA applications are handled in the gel and entangled polymer chapters (Chapters 5 and 8). A trouble-shooting guide for CE operations is given in the Appendix.

We would like to thank all of the contributors for their support and cooperation. The opportunity to develop and complete this project with a team of excellent scientists was an exciting and fulfilling experience, and has led to, we believe, a significant addition to the CE field. We would also like to thank those researchers who, although not directly contributing to this text, have helped develop the CE technology that forms the foundations of this effort.

Paul D. Grossman
Joel C. Colburn

BACKGROUND

CONCEPTS

Factors Affecting the Performance of Capillary Electrophoresis Separations: Joule Heating, Electroosmosis, and Zone Dispersion

Paul D. Grossman

I. Introduction

Joule heating, electroosmosis, and zone dispersion are important factors to consider when performing any electrophoretic separation, whether it is performed in a traditional slab, tube format, or in a microcapillary. However, because of the high electrical fields, the large surface-area-to-volume ratio, and the low viscosity of the supporting electrolyte typical of capillary electrophoresis (CE) separations, these factors become particularly important. The proper control of these effects will dictate the ultimate separation performance achieved. Because they are fundamental to the performance of all CE separations, the concepts presented here will be referred to throughout the chapters in this volume. This chapter introduces these concepts in a quantitative way in order to develop a more rigorous understanding of the relationships among the parameters that control the performance of CE separations, to lead to a more effective utilization of this powerful analytical technique.

II. Joule Heating Effects

A major limitation on the speed, resolution, and scale of electrophoretic separations is the ability to dissipate the Joule heat that is generated as a result of the electric current passing through the electrophoresis buffer. This Joule heating and the resulting temperature gradient can negatively affect the quality of the separation in a number of ways. First, if the temperature gradients are steep enough, density gradients in the electrophoresis buffer can be induced, which in turn can cause natural convection. Any such convection would serve to remix separated sample zones, severely reducing separation performance. Second, even if the temperature gradients are not large enough to cause natural convection, because the electrophoretic mobility, μ, is a strong function of temperature, separation performance can be compromised by introducing a spatial dependence on μ. This spatial dependence of mobility can cause a deformation in the migrating zones, leading to a reduced separation performance. Third, if the average temperature of the buffer becomes too high, the structural integrity of the sample analyte may be compromised.

Efficient heat dissipation is the primary reason for performing electrophoretic separations in a capillary format. The efficient removal of Joule heat allows CE to use high electrical fields, to perform separations without a rigid stabilizing medium, and to achieve very low band dispersion.

A. Natural Convection

Natural convection (or free convection) is the bulk flow of fluid due to density differences between neighboring regions of a fluid phase. These density differences can be caused by differences in either composition or temperature. Here, we are interested in density differences that result from differences in temperature.

A parameter that can be used to correlate the effects of natural convection is the dimensionless Rayleigh number, Ra (Cussler, 1984), defined as

$$\mathrm{Ra} = \frac{R_1^4 g}{\eta \alpha} \left(\frac{\Delta \rho}{\Delta r} \right) \qquad (1)$$

where R_1 is the radius of the electrophoresis chamber (in this case a tube), g is the acceleration of gravity, η is the viscosity of the medium, α is the thermal diffusivity of the medium, and $\Delta\rho/\Delta r$ is the density change per unit of radial distance caused by heating. According to theoretical predictions, if Ra $\ll 1$, then convective flow will be negligibly small, and if Ra is of the order of 1, natural convection will become significant. Thus, the Rayleigh number provides a quantitative criterion that can be used to determine whether or not natural convection will occur.

A number of approaches have been used to eliminate natural convection in electrophoretic processes. By far the most common approach has been to perform the electrophoresis in a rigid gel rather than in free solution. This makes the medium viscosity very high, thus reducing the value of Ra. Typical gel matrices include agarose, a glucose polymer, and polyacrylamide. It is this anticonvective property of gels, rather than their sieving properties, that first prompted their use in electrophoresis. Although gel electrophoresis eliminates the problem of convection, the spatial temperature gradient remains. Thus, only low electrical fields can be used (typically 1–10 V/cm). A drawback of using gels is that it makes the scale-up of such processes difficult, and it requires that a new gel matrix be prepared before each electrophoresis experiment, resulting in a lack of reproducibility.

Historically, the first method used to reduce temperature-induced natural convection was to superimpose a composition-induced density gradient. Typically, this was done using sucrose. In a vertical tube, the composition-induced density gradient serves to stabilize any convection caused by temperature gradients. Furthermore, these separations were performed at 4°C, where $d\rho/dT$ for water is a minimum. In a horizontal tube, convection can be hindered by rotation of the tube around its longitudinal axis (Kolin, 1954; Hjertén, 1967). This rotation creates a centrifugally induced radial density gradient, which serves to stabilize the convection caused by a thermal temperature gradient.

Another approach that has been attempted to reduce natural convection in electrophoretic separations is to perform the separations in outer space. Performing the separation in a microgravity environment reduces the value of g in Eq. (1). This approach is thought to be especially promising for large-scale preparative separations, because it eliminates the need for a gel. However, for obvious practical and economic reasons, this approach is not in common use.

Capillary electrophoresis represents a new and powerful alternative solution to the problem of minimizing natural convection in electrophoresis. Performing the electrophoresis in a microcapillary makes R_1 small in Eq. (1), thus reducing the value Ra. As seen from Eq. (1), this is a particularly powerful approach because of the fourth-power dependence of Ra on R_1. Capillary electrophoresis allows the use of electric fields one to two orders of magnitude higher than those allowable using traditional electrophoresis formats, without the need for a stabilizing gel, resulting in significantly shorter separation times. However, because of the very small sample volumes involved (typically on the order of *nano*liters), capillary electrophoresis is practical only as an analytical technique.

In order to quantitatively examine the effects of Joule heating, we need to know how the temperature varies in the electrophoresis medium as a function of the various operating parameters: electric field strength, buffer properties, and properties of the capillary and the environment surrounding the capillary.

B. Intracapillary Temperature Profile

To quantitatively determine the temperature gradients within the capillary caused by Joule heating, one must solve the energy-balance equation (Fahien, 1983). This equation simply states that the rate of energy accumulation in any control volume is equal to the difference between the rate at which thermal energy enters and leaves the volume. In cylindrical coordinates, this is expressed by the relation

$$\frac{1}{r}\frac{d}{dr}(rq_r) = \text{Se} \tag{2}$$

where q_r is the energy flux at the radial position r and Se is the rate of power generation within the control volume per unit volume. For the case of Joule heating, Se is given by

$$\text{Se} = \frac{I^2}{\kappa_e} \tag{3}$$

where I is the current density and κ_e is the electrical conductivity of the electrophoresis buffer. In this derivation, in order to simplify the mathematical analysis and arrive at a more practically useful expression, we shall assume that κ_e is constant throughout the conducting medium. This is not to say that the absolute value of κ_e is constant, but only that we can neglect the variation of κ_e across the capillary. It has been demonstrated that for typical situations encountered in CE, this assumption introduces a negligible error (Jones and Grushka, 1989). A schematic diagram of the capillary cross-section is shown in Fig.1.

Integrating Eq. (2), using the boundary condition that q_r is not infinite at

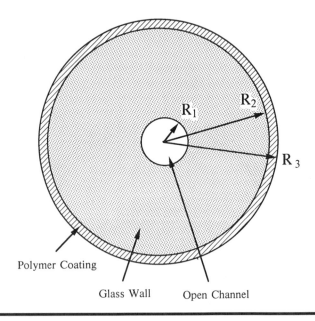

Polymer Coating

Glass Wall   Open Channel

Figure 1   Schematic illustration of capillary cross-section, where R_1, R_2, and R_3 are the radii of the open channel, the glass wall, and the polymer coating, respectively.

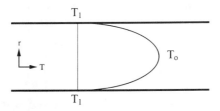

Figure 2   Parabolic temperature profile within the capillary. T_1 is the temperature at R_1, and T_0 is the temperature at the capillary center-line.

$r = 0$, results in the equation

$$q_r = \frac{Ser}{2} \tag{4}$$

which states that the heat flux increases linearly with r. Next, in order to express Eq. (4) in terms of T rather than q_r, we use Fourier's law of heat conduction, which states that

$$q_r = -k_b \frac{dT}{dr} \tag{5}$$

where k_b is the thermal conductivity of the buffer. Thus, Eq. (4) becomes

$$-k_b \frac{dT}{dr} = \frac{Ser}{2} \tag{6}$$

If k_b is assumed to be constant across the capillary, then Eq. (6) can be easily integrated using the boundary condition that at $r = R_1$, $T = T_1$, resulting in the expression

$$T(r) - T_1 = \frac{SeR_1^2}{4k_b}\left(1 - \left(\frac{r}{R_1}\right)^2\right) \tag{7}$$

Figure 2 shows an example of the parabolic temperature profile calculated from Eq. (7). We shall apply this profile later when we discuss its effect on zone dispersion. To compute the total temperature difference across the buffer, $T_0 - T_1$, we can simply evaluate Eq. (7) at $r = 0$, giving

$$T_0 - T_1 = \frac{SeR_1^2}{4k_b} \tag{8}$$

where T_0 is the temperature at the capillary center-line. Note that this expression says nothing about the overall magnitude of the temperature rise within the capillary relative to the temperature of the surroundings, T_a. It describes only the difference between the temperature at the capillary center-line and that at the inside wall.

C. Overall Temperature Rise

Next, we want to determine the overall temperature rise within the electrophoresis buffer under a given set of operating conditions. In order to calculate the total temperature rise, $T_0 - T_a$, where T_a is the ambient temperature of the surrounding medium, one must solve Eq. (2) inside the tube wall, in the polymer coating, and in the surrounding air. Following the same procedure as before, inside the glass wall

$$T_1 - T_2 = \frac{SeR_1^2}{2k_g} \ln\left(\frac{R_2}{R_1}\right) \tag{9}$$

and within the polymer coating,

$$T_2 - T_3 = \frac{SeR_1^2}{2k_p} \ln\left(\frac{R_3}{R_2}\right) \tag{10}$$

where T_2 and T_3 are the temperatures at positions R_2 and R_3 (see Fig. 1) and k_g and k_p are the thermal conductivities of the glass and the polymer coating respectively. To relate T_3 to T_a, we use the fact that the total amount of energy leaving the capillary at the capillary–air interface must equal the rate of heat generation within the capillary, Se_{tot}. Therefore, in terms of the capillary–air heat transfer coefficient, h,

$$hA\,(T_3 - T_a) = Se_{tot} \tag{11}$$

where A is the outside surface area of the capillary. Thus, substituting for A and the volume of the capillary,

$$T_3 - T_a = \frac{SeR_1^2}{2R_3h} \tag{12}$$

Therefore, to calculate $T_0 - T_a$, we simply add Eqs. (8) through (10) and (12). The resulting expression is

$$T_0 - T_a = \frac{SeR_1^2}{2}\left(\frac{1}{2k_b} + \frac{1}{k_g}\ln\left(\frac{R_2}{R_1}\right) + \frac{1}{k_p}\ln\left(\frac{R_3}{R_2}\right) + \frac{1}{R_3h}\right) \tag{13}$$

Equation (13) is the relation we have been looking for. It relates the temperature rise inside the capillary to the rate of power generation, and thus electric field strength, as well as to properties of the capillary, the buffer, and the surrounding medium.

It is important to realize that in most cases of practical interest, the magnitude of the overall temperature rise, $T_0 - T_a$, is dominated by the last term in Eq. (13). Thus,

$$T_0 - T_a \approx \frac{SeR_1^2}{2R_3 h} \tag{14}$$

This is because the resistance to heat transfer between the outside wall of the capillary and the surroundings is typically much larger than any of the other resistances.

D. Application of Temperature Relationships

Before we can apply Eqs. (7) and (13) to evaluate temperature profiles in real systems, we must first determine the values of the relevant parameters and develop a workable calculational scheme.

1. Evaluation of the Parameters in Equation 13

Several factors complicate the direct application of Eq. (13). These include the temperature dependence of the thermal and electrical conductivity of the buffer, the strong composition dependence of the electrical conductivity of the buffer, and the evaluation of the capillary–air heat-transfer coefficient.

First, we must determine the temperature dependence of the thermal conductivity of the buffer medium, k_b. We assume that the thermal conductivity is independent of the buffer ions present and is simply given by the thermal conductivity of water. Applying a second-order polynomial fit to published data can generate a function relating k_b to T. This function is given by

$$k_b(T) = 0.5605 + 1.998 \cdot 10^{-3} T - 7.765 \cdot 10^{-6} T^2 \tag{15}$$

where the units of T are °C and for k_b are $Wm^{-1} K^{-1}$. The thermal conductivities of the glass and the polymer coating are assumed to be independent of temperature over the range of interest and have values of 1.50 $Wm^{-1} K^{-1}$ and 1.55 $Wm^{-1} K^{-1}$ respectively. Equation (15) is valid over the range of 15 to 100°C at atmospheric pressure.

Next, we need to determine the electrical conductivity, κ_e, of the buffer. As stated before, κ_e will be both a strong function of both temperature and buffer composition. Therefore, not only does κ_e need to be measured for each buffer system, but $\Delta\kappa_e/\Delta T$ must also be measured. Fortunately, CE systems provide a simple way to measure both of these parameters. The definition of κ_e (Atkins, 1978) states that

$$\kappa_e = \frac{L_{tot}}{RS} \tag{16}$$

where R is the electrical resistance across the capillary, L_{tot} is the total length of the capillary, and S is the cross-sectional area of the capillary. This can be easily related to parameters that are typically measured in a CE experiment using Ohm's law, resulting

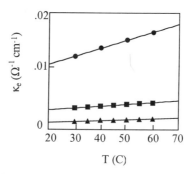

Figure 3   Measured values of the electrical conductivity, κ_e, of three different buffer systems: ($\bullet$) 100 mM sodium phosphate, pH 7.0; ($\blacksquare$) 50 mM sodium citrate, pH 2.5; ($\blacktriangle$) 20 mM [cyclohexylamino] propanesulfonic acid (CAPS), pH 11.0.

in the expression

$$\kappa_e = \frac{L_{tot}\,i}{SV}$$

(17)

where i is the measured current through the capillary and V is the voltage applied across the capillary. Thus using easily measured parameters, κ_e can be quickly determined for any buffer. Furthermore, to find the temperature dependence of κ_e, one simply has to make the above measurements at a number of different ambient temperatures. (When performing these measurements, it is important to use a low field strength in order to reduce the temperature rise due to Joule heating.) Examples of measurements of κ_e as a function of temperature for three different buffers are presented in Fig. 3. As can be seen, there is quite a bit of variation in both κ_e and $\Delta\kappa_e/\Delta T$ among these different buffer systems.

Finally, before we can apply Eq. (13) to the calculation of the overall temperature rise, we must obtain an estimate for the heat-transfer coefficient, h, between the capillary outer wall and the surroundings. The value of h depends strongly on the nature and the degree of motion of the surrounding medium and can vary over orders of magnitude. Here, because it is the most common situation, we consider the case of a capillary surrounded by agitated air. Many empirical correlations exist to estimate h for the case of gas flowing over a tube. One such correlation is given by (Chapman, 1974)

$$\mathrm{Nu} = 0.615\,\mathrm{Re}^{0.466}$$

(18)

where Nu is the Nusselt number, a dimensionless heat-transfer coefficient, and Re is the Reynolds number. The definitions of Re and Nu are

$$Nu = \frac{hD}{k_{air}} \qquad (19)$$

where D is the outside diameter of the capillary and k_{air} is the thermal conductivity of the surrounding air, and

$$Re = \frac{Dv_{air}\rho_{air}}{\eta_{air}} \qquad (20)$$

where v_{air} is the velocity of the air over the tube and ρ_{air} and η_{air} are the density and viscosity of the air, respectively. Equation (18) is valid for $40 < Re < 4000$. Assuming that the velocity of the air over the capillary is 2.0 m/sec and that the air temperature is 25 °C, Eq. (18) predicts a value for h of 263 W/m^2 °C. Note that this value for h is considerably smaller than that assumed in previous reports (Jones and Grushka, 1989), where h was assumed to be 10,000 W/m^2 °C. This is because of the poor heat-transfer characteristics of air. The value of 10,000 W/m^2 °C would be typical for a liquid heat-transfer medium.

Because this value of h is so important for the determination of $T_0 - T_a$, in the forthcoming calculations, we *fit* a value of h to our experimental findings. This is done by comparing a calculated value of the current density, I_{calc}, to the measured value, I_{meas}, where

$$I_{calc}(T) = \kappa_e(T)\, E \qquad (21)$$

and

$$I_{meas}(T) = \frac{i}{S} \qquad (22)$$

where E is the electric field strength. Thus, from I_{meas} and Eqs. (13) and (17), one can obtain an estimate for $T_0 - T_a$, which is a function of h. Next, knowing T_0, using Eq. (21), one can calculate I_{calc}. Then I_{meas} can be compared to I_{calc}. Then, h can be adjusted in Eq. (13) to force $I_{calc} = I_{meas}$. That is, $I_{calc} > I_{meas}$ indicates that h is too small, and $I_{calc} < I_{meas}$ indicates that h is too large. Figure 4 shows plots of I_{calc} and I_{meas} as a function of electrical field strength for three different buffers, using a best-fit value for h. The average value of h for the three buffer systems is 49 W/m^2 °C, where % relative standard deviation (RSD) = 6%. This value is a factor of 5 lower than that predicted from Eq. (18). This is probably because, in the experimental apparatus, a portion of the capillary length was not exposed to the flowing air stream, as well as because of uncertainties in the correlation itself. It should be mentioned that once a value of h has

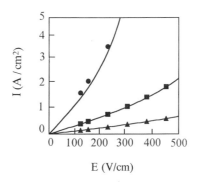

Figure 4 Comparison of measured values of the current density (points), I, with calculated values (solid curves), for the three buffer systems in Fig. 3. The solid curves were generated using Eqs. (13) and (20). The curves were fit to the experimental data by adjusting the value of h in Eq. (13). The average value of h was 49 W/m^2 °C. The parameter values used in the evaluation of Eqs. (13) and (20) are $R_1 = 25$ μm, $R_2 = 173$ μm, and $R_3 = 187$ μm. Values for κ_e as a function of temperature were computed using a least-squares fit of the data in Fig. 3. $k_b(T)$ was computed using Eq. (14).

been determined for a given CE apparatus, it can be used for any buffer system or field strength.

2. Iterative Calculational Scheme to Evaluate Equation 13

Because of the temperature dependence of κ_e and k_b, Eq. (13) must be solved iteratively. A schematic diagram of the iteration protocol is given in Fig. 5. First, a guess for T_0 must be made. A reasonable guess is that $T_0 = T_a$. Then, initial values for κ_e and k_b are evaluated. Next, having values for κ_e and k_b, Eq. (13) is used to compute a new value for T_0. Then, if $T_0(\text{new}) \neq T_0(\text{old})$, κ_e and k_b are recalculated, this time using $T_0(\text{new})$. This process is repeated until $T_0(\text{new}) = T_0(\text{old})$. In practice, the iteration converges fairly rapidly.

3. Calculated Overall Temperature Rise and Temperature Profiles

Figure 6A shows calculated values of $T_0 - T_a$ as a function of the electrical field strength, whereas Fig. 6B plots $T_0 - T_1$ as a function of electrical field strength. Both of these graphs show results using three different buffer systems and assume that $h = 49$ W/m^2 °C. Several aspects of these plots need to be considered.

Comparing Figs. 6A and 6B, it is apparent that even though $T_0 - T_a$ can be quite large, $T_0 - T_1$ is typically small. This is important because the temperature rise across the electrophoresis chamber can negatively influence the efficiency of the separation or cause the onset of convective mixing. The relationship between the temperature gradient across the capillary and the separation efficiency will be discussed further in a following section of this chapter.

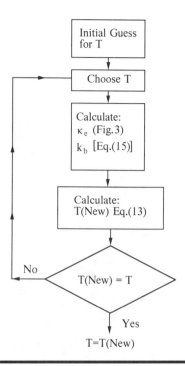

Figure 5 Diagram of iterative scheme used to evaluate Eq. (13).

It is also clear from Fig. 6A,B that there is a large difference between the behavior of different buffers. For example, at 300 V/cm, the overall temperature difference, $T_0 - T_a$, for the 20 mM [cyclohexylamino]propanesulfonic acid (CAPS) buffer is only 4.6°C, whereas for the 100 mM phosphate buffer under the same electrical field, $T_0 - T_a$ is 63°C. Thus the overall temperature rise can vary by over an order of magnitude between commonly used buffers. This highlights the fact that proper choice of buffer is an important factor controlling heating effects. These large values of $T_0 - T_a$ are largely a result of poor heat transfer between the capillary outer wall and the surrounding air. In this situation practically all the heat-transfer resistance is owing to h, thus $T_0 - T_a$ will be roughly proportional to h. These large values of $T_0 - T_a$ could potentially cause denaturation of analyte molecules and could affect the value of their electrophoretic mobilities. In practice, if one finds that the electrophoretic mobility of an analyte in free solution is a function of the electrical field strength, this may be caused by an increase in the temperature of the electrophoresis buffer due to joule heating. In free solution, the electrophoretic mobility increases approximately 2%/°C, primarily owing to the decreased viscosity of the aqueous buffer as the temperature is increased.

Finally, it is instructive to relate these calculated temperature gradients back to the Rayleigh number and the likelihood for convective mixing. As an example, let us compute the value of Ra for the case of the 50 mM citrate buffer at an electric field of

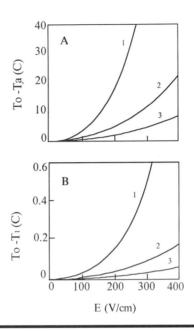

Figure 6 Calculated curves showing $T_0 - T_a$ and $T_0 - T_1$ as a function of electrical field strength for 3 different buffers. The curves were generated using Eqs. (13) and (17). The buffers used were the same as those in Fig. 3: (1) 100 mM sodium phosphate, pH 7.0; (2) 50 mM sodium citrate, pH 2.5; (3) 20 mM [cyclohexylamino] propanesulfonic acid (CAPS), pH 11.0.

400 V/cm. In this case, from Eq. (7), $T_0 - T_1 = 0.182°C$ (e.g., $\Delta T/\Delta r \cdot 7280°C/m$) and $\Delta \rho / \Delta T(\text{water}) \approx 0.221$ kg/m^3 °C. Recognizing that $\Delta \rho / \Delta r = \Delta T / \Delta r \cdot \Delta \rho / \Delta T$, Eq. (1) predicts a value for Ra of $4.3 \cdot 10^{-5}$. Thus, convective mixing is not an issue in this case.

III. Electroosmosis

Electroosmosis (EO) is the bulk flow of liquid due to the effect of the electric field on counterions adjacent to the negatively charged capillary wall. Because the wall of the fused-silica capillary is negatively charged at most pH conditions, there is a build-up of positive counterions in the solution adjacent to the capillary wall. When an electric field is applied, this layer of positive charge is drawn toward the negative electrode, resulting in the bulk flow of liquid toward that electrode.

Electroosmosis is present to some degree in all electrophoresis formats, but it is particularly important in CE. This is because of the high surface-area-to-volume ratio, the low viscosity of the electrophoresis medium, and the very high electric fields used in CE. A brief theoretical discussion follows of the mechanisms underlying EO and practical methods that have been developed to control it.

A. The Electrical Double Layer

The first step in describing EO is to quantitatively describe the charge density of counterions in solution near the capillary wall. This counter-ion-rich region is called the electrical double layer. The following treatment follows closely that of Hiemenz (1986).

The charge density of ions in solution, $\rho_e(x)$, where x is the distance from the capillary wall, is related to the ion concentration by the expression,

$$\rho_e(x) = \sum_i z_i e n_i(x) \tag{23}$$

where z_i is the valence of each ion, e is the charge on an electron, $n_i(x)$ is the number of ions of type i per unit volume near the capillary surface, and the sum is over all ions in the solution. In this analysis we will assume that n_i is a function of x only. Thus, in order to find $\rho_e(x)$, we must find an expression for $n_i(x)$.

Debye and Hükel proposed that the number density of ions in solution near a charged surface is determined by the competition between electrostatic attraction (or repulsion) of the ions for the charged interface and the randomizing influence of Brownian motion. This relationship is expressed in terms of a Boltzmann distribution, where

$$\frac{n_i(x)}{n_{i0}} = \exp\left(\frac{-z_i e \psi(x)}{kT}\right) \tag{24}$$

where n_{i0} is the number of ions per unit volume at a point of zero potential, $\psi(x)$ is the electrical potential at position x, k is Boltzmann's constant and T is the absolute temperature. Combining Eqs. (23) and (24) gives $\rho_e(x)$ as a function of the properties of the ions in solution and $\psi(x)$,

$$\rho_e(x) = \sum_i z_i e n_{i0} \exp\left(\frac{-z_i e \psi(x)}{kT}\right) \tag{25}$$

At this point an important assumption is made in order to simplify the further analysis. This assumption, known as the Debye-Hückel approximation, states that if $z_i e \psi(x) < kT$, the exponential term in Eq. (25) may be expanded into a power series truncated after the first-order terms [i.e., $e^x \approx (1 - x)$]. Inserting this approximate expression for the exponential term into Eq. (25) gives

$$\rho_e(x) = \sum_i \left(z_i e n_{i0} - \frac{z_i^2 e^2 n_{i0} \psi(x)}{kT} \right) \tag{26}$$

Because of the requirement for electroneutrality, the terms $\sum z_i e n_{i0}$ cancel from the above summation, so that Eq. (26) reduces to

$$\rho_e(x) = \sum_i \frac{z_i^2 e^2 n_{i0}}{kT} \psi(x) \tag{27}$$

To find an explicit expression for $\rho_e(x)$, we need another expression relating $\psi(x)$ to $\rho_e(x)$. This equation is Poisson's equation, which gives the electrical potential as a function of the charge density. Because we are considering only the variation of ρ_e in the x direction, Poisson's equation reduces to

$$\frac{d^2\psi}{dx^2} = \frac{\rho_e(x)}{\varepsilon} \tag{28}$$

where ε is the electrical permittivity of the solvent. (The permittivity of a material is the constant, ε, in Coulomb's inverse square law, $F = q_1 q_2 / 4\pi\varepsilon r^2$, where F is the force between charges q_1 and q_2 separated by a distance r. The value of ε is given by the expression $\varepsilon = \varepsilon_0 \varepsilon_r$ where ε_0 is the permittivity of a vacuum, $8.854 \cdot 10^{-12}$ $C^2\,N^{-1}$ m^{-2}, and ε_r is the relative permittivity, or dielectric constant, of the medium; 78.54 for water at 25°C.) Thus, substituting Eq. (27) into Eq. (28) gives

$$\frac{d^2\psi}{dx^2} = \sum_i \frac{z_i^2 e^2 n_{i0}}{\varepsilon kT} \psi(x) \tag{29}$$

resulting in an explicit expression for $\psi(x)$. At this point, the grouping of constants in the sum in Eq. (29) are combined into a single constant, κ^2, where

$$\kappa^2 = e^2 \sum_i \frac{z_i^2 n_{i0}}{\varepsilon kT} \tag{30}$$

Thus, Eq. (29) becomes simply

$$\frac{d^2\psi}{dx^2} = \kappa^2 \psi(x) \tag{31}$$

The physical significance of κ^2 will be discussed shortly. Equation (31) can now be easily solved for $\psi(x)$ by integrating twice, using the boundary conditions: (1) $\psi = \psi_0$ when $x = 0$; (2) $\psi = 0$ when $x = \infty$, where ψ_0 is the electrical potential at the capillary–solution interface. The result of these integrations is

$$\psi(x) = \psi_0 \exp(-\kappa x) \tag{32}$$

Thus, as one moves away from the capillary wall, the electrical potential falls off exponentially, with a decay constant of κ. See Fig. 7. Finally, we are at the point where

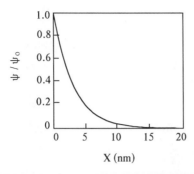

Figure 7   Plot of the electrical potential, ψ, as a fraction of the surface potential, ψ_0, as a function of the distance from the capillary wall, x. This curve was generated using Eq. (31) where a value of $3.29 \cdot 10^8$ m was used for κ (see Table I).

we can find an explicit expression for $\rho_e(x)$. Substituting Eq. (32) into Eq. (27) we obtain,

$$\rho_e(x) = \varepsilon \kappa^2 \psi_0 \exp(-\kappa x)$$

(33)

Thus given a value for κ, a property of the electrolyte solution, and a value for ψ_0, a property of the capillary wall, one can compute the charge density, $\rho_e(x)$, as a function of distance from the capillary wall.

   At this point, it is important to make some observations regarding Eq. (33). First, with respect to the assumption of low electrical potential at the solution – capillary interface, we have assumed in the derivation of Eq. (33) that $ze\psi < kT$, which implies that for a monovalent species,

$$\psi_0 < \frac{kT}{\varepsilon} = 25.7 \text{ mV}$$

(34)

assuming $T = 25°C$. Thus, to be rigorously correct, Eq. (33) is valid only for surface potentials less than approximately 25 mV. Later we shall check to see if this is a reasonable assumption for the case of typical capillary buffer systems used in CE.

   Next, we should examine the physical significance of the constant κ in Eq. (33). From Eq. (32) we can see that for the term in the exponent to be dimensionless, κ must have the dimensions of inverse length. Therefore, κ^{-1} has the dimensions of length. Writing Eq. (32) in terms of κ^{-1},

$$\psi(x) = \psi_0 \exp\left(-\frac{x}{\kappa^{-1}}\right)$$

(35)

From Eq. (35) it can be seen that when $x = \kappa^{-1}$, $\psi = \psi_0/e$. For this reason κ^{-1} is called the double-layer "thickness." See Figure 7. Because κ^{-1} is such an important parameter, it is helpful to express it in terms of ionic concentrations rather than number densities, as is done in Eq. (30). In terms of ionic concentrations in moles per liter, C_i, assuming that n_i has units of (number of ions)/m³,

$$n_i = 1000 \, N_A \, C_i \tag{36}$$

where N_A is Avogadro's number. Thus, Eq. (30) becomes

$$\kappa = \left(\frac{1000 N_A e^2}{\varepsilon k T} \sum_i z_i^2 C_i \right)^{1/2} \tag{37}$$

Note that the summation in Eq. (37) is equal to twice the ionic strength of the solution. Hiemenz (1986) provides a convenient table of values of κ based on Eq. (37) for aqueous electrolyte solutions at 25°C as a function of electrolyte concentration and valence. This table is reproduced in Table I.

Finally, it is sometimes helpful to express Eq. (33) in terms of surface charge density, σ^*, rather than surface potential, ψ_0. Based on the requirement for electroneutrality, the total charge on the capillary wall must equal the total charge in the solution, therefore

$$\sigma^* = \varepsilon \int_0^\infty \rho(x) dx \tag{38}$$

Table I Values[a] of the Electrical Double-Layer Thickness, κ^{-1}

Molarity of electrolyte	Symmetrical electrolyte		Asymmetrical electrolyte	
	$Z^+ : Z^-$	κ^{-1} (m) · 10^9	$Z^+ : Z^-$	κ^{-1} (m) · 10^9
0.001	1 : 1	9.61	1 : 2, 2 : 1	5.56
	2 : 2	4.81	3 : 1, 1 : 3	3.93
	3 : 3	3.20	2 : 3, 3 : 2	2.49
0.01	1 : 1	3.04	1 : 2, 2 : 1	1.76
	2 : 2	1.52	1 : 3, 3 : 1	1.24
	3 : 3	1.01	2 : 3, 3 : 2	0.787
0.1	1 : 1	0.961	1 : 2, 2 : 1	0.556
	2 : 2	0.481	1 : 3, 3 : 1	0.393
	3 : 3	0.320	2 : 3, 3 : 2	0.249

[a] For different electrolyte concentrations and valences for aqueous solutions at 20°C. Values are calculated using Eq. (36).

Combining Eqs. (28) and (38) results in an expression relating ψ and σ^*,

$$\sigma^* = \varepsilon \int_0^\infty \frac{d^2\psi}{dx^2}\, dx \tag{39}$$

Performing the integrations using the boundary conditions: (1) $x = \infty$ when $d\psi/dx = 0$; (2) $x = 0$ when $\psi = \psi_0$, gives

$$\sigma^* = \varepsilon \kappa \psi_0 \tag{40}$$

Thus, Eq. (40) provides a relationship between the surface charge density and the electrical potential at the surface of the capillary in a given electrolyte solution. Therefore, in terms of surface charge density, Eq. (33) becomes

$$\rho_e(x) = \kappa \sigma^* \exp(-\kappa x)$$

$$\tag{41}$$

Now that we have a quantitative relationship describing $\rho_e(x)$, we can determine the electroosmotic velocity profile.

B. Electroosmotic Velocity Profile

In this section we derive the velocity profile for electroosmotic flow. The nature of this profile is one of the key reasons for the very high separation efficiencies (or low dispersion) possible in CE separations. This problem was first solved by Rice and Whitehead (1965) for flow in cylindrical capillaries. This presentation will follow closely that of Rice. However, to improve the clarity of the treatment, we shall perform the derivation in rectangular rather than cylindrical coordinates. As we shall see, this will not affect the final conclusions of the analysis.

The equation which relates the forces on a fluid to its velocity profile is the equation of motion (Fahien, 1983). The equation of motion for a Newtonian electrolyte solution, flowing in one dimension across a flat surface, under the influence of an electric field in the direction of flow, is

$$\eta \frac{d^2 v_z(x)}{dx^2} E \rho_e(x) \tag{42}$$

where η is the viscosity of the solution, v_z is the fluid velocity in the z direction and E is the electrical field strength. To solve Eq. (42) for $v_z(x)$, we simply integrate Eq. (42) twice using the boundary conditions: (1) $dv/dx = 0$ at $x = \infty$; (2) $v_z(x) = 0$ at $x = 0$. The result is

$$v_z(x) = \frac{-E\varepsilon\psi_0}{\eta}(1 - \exp(-\kappa x)) \tag{43}$$

The key feature of Eq. (43) is the fact that for typical values of κ and x used in CE, the exponential term quickly vanishes. At this point we can calculate how far away from the capillary wall the fluid velocity reaches its constant value of v_∞. Substituting Eq. (40) into Eq. (43) gives

$$v_z(x) = \frac{-E\sigma^*}{\kappa\eta}(1 - \exp(-\kappa x)) \qquad (44)$$

If we assume that the wall of a fused-silica capillary has a charge density of 0.01 C/m^2 (Churaev et al., 1981) and that $\kappa = 3.29 \cdot 10^8$ m^{-1} (see Table I), we can calculate $v_z(x)$. In Fig. 8, Eq. (44) is plotted for the case where $E = 30{,}000$ V/m and $\eta = 0.001$ N sec/m^2. As can be seen in Fig. 8, the velocity does indeed reach a constant value very close to the capillary wall. In fact, within 14 nm of the capillary wall, $v_z(x) = 0.99v_\infty$. Therefore, in a 50-μm inside diameter (i.d.) capillary, $v_z(x) = v_\infty$ across 99.95% of the capillary cross-section. Thus the assumption of a flat, radially independent velocity profile appears to be justified.

Therefore, given that the exponential term in Eq. (43) can be neglected, Eq. (43) reduces to

$$v_z(x) \approx v_\infty = \frac{-E\varepsilon\psi_0}{\eta} \qquad (45)$$

where v_∞ is the velocity of the fluid far away from the influence of the capillary wall. In terms of the electroosmotic mobility, i.e., the electroosmotic velocity per unit elec-

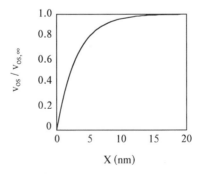

Figure 8   Calculated values of the electroosmotic velocity, v_{os}, as a function of the distance from the capillary wall, x, where $v_{os,\infty}$ is the fluid velocity in the bulk solution. This curve was evaluated using Eq. (45) where $E = 30{,}000$ V/m, $\kappa = 3.29 \cdot 10^8$ m, and $\sigma^* = 0.01$ C/m^2. From Churaev et al. (1981).

trical field strength, Eq. (45) becomes

$$\mu_{os} = \frac{-\varepsilon \psi_0}{\eta}$$

(46)

The striking feature of Eq. (45) is that it says that the velocity of the fluid is *not a function of radial position*. This is in contrast to the case of pressure-driven flow where the well-known parabolic velocity profile is present. This "flat" velocity profile makes possible the very low dispersion found in CE separations. Clearly, because v_z is independent of the distance from the solid surface, it makes no difference that we performed the analysis in rectangular coordinates.

At this point, we are in a position to perform some calculations to help gain a quantitative feel for these effects. For example, how do predicted values of the electroosmotic mobility, μ_{os}, compare with typical measured values? Substituting Eq. (40) into Eq. (46) gives

$$\mu_{os} = \frac{\sigma^*}{\kappa \eta}$$

(47)

Using the same values for σ, κ, and η as in the previous example gives us a value for μ_{os} of $6.1 \cdot 10^{-8}$ m²/V · sec. Given that the value for σ was measured at a pH of 7.1, it can be seen from Fig. 9 that the calculated value of μ_{os} agrees with experimental observations.

Next, we can see if the potential at the surface of the capillary, ψ_0, lies within the limits of the Debye-Hückel approximation. Given experimental values for μ_{os} and Eq. (46), we can obtain a direct measure of ψ_0,

$$\psi_0 = \frac{v_\infty \eta}{E \varepsilon} = \frac{\mu_{os} \eta}{\varepsilon}$$

(48)

Given that the electrical permittivity of water at 25°C is $6.95 \cdot 10^{-10}$ C²/J · m, according to Eq. (48), $\psi_0 = 86$ mV. Although this is larger than the 25-mV limit stated for the Debye-Hückel approximation, the error introduced by this slightly higher value of ψ_0 does not appreciably affect the results of this analysis.

Finally, we can find out how the magnitude of the volumetric flow rate caused by electroosmosis compares to that caused by a pressure gradient. The volumetric flow rate due to electroosmosis, Q_{os}, is related to the electroosmotic mobility by

$$Q_{os} = v_\infty S = \mu_{os} E S$$

(49)

where S is the cross-sectional area of the capillary. Using the same values as before for μ_{os} and E, Eq. (49) gives a value for Q_{os} of $3.53 \cdot 10^{-12}$ m³/sec or 3.53 nl/sec. To calculate the volumetric flow rate due to a pressure gradient, we can use the Poiseuille

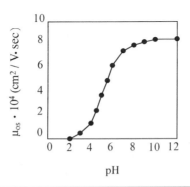

Figure 9 Dependence of electroosmotic mobility, μ_{os}, on buffer pH for the case of a fused-silica capillary that has been prewashed with 1.0 M NaOH. Reprinted with permission from Applied Biosystems (1990).

equation (Fahien, 1983),

$$Q = \frac{\pi \, \Delta P \, R_1^4}{8 \, L_{tot} \, \eta} \tag{50}$$

where ΔP is the pressure drop across the capillary and L_{tot} is its length. Thus, to generate a flow of $3.53 \cdot 10^{-12}$ m^3/sec would require a $\Delta P/L$ of $2.3 \cdot 10^4$ N/m^2/m or 3.4 psi/m. Thus, for this example, with regard to volumetric flow rate, an electrical potential of 30,000 V/m is comparable to a pressure gradient of 3.4 psi/m.

C. Control of Electroosmosis

For many applications it is desirable to be able to manipulate the magnitude of the electroosmotic flow in order to optimize separation performance. Many studies have been conducted describing various methods that can be used to control electroosmotic flow.

In order to control electroosmosis, it is clear from Eqs. (46) or (47) that one must control either the charge density on the capillary wall, the double-layer thickness, or the viscosity of the solution adjacent to the capillary wall. This can be clearly seen if we express Eq. (47) in terms of the double-layer thickness, κ^{-1},

$$\mu_{os} = \frac{\sigma^* \, \kappa^{-1} \, (\varepsilon, C_i)}{\eta} \tag{51}$$

where the dependence of κ^{-1} on ε and C_i has been indicated. Therefore, each of the techniques that follow acts by affecting one or the other of these parameters.

Two approaches have been used to control μ_{os} by reducing the double-layer thickness, κ^{-1}. The first simply uses an increased concentration of electrolytes in the electrophoresis buffer. As can be seen in Eq. (37), this will serve to reduce κ^{-1}. Detailed studies of the effect of varying NaCl concentration on μ_{os} have been reported (Fujiwara and Honda, 1986). A drawback to this approach is that the increased ion concentration will increase the amount of Joule heating within the capillary, thus potentially affecting separation performance. A second approach is to decrease κ^{-1} by decreasing the permittivity of the buffer by the addition of simple organic solvents. Again from Eq. (37), it can be seen that as ε is decreased, κ^{-1} will become smaller. Organic solvents that have been investigated include acetonitrile and methanol (Fujiwara and Honda, 1987). A potential drawback to using organic solvents in the electrophoresis buffer is the large ultraviolet (UV) absorbance background associated with these solvents, which could negatively affect detection sensitivity while using a UV-absorbance detector.

Three main approaches have been examined to control μ_{os} by influencing the surface charge density on the capillary wall, σ^*. The first uses physically adsorbed small cationic molecules to neutralize the charge on the capillary wall. Molecules that have been used for this purpose include cetyltrimethylammonium bromide (Altria and Simpson, 1986), tetradecyltrimethylammonium bromide (Huang *et al.,* 1989), putrescine (Lauer and McManigill, 1986a), and *s*-benzylthiouronium chloride (Altria and Simpson, 1987). Using multivalent ions, one can even reverse the *direction* of the electroosmotic flow using this method (Wiktorowicz and Colburn, 1990). A drawback to this approach is that the cations can potentially bind to the analyte molecule, changing its net charge and thus its electrophoretic mobility. A second approach to influence σ^* is to covalently block the charged silanol groups on the capillary surface. Chemical derivitizing agents that have been used for this purpose include trimethylchlorosilane (Jorgenson and Lukacs, 1983) and (γ-methacryloxypropyl)-trimethoxy silane (Hjertén, 1985). A study of the effect of a number of silanating reagents is given by McCormick (1988). A drawback to this approach is that over time, the covalent bond can hydrolyze, thus allowing the blocking agent to leach off the surface of the capillary, contaminating the buffer and causing μ_{os} to change over time. A third approach used to manipulate σ^* is simply to titrate the charge on the capillary surface. The point of zero charge for fused silica has been found to be approximately at a pH of 2.0 (Churaev *et al.,* 1981). As can be seen from the titration curve in Fig. 9, this does indeed appear to be the case. An obvious drawback of this approach is that if, at the desired pH, the analyte molecule is oppositely charged, a significant amount of interaction between the analyte and the wall can result. However, this method has been effectively exploited for peptide separations at a pH of 2.5 without any apparent wall sticking (Grossman *et al.,* 1989).

The last parameter that can be used to alter μ_{os} according to Eq. (51) is η, the solution viscosity. By adding to the buffer a polymer which adsorbs to the capillary wall, one is able to greatly increase the effective viscosity of the buffer near the capillary–buffer interface. The use of a number of different noncovalently bound polymers for this purpose has been investigated by Herrin *et al.* (1987).

As can be seen from this discussion, there are a large number of alternative approaches to solving the problem of controlling electroosmotic flow. The proper choice of protocol will depend on the specifics of the particular application under consideration.

IV. Zone Dispersion and Resolution

As in chromatography, the quantitative measure of separation performance in CE is the resolution, Res. The resolution is defined as the distance between the centers of two zones divided by the average width of each zone, or

$$\text{Res} \equiv \frac{x_2 - x_1}{\frac{1}{2}(w_1 + w_2)}$$

(52)

where x_2 and x_1 are the positions of the zone centers for species 1 and 2, respectively, and w_1 and w_2 are their band widths measured at the base of the peak (see Fig. 10) (Littlewood, 1970). When Res = 1.5, the separation of two bands is essentially complete, and the peaks are said to be resolved to "base line."

The average position of a band, x_i, is determined by its net rate of migration, whereas the peak width, w_i, is determined by diffusion and other dispersive phenomena. The fact that CE is such a low-dispersion technique provides the basis for much

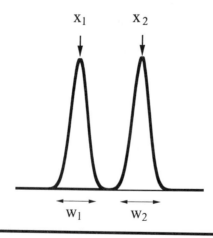

Figure 10 Schematic illustration showing the parameters used to compute the resolution using Eq. (51).

of its exceptional resolving capability. In this section we discuss the factors that control the peak width in CE separations and show how these factors affect the ultimate separation performance achievable by CE.

A. Diffusion, Dispersion, and Theoretical Plates

1. Concentration Distribution

In order to fully describe the peaks resulting from a CE separation, we need to develop an expression that describes the concentration of a solute as a function of its position in the capillary and time. Such an equation can be derived starting from the convective diffusion equation (Fahien, 1983)

$$\frac{\partial C_i(x,t)}{\partial t} = \frac{D_i \, \partial^2 C_i(x,t)}{\partial x^2} - v_{z,i} \frac{\partial C_i(x,t)}{\partial x} \tag{53}$$

where C_i is the concentration, D_i is the diffusion coefficient, and $v_{z,i}$ is the net migration velocity of species i. The convective diffusion equation is simply a mass balance, which states that C_i changes as a result of molecules entering and leaving a control volume by either diffusion or convection. Note that in the case of zone electrophoresis, $v_{z,i} = \mu_{app}E$, where μ_{app} is the apparent electrophoretic mobility of species i and E is the electric field strength. The apparent mobility, μ_{app}, is defined as the vector sum of the solute electrophoretic mobility, μ, and the electroosmotic mobility, μ_{os}. Equation (53) assumes that the only dispersive process is molecular diffusion, and that the diffusion coefficient is not a function of solute concentration.

Equation (53) can be solved using the method of Fourier transforms using the boundary conditions: (1) $t > 0$, $x = \infty$, $C = 0$; (2) $t > 0$, $x = 0$, $dC/dx = 0$. The first boundary condition states that far from the initial pulse, the solute concentration is zero, whereas the second boundary condition states that because diffusion occurs at the same speed in both directions, the pulse remains symmetrical throughout the separation. The initial condition is that at $t = 0$, $C = (M/S)\delta(x)$, where S is the cross-sectional area of the capillary, M is the initial mass of solute and $\delta(x)$ is the Dirac delta function. This states that the pulse is initially present as an infinitely thin zone. The solution to Eq. (53) is

$$C_i(x,t) = \frac{M}{S\sqrt{4\pi D_i t}} \exp\left\{ -\left(\frac{(x - \mu_{app}Et)^2}{4D_i t} \right) \right\} \tag{54}$$

Note that this is the expression that we have been looking for, which relates the concentration of solute i to its position in the capillary and time.

2. Resolution and Number of Theoretical Plates

It can be seen that Eq. (54) is in the form of a Gaussian, or normal distribution,

$$P(x) = \frac{1}{\sigma\sqrt{2\pi}} \exp\left(\frac{-(x - x_{avg})^2}{2\sigma^2}\right) \tag{55}$$

where $P(x)$ is a probability, x_{avg} is the average value of the variable x, and σ^2 is the variance of the distribution (Maisel, 1971). Comparing Eqs. (54) and (55), it can be seen that the average position of the migrating zone is given by

$$x_{avg} = \mu_{app} E t \tag{56}$$

and that the variance, σ^2, is given by

$$\sigma_D^2 = 2D_i t \tag{57}$$

where the subscript D indicates that this is the variance resulting from diffusion only. These are important conclusions. They show that the average position of a peak as well as its width can be expressed in terms of simple expressions. As pointed out by Jorgenson and Lukacs (1981), σ_D can be expressed in terms of the electrical field strength by recognizing that, in Eq. (57),

$$t = \frac{L}{v_z} = \frac{L^2}{\mu_{app} V} = \frac{L}{\mu_{app} E} \tag{58}$$

where μ_{app} is the apparent electrophoretic mobility, i.e., $\mu_{app} = \mu + \mu_{os}$, and L is the total distance migrated by the zone. Making this substitution, Eq. (57) becomes

$$\sigma_D^2 = \frac{2 D_i L}{\mu_{app} E} \tag{59}$$

By substituting Eqs. (56) and (57) into Eq. (52), we can derive an expression for the resolution, Res. If we assume that $w_i = 4\sigma_i$,

$$\text{Res} = \frac{\Delta\mu_{app} E\sqrt{t}}{2\sqrt{2}\left(\sqrt{D_i} + \sqrt{D_j}\right)} \tag{60}$$

where $\Delta\mu_{app}$ is the difference between the apparent electrophoretic mobilities of species i and j. It should be remembered that Eq. (60) assumes that the only dispersive influence is that of molecular diffusion.

However, Eq. (60) can be further generalized if we combine it with the definition for the number of theoretical plates, N, where N is defined as

$$N = \frac{L^2}{\sigma_T^2} \tag{61}$$

where σ_T^2 is the total variance of the zone, including contributions from all dispersive phenomena. For the diffusion-only case, by combining Eqs. (59) and (61), we see that

$$N = \frac{\mu_{app,avg}^2 E^2 t}{2D_{i,avg}} \tag{62}$$

where $D_{i,avg}$ is an average diffusion coefficient and $\mu_{app,avg}$ is the average apparent electrophoretic mobility. If we express Eq. (60) in terms of an average diffusion coefficient, Eq. (60) becomes

$$Res = \frac{\Delta\mu_{app} E \sqrt{t}}{4\sqrt{2D_{i,avg}}} \tag{63}$$

By comparing Eq. (62) and (63) it can be seen that

$$Res = \frac{1}{4} \frac{\Delta\mu_{app}}{\mu_{app,avg}} N^{1/2} \tag{64}$$

Equation (64) shows that the overall resolution is made up of a selectivity term given by $\Delta\mu_{app}/\mu_{app,avg}$, and an efficiency term, N. The selectivity term is determined by the properties of the analytes and buffer, whereas the efficiency term is dictated primarily by the properties of the instrumentation. Furthermore, Eq. (64) is completely general because the definition of N can include contributions from all dispersive phenomena. In the following section, we derive relationships that will allow us to estimate the value of σ_T^2.

Another measure of separation efficiency, in addition to σ_T^2 and N, is the height of a theoretical plate, H. H is defined as

$$H = \frac{L}{N} = \frac{\sigma_T^2}{L} \tag{65}$$

H is a more convenient measure of efficiency than N because, like σ^2, individual contributions to H can be easily combined to give an overall value.

It is important to recognize that it is misleading to discuss electrophoretic separations in terms of plates. The concept of a theoretical plate implies that a separation is accomplished as a result of an *equilibrium* partitioning between two immiscible phases. This is certainly not true in the case of electrophoresis, where the separation is accomplished through differences in *rates* of transport. Thus, in the context of electrophoretic separations, theoretical plates is simply a convenient way to describe the shape of a Gaussian curve.

3. Measurement of N

A number of methods are available for the determination of N from chromatographic or electrophoretic data (Littlewood, 1970). All are based on measurements of the standard deviation and average position of a Gaussian peak followed by the application of Eq. (61). The differences in the various methods lie in the way σ is determined from the experimental data. A particularly useful method for use with an electronic integrator is (James and Martin, 1952)

$$N = 2\pi\left(\frac{h_p L}{A_p}\right) \tag{66}$$

where A_p is the area of the peak, h_p is the height of the peak, and L is the distance traveled by the peak. To make the term $(h_p L)/A_p$ dimensionless, both L and A_p must have units of distance or time. This expression is particularly handy, because most electronic integrators report the quantity h_p/A_p directly.

B. Additional Factors Contributing to Zone Dispersion

Up to this point we have considered molecular diffusion as the only mechanism of dispersion. If this were true, according to Eq. (62), for a macromolecule run under typical free-solution CE conditions, one would expect N to be on the order of 10^6. In fact, such high plate-counts are rarely achieved in practice. This is because a number of additional factors contribute to band dispersion in CE. An experimental study showing the influence of various factors on the total variance has been published by Jones *et al.* (1990). Giddings (1965) has shown that the individual contributions to the total peak variance can be summed to give a total variance, σ_T^2. Therefore

$$\sigma_T^2 = \sum_n \sigma_n^2 \tag{67}$$

where the sum is taken over all n contributions to the variance. This conclusion is based on the assumption that all dispersive processes can be considered independent, random processes. Because the variances of individual contributions can be combined in this way, σ^2 is a convenient parameter to use to quantitatively describe peak broadening.

In this section we discuss these nondiffusional contributions to zone broadening and develop quantitative models to describe them. Furthermore, we will show that these relationships can be used to help determine the optimal operating conditions for a given separation.

1. Variance Caused by Parabolic Temperature Profile ($\sigma_{\Delta T}^2$)

a. Evaluation of $\sigma_{\Delta T}^2$ As we have seen in Section I, the Joule heating caused by the passage of current through the electrophoresis buffer results in a parabolic temperature profile within the capillary. Because the electrophoretic mobility is a strong function of temperature, the parabolic temperature profile results in a parabolic velocity profile across the capillary. In this section, we develop the relationship that describes the manner in which this temperature-induced velocity profile affects band dispersion in CE. This problem has been studied by many authors (Grushka *et al.*, 1989; Hjertén, 1967; Coxon and Binder, 1974; Virtanen, 1974); here we follow the treatment of Virtanen because it provides clear physical insight into the relevant process, while avoiding unnecessary complexity.

Virtanen developed an expression for the contribution of the parabolic temperature profile to σ_T^2, $\sigma_{\Delta T}^2$, by using an analogy with Taylor dispersion (Taylor, 1953). Taylor dispersion is that caused by the parabolic velocity profile present in laminar flow in cylindrical tubes. Qualitatively, Taylor dispersion can be explained as follows. Because of the parabolic velocity profile, solutes in the center of the capillary flow faster than those near the wall. Therefore, if a solute were initially evenly distributed across the capillary cross-section, over time, the solutes in the middle of the capillary would have moved farther down the length of the tube. In addition, counteracting this distorting influence of the parabolic velocity profile, radial diffusion causes solutes present in the front of the profile to diffuse radially into regions of lower concentration at the edges. Thus, if the rate of radial diffusion is fast relative to the longitudinal convection, dispersion because of the parabolic velocity profile is nullified (see Fig. 11.)

Taylor determined that the dispersion caused by a parabolic velocity profile is given by

$$\sigma_{\text{Taylor}}^2 = \frac{R_1^2 v_{z,\text{avg}}^2 t}{24 D_i} \tag{68}$$

where $v_{z,\text{avg}}$ is the average linear velocity of the solute across the capillary cross-section. This important result will be applied to describe a number of dispersive phenomena in following sections. For laminar flow in a tube, the parabolic velocity profile is given by the expression (Fahien, 1983)

$$v_z(r) = \frac{\Delta P R_1^2}{4 L_{\text{tot}} \eta} \left(1 - \left(\frac{r}{R_1} \right)^2 \right) \tag{69}$$

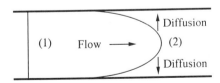

Figure 11   Diagram showing mechanism of Taylor dispersion. An initially uniform concentration profile (1) is distorted by viscous flow into a paraboloid (2). In addition, solutes located at the tip of the parabola diffuse into less-concentrated regions near the tube wall, serving to wash out the parabolic profile.

where ΔP is the pressure drop across the capillary, and η is the fluid viscosity. The average velocity across the capillary is then given by

$$v_{z,\text{avg}} = \frac{1}{S}\int_S v_z \, dS = \frac{1}{\pi R_1^2}\int_0^{R_1} v_z(r) 2\pi r \, dr = \frac{\Delta P R_1^2}{8L\eta} \tag{70}$$

where S is the cross-sectional area of the capillary. Equation (70) is the Poiseuille equation. This pressure-induced parabolic velocity profile can be compared to the temperature profile given by Eq. (7) in Section I. However, before we can make a direct analogy between the two profiles, we must provide a relationship between the temperature and the electrophoretic velocity. This is done by defining a temperature coefficient of the electrophoretic mobility, Ω_T, as

$$\Omega_T = \frac{\mu(T) - \mu(T_1)}{\mu(T_1)(T - T_1)} \tag{71}$$

where $\mu(T)$ and $\mu(T_1)$ are the electrophoretic mobility at temperatures T and T_1 respectively. Because the temperature dependence of μ is primarily the result of the influence of temperature on the viscosity of the solvent, in aqueous solution, Ω_T is approximately 0.02 K^{-1}. Rearranging Eq. (71) and substituting Eq. (7) for $T - T_1$ results in the expression

$$\mu(T) - \mu(T_1) = \frac{\Omega_T \mu(T_1) S e R_1^2}{4k_b}\left(1 - \left(\frac{r}{R_1}\right)^2\right) \tag{72}$$

Next, if we multiply Eq. (72) by E in order to express Eq. (72) in terms of electrophoretic velocities rather than mobilities and define a velocity increment, Δv, as $v_z(T) - v_z(T_1)$, Eq. (72) becomes

$$\Delta v = \frac{E \, \Omega_T \mu_{\text{app}} S e R_1^2}{4k_b}\left(1 - \left(\frac{r}{R_1}\right)^2\right) \tag{73}$$

Also, because the absolute value of μ does not vary much across the capillary, $\mu(T_1)$ has been replaced by the average apparent mobility, μ_{app}. At this point the analogy between the parabolic velocity profiles caused by laminar flow and by the parabolic temperature profile is complete. It is clear that Eqs. (69) and (73) are analogous. Finally, in order to apply Eq. (73) to the temperature-induced parabolic velocity profile, we must calculate the average value for Δv. Again, by analogy with Eq. (69), it is clear that

$$\Delta v_{\text{avg}} = \frac{\Omega_T E \mu S e R_1^2}{8k_b} \tag{74}$$

Substituting Δv_{avg} into Eq. (68) gives

$$\sigma_{\Delta T}^2 = \frac{R_1^6 E^6 \kappa_e^2 \Omega_T^2 \mu_{\text{app}}^2}{1536 D_i k_b^2} t$$

(75)

where we have used the fact that $Se = E^2 \kappa_e$. The extremely strong dependence of $\sigma_{\Delta T}^2$ on R_1 and E should be noted. This again highlights the motivation for performing electrophoretic separations in narrow-bore capillaries.

b. Optimal Field Strength Equation (59) shows that as E increases, σ_D^2 decreases. This is contrasted with the field-strength dependence of $\sigma_{\Delta T}^2$ as given by Eq. (75), where as E is increased, the variance increases. Thus, as pointed out by Lauer and McManigill (1986b), there must exist an optimal value of E that minimizes the combined variance, $\sigma_D^2 + \sigma_{\Delta T}^2$. An example of the field dependence of H_D, $H_{\Delta T}$, and $H_D + H_{\Delta T}$, is shown in Fig. 12. Foret *et al.* (1988) showed that an analytical expression for the optimal field strength can be derived by simply taking the derivative of the equation for $\sigma_D^2 + \sigma_{\Delta T}^2$ with respect to E, setting the resulting expression equal to zero and solving for E_{opt}. The resulting expression for E_{opt} is

$$E_{\text{opt}} = \frac{2.92}{R_1} \left(\frac{D_i k_b}{\mu_{\text{app}} \kappa_e \Omega_T} \right)^{1/3}$$

(76)

It should be mentioned that in some cases, when h is small, when $E = E_{\text{opt}}$, the overall temperature rise, $T_0 - T_a$, may be prohibitively large.

2. Variance Caused by Finite Injection Volume (σ_I^2)

In deriving Eq. (54), we assumed that the sample was initially present as an infinitely thin zone. Obviously, this is an unrealistic assumption. In fact, the injection zone has a finite width whose magnitude can greatly influence the overall band dispersion.

Many authors have used the expression of Sternberg (1966) to describe the variance caused by a finite injection volume,

$$\sigma_I^2 = \frac{l^2}{12}$$

(77)

where l is the length of the sample slug. However, this expression assumes that the sample solution enters the capillary as a rectangular plug. Whereas this is a reasonable assumption if the sample is applied to the capillary using an injection loop, it is not

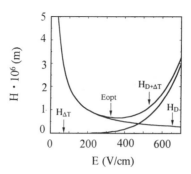

Figure 12   Plate-height contribution due to diffusion, H_D, parabolic temperature profile, $H_{\Delta T}$, and the sum of the two effects, $H_{D+\Delta T}$ as a function of electrical field strength. The field strength at which $H_{D+\Delta T}$ is a minimum, E_{opt}, is also indicated. These curves were generated using Eqs. (58) and (75) using the parameter values: $D = 1 \cdot 10^{-10}$ m^2/sec, $\eta = 0.001$ N $\cdot$ sec/m^2, $\mu_{app} = 1 \cdot 10^{-8}$ m^2/V $\cdot$ sec, $R_1 = 50 \,\mu$m, $\kappa_e = 1.25 \,\Omega^{-1}$ m^{-1}, and $k_b = 0.6$ Wm^{-1} K^{-1}.

valid if the sample is injected directly into the capillary. In this section we derive a quantitative expression that can be used to estimate the contribution that the injection slug, σ_I^2, makes to the overall variance.

a. Hydrodynamic Injection   In hydrodynamic injection, sample is introduced into the capillary by applying a pressure difference across the capillary while one end of the capillary is immersed in the sample solution. In this case, an expression for σ_I^2 can be directly evaluated by again applying the expression for Taylor dispersion. Thus,

$$\sigma_{I,HD}^2 = \frac{R_1^2 v_{inj}^2}{24 D_i} t_{inj} \tag{78}$$

where t_{inj} is the injection time, v_{inj} is the average injection velocity, and the subscript HD indicates hydrodynamic injection. Using the Poiseuille equation, Eq. (70), one can express Eq. (78) in terms of the applied pressure drop rather than the average velocity. In this case, Eq. (78) becomes

$$\sigma_{I,HD}^2 = \frac{R_1^6 \Delta P_{inj}^2}{1536 L_{tot}^2 \, \eta^2 D_i} t_{inj} \tag{79}$$

where L_{tot} is the total length of the capillary. Note that because t_{inj} is inversely proportional to ΔP_{inj}, Eq. (79) implies that to minimize σ_I^2, one should perform injections using the lowest possible pressure difference. However, at some point, if ΔP_{inj} becomes too small, σ_I^2 will again begin to increase because of the effects of molecular diffusion. This implies that there is an optimal injection pressure, or injection velocity, which will minimize σ_I^2. If we include a longitudinal diffusion contribution to σ_I^2, Eq. (78) becomes

$$\sigma_{I,HD}^2 = \frac{R_1^2 v_{inj}^2 t_{inj}}{24 D_i} + 2 D_i t_{inj} \tag{80}$$

In order to find the optimal value for v_{inj}, we must determine the minimal value of σ_I^2 that satisfies the constraint of constant injection volume, i.e., that $v_{inj} t_{inj} = l$, where l is the length of the injection slug. This optimal value of the injection velocity, $v_{inj,opt}$, can be obtained by substituting $t_{inj} = l/v_{inj}$ into Eq. (80), taking the derivative with respect to v_{inj} and setting the resulting expression equal to zero. Then, solving for $v_{inj,opt}$,

$$v_{inj,\,opt} = 6.93 \frac{D_i}{R_1} \tag{81}$$

Or, in terms of ΔP,

$$\Delta P_{inj,\,opt} = 55.4 \frac{D_i L_{tot} \eta}{R_1^3} \tag{82}$$

A plot of H vs. t_{inj} for a constant injection volume of 1% of the total column volume is shown in Fig. 13. It can be seen that for macromolecules in small capillaries, $v_{inj,opt}$ will be prohibitively small, requiring very low injection pressures, which would result

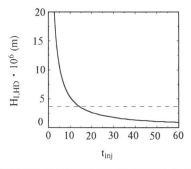

Figure 13   Plate-height contribution due to hydrodynamic injection as a function of injection time, assuming a total sample volume of 1% of the total effective column volume. The dashed line indicates the value of $H_{I,HD}$ calculated from Eq. (78), which assumes that the sample solution enters the capillary as a rectangular plug. Calculations were performed using Eq. (81) where the relevant parameter values are the same as those in Fig. 12.

in long injection times and would be hard to adequately control. It is also clear from Fig. 13 that by operating away from the optimal injection velocity, a large amount of band dispersion is introduced. However, it can also be seen from Fig. 13 that, once t_{inj} is larger than some critical value, there is little to be gained by further increasing t_{inj}. A good rule of thumb is to choose a t_{inj} where $\sigma^2_{I,HD} \approx l^2/12$.

b. Electrokinetic Injection   In electrokinetic injection, sample is drawn into the capillary by a combination of electrophoresis and electroosmosis. Thus, the amount of material that is loaded is proportional to the apparent electrophoretic mobility of each solute, μ_{app}. To determine the variance caused by electrokinetic injection, we turn to Eq. (75), where, if we substitute t_{inj} for t,

$$\sigma^2_{I,EK} = \frac{R_1^6 E^6 \kappa_e^2 \Omega_T^2 \mu_{app}^2}{1536 D_i k_b^2} t_{inj} \tag{83}$$

As was the case for hydrodynamic injection, there exists an optimal value of the electrical field strength that will minimize $\sigma^2_{I,EK}$, where E_{opt} is given by Eq. (76).

It is instructive to compare the dispersion caused by hydrodynamic injection and electrokinetic injection. As we showed before, using the sample parameters in Fig. 12, $E_{opt} = 36,300$ V/m. Using this field, from Eq. (65) and (83), $H_{I,EK} = 2.7 \cdot 10^{-6}$ m. At this field strength, $t_{inj} = 13.8$ sec. We can compare this value of $H_{I,EK}$ to $H_{I,HD}$ using the same injection time. In this case, $H_{I,HD} = 3.78 \cdot 10^{-6}$ m and $\Delta P_{inj} = 0.13$ psi. However, if ΔP_{inj} is increased to 3 psi, keeping the injection volume constant, $t_{inj} = 0.6$ sec and $H_{I,HD} = 8.68 \cdot 10^{-5}$ m, a factor of 20 worse.

Electrokinetic injection has one major drawback. Because $\sigma^2_{I,EK}$ is a function of a large number of parameters that can be hard to control, it is difficult to obtain good day-to-day injection reproducibility. The parameter that is particularly difficult to control between different capillaries is the electroosmotic mobility, which affects μ_{app} in Eq. (83).

3. Variance Caused by Electroosmotic Flow and Solute–Wall Interactions (σ^2_W)

The interaction of analytes with the capillary wall can have a negative influence on the overall efficiency of a CE separation. Even very small degrees of interaction can dramatically increase σ^2_T. Martin et al. (1985) has developed an expression, based on the Aris generalized dispersion theory (Aris, 1959), relating σ^2_W to the degree of solute–wall interaction. In this treatment,

$$\sigma^2_W = \frac{C_m R_1^2 v_z^2}{D_i} t \tag{84}$$

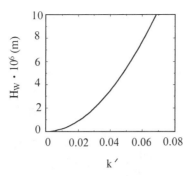

Figure 14   Plate-height contribution due to solute–wall interactions as a function of the solute-wall capacity factor, k'. Calculations were performed using Eq. (85) where $E = E_{opt} = 36,300$ V/m, and the other relevant parameters are the same as those in Figs. 8 and 12.

where

$$C_m = \frac{4 + (4n + 16)k' + (n^2 + 10n + 20)k'^2}{4(n + 2)(n + 4)(1 + k')^2} \tag{85}$$

where k' is the solute capacity factor and $n = R_1/\kappa^{-1} + 3/2$ for $R_1/\kappa^{-1} \gg 1$, where κ^{-1} is the thickness of the electrical double layer adjacent to the capillary wall [see Eq. (37)].

For the case in which $k' = 0$, i.e., no solute–wall interaction, Eq. (84) provides the contribution to σ_T^2 of electroosmosis alone, σ_{os}^2. Setting $k' = 0$ and assuming $\kappa = 3.29 \cdot 10^8$ m^{-1}, and $E = E_{opt} = 36,300$ V/m, $R_1 = 50 \cdot 10^{-6}$ m, $\mu_{app} = 1 \cdot 10^{-8}$ m^2/V s, $D_i = 1 \cdot 10^{-10}$ m^2/s, and $L = 0.5$ m, Eq. (84) says that $\sigma_{os}^2 = 1.67 \cdot 10^{-11}$ m^2 or $H_{os} = 3.34 \cdot 10^{-11}$ m. Thus, as we might have expected based on the flat electroosmotic velocity profile, pure electroosmosis does not significantly contribute to band broadening.

Next, we can see the effect of a finite value of k' on the magnitude of σ_T^2. Figure 14 shows the dramatic impact of k' on H_w, the plate-height contribution of solute–wall interaction. Note that this analysis does not take into account the influence of k' on migration velocity. It can be seen that even for small values of k', σ_w^2 can be significant. Therefore, to achieve high-efficiency CE separations, it is critically important to eliminate any solute–wall interactions. This can usually be accomplished by a proper choice of buffer pH.

4. Variance Caused by Siphoning ($\sigma_{\Delta h}^2$)

Unwanted hydrodynamic flow caused by siphoning, due to a height difference between the ends of the electrophoresis capillary, can result in a devastating loss of separation efficiency. Even seemingly small differences in height can result in severe band broadening. This can be clearly seen by again turning to the expression describing Taylor dispersion. If we define $\sigma_{\Delta h}^2$ as the contribution of the siphon-induced variance to the total variance, we can see from Eq. (79) that,

$$\sigma_{\Delta h}^2 = \frac{R_1^6 \Delta P^2}{1536 L_{tot}^2 \, \eta^2 D} \tag{86}$$

In the case of siphoning, the pressure difference, ΔP, is caused by the height difference between the ends of the capillary tube, Δh. For water at 39°C, $\Delta P = 98.1$ N/m^2 per cm of height difference. Substituting this value of ΔP into Eq. (86) results in the expression

$$\sigma_{\Delta h}^2 = 6.26 \frac{R_1^6 \, \Delta h \text{ (cm)}}{L_{tot}^2 \, \eta^2 D} t \tag{87}$$

where Δh is measured in cm. From Eq. (87) it can be seen that, using parameter values from Fig. 12 and assuming $L_{tot} = 0.75$ m, $L = 0.5$ m, and $E = 36{,}300$ V/m, for a 1-cm height difference, $H_{\Delta h} = 4.8 \cdot 10^{-6}$ m. Therefore, to ensure that no siphoning occurs during a separation, it is important to be sure that the level of the buffer in each reservoir is equal.

5. Variance Caused by Conductivity Differences ($\sigma_{\Delta \kappa}^2$)

In zone electrophoresis, the concentration of the analyte ion is kept low relative to the surrounding buffer in order to keep the conductivity difference between the sample zone and the surrounding buffer as small as possible. This is done to ensure that the field strength across the capillary is constant and that each sample ion migrates independently at a rate governed by its electrophoretic mobility. Alternatively, if the sample ion contributes significantly to the conductivity of the sample zone, the migration of the analyte will become coupled to the migration of the surrounding buffer ions. This form of electrophoresis is known as isotachophoresis. In this chapter, we are concerned only with zone electrophoresis. However, even if the analyte ion only slightly contributes to the overall conductivity of the sample zone, it can still result in significant band broadening.

Qualitatively, the mechanism of band broadening is as follows (see Fig. 15). If $\kappa_{e,A} < \kappa_{e,B}$, a solute molecule entering region B to the right will migrate slower than it did in zone A, because the field strength in zone B is lower than that in zone A. Thus, the migrating zone A will overtake it, resulting in a sharpening of the A–B interface. Conversely, if the solute enters region B to the left, it will be left behind the migrating zone, resulting in tailing. Hjertén (1990) has derived a simple approximate expression

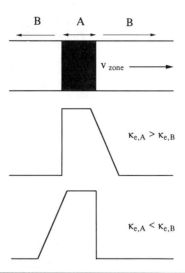

Figure 15   Schematic illustration showing the mechanism of band broadening due to electrical conductivity differences between the sample zone and the running buffer. When $\kappa_{e,A} > \kappa_{e,B}$, solutes diffusing out of the sample zone (A) to the right, will be propelled ahead of the sample zone, whereas solutes diffusing to the left will be propelled back into the sample zone (A). Thus the left boundary of the zone will sharpen, while the right boundary will broaden. The situation is reversed when $\kappa_{e,A} < \kappa_{e,B}$.

to describe conductivity-induced band broadening quantitatively,

$$\sigma^2_{\Delta\kappa} = \frac{(\mu_{app} Et)^2}{16} \left(\frac{\Delta\kappa_e}{\kappa_e}\right)^2 \qquad (88)$$

where $\Delta\kappa_e$ is the conductivity difference between the buffer and the migrating zone, and κ_e is the conductivity of the buffer. Thus, to have a minimum of band broadening, one should make $\Delta\kappa_e$ as small as possible. Typically, one wants to make $\Delta\kappa_e/\kappa_e$ less than 0.5% (Lauer and McManigill, 1986b). The $\Delta\kappa_e$ can be related to the properties of the buffer ions and the analyte using the expression

$$\Delta\kappa_e = \frac{zCF}{\mu_s}(\mu_A - \mu_s)(\mu_B - \mu_s) \qquad (89)$$

where z is the valency of the sample ion, C is the molar concentration of the sample, F is Faraday's constant, and μ_s, μ_A, and μ_B are the mobilities of the sample ion, the coion, and the counterion, respectively. Therefore, to minimize $\Delta\kappa_e$, one should choose a buffer whose ions have mobilities close to those of the sample ion and use a low sample concentration.

It should be pointed out that, in a rigorous sense, zone broadening due to conductivity differences is not a true dispersive phenomenon. For all true dispersive effects, the variance, σ, is proportional to $t^{1/2}$. This is not the case in Eq. (88), whereas it is for all the other dispersive phenomena discussed in this chapter. Therefore, it is not rigorously correct to include $\sigma^2_{\Delta\kappa}$ in the sum in Eq. (66).

6. Variance Caused by Finite Detection Volume (σ^2_{DET})

In any detection scheme, a finite volume of fluid is in contact with the sensing element at any given time, because it is impossible to interrogate an infinitely small volume. Therefore, the output of the detector represents a signal that is averaged over a finite volume. It has been pointed out by Said (1959) that the variance introduced in this manner, σ^2_{DET}, is the same as that from a finite injection slug. Thus

$$\sigma^2_{DET} = \frac{l_d^2}{12} \tag{90}$$

where l_d is the length of the detector cell. For typical CE applications, this contribution to the total variance is minimal. In the case of UV-absorbance detection, if the illuminating light beam is 0.5 mm in diameter, $H_{DET} = 4.16 \cdot 10^{-8}$ m, if $L = 0.5$ m.

7. Total Variance (σ^2_T)

In an "ideal" experiment, only the influence of the parabolic temperature profile, finite injection volume, and diffusion on σ^2_T need to be considered. Thus,

$$\sigma^2_T = \sigma^2_D + \sigma^2_{\Delta T} + \sigma^2_I \tag{91}$$

Therefore, in terms of theoretical plates, using the expressions from this chapter,

$$H_T = \frac{2D_i}{L}t + \frac{R_1^6 E^6 \kappa_e^2 \Omega_T^2 \mu_{app}^2}{1536 D_i k_b^2 L}t + \frac{R_1^6 \Delta P_{inj}^2}{1536 L_{tot}^2 \eta^2 D_i L}t_{inj} \tag{92}$$

where we have assumed that hydrodynamic injection is used.

At this point we can use Eq. (92) to estimate the value of H_T for a given set of experimental parameters and compare these results with experimental findings. Using the example parameters provided in Table II, we see that, according to Eq. (92), $H_D = 2.0\ \mu m$, $H_{\Delta T} = 0.00022\ \mu m$, and $H_{I,HD} = 1.86\ \mu m$. Therefore, $H_T = 3.86\ \mu m/plate$, or $N = 259,100$ plates/m. Note that because of the small capillary diameter, $H_{\Delta T}$ is negligible. This value for H_T is comparable to recently published experimental values for small proteins (Jones *et al.*, 1990). Thus, given that $N = 259,100$ plates/m, using Eq. (64), we can see that to achieve a resolution of 1.0, given an L of 50 cm, $\Delta\mu/\mu_{avg} \approx$

Table II   **Parameters Used in the Calculation of Typical Plate-Height Contributions**

Parameter	Value	Units
η	0.001 (1cp)	N sec m^{-2}
D_i	$3 \cdot 10^{-10}$	m^2 sec^{-1}
μ_{app}	$1 \cdot 10^{-8}$	m^2 V^{-1} sec^{-1}
R_1	$25 \cdot 10^{-6}$	m
E	30,000	V m^{-1}
κ_e	1.25	Ω^{-1} m^{-1}
Ω_T	0.02	K^{-1}
k_b	0.6	W m^{-1} K^{-1}
L	0.5	m
L_{tot}	0.75	m
ΔP_{inj}	$2.07 \cdot 10^{-4}$ (3 psi)	N m^{-2}
Injection volume	1% of effective column volume	m^3
t_{inj}	2.3	sec

1.0%. Next, we can see the dramatic impact of increasing the internal diameter of the capillary. If R_1 goes from 25 μm to 37.5 μm, $H_D = 2.0$ μm, $H_{\Delta T} = 0.0076$ μm, and $H_{I,HD} = 28.4$ μm ($t_{inj} = 1.03$ sec). Thus, $H_T = 30.4$ μm or $N = 32{,}870$ plates/m. Thus, by increasing the capillary diameter by a factor of 1.5, the plate-height contribution due to hydrodynamic injection increases by a factor of 15, assuming that the injection pressure remains the same, resulting in an increase in H_T of a factor of eight.

Equation (92) is an important result. It shows how the various parameters in a CE experiment affect the ultimate separation performance. These relationships among the various parameters and their impact on separation performance are complex and not at all intuitive. Whereas Eq. (92) can be useful for designing experiments, it must always be remembered that it does not include the influence of effects such as siphoning, sample–wall interactions and conductivity differences. In fact, Eq. (92) provides a "best case" estimate of the performance that can be expected under a given set of experimental conditions. Therefore Eq. (92) can serve as a benchmark for experimental results.

C. Quantitative Comparison of Capillary and Traditional Electrophoresis Formats

We are now in a position to compare the theoretical performance of capillary electrophoresis with that of traditional electrophoresis formats. This comparison most clearly demonstrates the potential of CE. The comparison will be made between a 50-μm i.d. microcapillary and a 1-mm i.d. tube format. In each case, the separation distance, L, will be 1 m, and the field strength will be the optimal field strength for each case, E_{opt}, as calculated by Eq. (76). This is the key aspect of this analysis—to make a

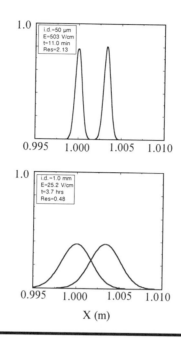

Figure 16   Comparison of separation performance between CE and a traditional tube format using the optimal field strength for each.

valid comparison of the performance of each format, we must operate each under its optimal operating conditions; otherwise, the comparison is meaningless. The two analytes will have mobilities of $3.00 \cdot 10^{-4}$ cm^2/V $\cdot$ sec and $3.01 \cdot 10^{-4}$ cm^2/V $\cdot$ sec, respectively, and a diffusion coefficient of $1 \cdot 10^{-6}$ cm^2/sec. Values of κ_e, k_b, and Ω_T are given in Table II. Figure 16 shows the results of this analysis where Eq. (55) is plotted and σ^2 is given by the sum of $\sigma_{\Delta T}^2$, Eq. (75), and σ_D^2, Eq. (67) for each species in each format. Thus, for a given separation distance, the CE separation is over 20 times faster and provides a separation over 4 times better. This comparison clearly demonstrates the potential of CE separations compared to those achievable using traditional formats. It should also be remembered that this analysis is valid regardless of the separation mechanism responsible for the selectivity.

Appendix: List of Symbols

Symbol	Definition	SI units
A	Outside surface area of the capillary	m^2
A_p	Peak area	m^2
C	Concentration	mole m^{-3}
C_m	Constant in Eq. (85)	Dim.[a]
D	Outside diameter of capillary	m
Δh	Height difference between capillary ends	m
D_i	Diffusion coefficient	m^2 sec^{-1}
ΔP	Pressure drop	N m^{-2}
E	Electrical field strength	V m^{-1}
e	Charge of electron	C
g	Acceleration of gravity	m sec^{-2}
H	Height of a theoretical plate	m
h	Heat transfer coefficient between the outside surface of the capillary and the surroundings	W m^{-2} K^{-1}
h_p	Peak height	m
I	Current density	A m^{-2}
i	Current	A
k	Boltzmann constant	J K^{-1}
k'	Solute capacity factor	Dim.[a]
k_{air}	Thermal conductivity of air	W m^{-1} K^{-1}
k_b	Thermal conductivity of buffer	W m^{-1} K^{-1}
k_g	Thermal conductivity of glass wall	W m^{-1} K^{-1}
k_p	Thermal conductivity of polymer coating	W m^{-1} K^{-1}
L	Length to detector	m
l	Length of sample slug	m
l_d	Length of detector cell	m
L_{tot}	Total length of the capillary	m
M	Mass of solute	kg
N	Number of theoretical plates	Dim.[a]
N_A	Avogadro constant	mole^{-1}
n_i	Number density of ions	m^{-3}
Nu	Nusselt number	Dim.[a]
Q	Volumetric flow rate	m^3 sec^{-1}
q_r	Heat flux	J m^{-2} sec^{-1}
R	Electrical resistance	Ω
r	Radial distance	m
R_1	Inside radius of glass wall	m
R_2	Outside radius of glass wall	m
R_3	Outside radius of capillary	m

Symbol	Definition	SI units
Ra	Rayleigh number	Dim.[a]
Re	Reynolds number	Dim.[a]
Res	Resolution	Dim.[a]
S	Cross-sectional area	m^2
Se	Power generation per unit volume	$W\ m^{-3}$
Se_{tot}	Total power generation	W
T	Temperature	K
t	Time	sec
T_1	Temperature at radial position R_1	K
T_2	Temperature at radial position R_2	K
T_3	Temperature at radial position R_3	K
T_a	Temperature of the surroundings	K
t_{inj}	Injection time	s
T_0	Temperature at the capillary center-line	K
V	Voltage	V
v_{air}	Velocity of air	$m\ sec^{-1}$
v_z	Fluid velocity	$m\ sec^{-1}$
w	Zone width	m
x	Linear distance	m
z_i	Valence number	Dim.[a]
α	Thermal diffusivity	$m^2\ sec^{-1}$
ε	Electrical permittivity	$C^2 J^{-1}\ m^{-1}$
η	Viscosity	$N\ sec\ m^{-2}$
κ^{-1}	Double layer thickness	m
κ_e	Electrical conductivity	$\Omega^{-1}\ m^{-1}$
μ	Electrophoretic mobility	$m^2\ sec^{-1}\ V^{-1}$
μ_{app}	Apparent electrophoretic mobility	$m^2\ sec^{-1}\ V^{-1}$
μ_{os}	Electroosmotic mobility	$m^2\ sec^{-1}\ V^{-1}$
ρ	Density	$kg\ m^{-3}$
ρ_e	Charge density	$C\ m^{-3}$
σ^*	Surface charge density	$C\ m^{-2}$
σ^2	Variance of normal distribution	m^2
σ_D^2	Variance caused by diffusion	m^2
σ_{DET}^2	Variance caused by finite detector volume	m^2
$\sigma_{\Delta h}^2$	Variance caused by siphoning	m^2
$\sigma_{\Delta\kappa}^2$	Variance caused by conductivity differences	m^2
$\sigma_{\Delta T}^2$	Variance caused by temperature profile	m^2
σ_I^2	Variance caused by injection	m^2
σ_T^2	Total variance	m^2
σ_W^2	Variance caused by wall interaction	m^2
Ω_T	Temperature coefficient of electrophoretic mobility	K^{-1}
ψ	Electrical potential	V

[a]Dim., dimensionless.

References

Altria, K., and Simpson, C. (1986). *Anal. Proc.* **23**, 453–454.

Altria, K., and Simpson, C. (1987). *Chromatographia* **24**, 527–532.

Applied Biosystems (1990). "Application Note #26." Applied Biosystems, Foster City, California.

Aris, R. (1959). *J. R. Soc. London* **A252**, 538.

Atkins, P. W. (1978). "Physical Chemistry." W. H. Freeman, San Francisco.

Chapman, A. J. (1974). "Heat Transfer." Macmillan, New York.

Churaev, N. V., Sergeeva, I. P., Sobolev, V. D., and Derjaguin, B. V. (1981). *J. Colloid. Interfac. Sci.* **37**, 451–460.

Coxon, M., and Binder, M. J. (1974). *J. Chromatogr.* **101**, 1–16.

Cussler, E. L. (1984). "Diffusion: Mass Transfer in Fluid Systems." Cambridge University Press, Cambridge, England.

Fahien, R. W. (1983). "Fundamentals of Transport Phenomena." McGraw-Hill, New York.

Foret, F., Deml, M., and Bocek, P. (1988). *J. Chromatogr.* **452**, 601–613.

Fujiwara, S., and Honda, S. (1986). *Anal. Chem.* **58**, 1811–1814.

Fujiwara, S., and Honda, S. (1987). *Anal. Chem.* **59**, 487–490.

Giddings, J. C. (1965). "Dynamics of Chromatography." Marcel Dekker, New York.

Grossman, P. D., Colburn, J. C., and Lauer, H. H. (1989). *Anal. Biochem.* **58**, 1811–1814.

Grushka, E., McCormick, R. M., and Kirkland, J. J. (1989). *Anal. Chem.* **61**, 241–246.

Herrin, B., Shafer, S., van Alstine, J., Harris, J., and Snyder, R. (1987). *J. Colloid. Interfac. Sci.* **115**, 46–55.

Hiemenz, P. C. (1986). "Principles of Colloid and Surface Chemistry." Marcel Dekker, New York.

Hjertén, S. (1967). *Chromatogr. Rev.* **9**, 122–219.

Hjertén, S. (1967). "Free Zone Electrophoresis." Almqvist & Wicksells Boktryckeri, Uppsala, Sweden.

Hjertén, S. (1985). *J. Chromatogr.* **347**, 191–198.

Hjertén, S. (1990). *Electrophoresis* **11**, 665–690.

Huang, X., Luckey, J. A., Gordon, M. J., and Zare, R. N. (1989). *Anal. Chem.* **61**, 766–770.

James, A. T., and Martin, A. J. P. (1952). *Analyst* **77**, 915.

Jones, A. E., and Grushka, E. (1989). *J. Chromatogr.* **466**, 225.

Jones, H. K., Nguyen, N. T., and Smith, R. D. (1990). *J. Chromatogr.* **504**, 1–19.

Jorgenson, J. W., and Lukacs, K. D. (1981). *Anal. Chem.* **53**, 1298.

Jorgenson, J. W., and Lukacs, K. D. (1983). *Science* **222**, 266–272.

Kolin, A. (1954). *J. Appl. Phys.* **25**, 1442–1443.

Lauer, H. H., and McManigill, D. (1986a). *Anal. Chem.* **587**, 166–170.

Lauer, H. H., and McManigill, D. (1986b). *Trends Anal. Chem.* **5**, 11–15.

Littlewood, A. B. (1970). "Gas Chromatography." Academic Press, New York.

Maisel, L. (1971). "Probability, Statistics, and Random Processes." Simon and Schuster, New York.

Martin, M., Guiochon, G., Walbroehl, Y., and Jorgenson, J. W. (1985). *Anal. Chem.* **57**, 561–563.

McCormick, R. (1988). *Anal. Chem.* **60**, 2322–2328.

Rice, C. L., and Whitehead, R. (1965). *J. Phys. Chem.* **69**, 4017–4025.

Said, A. (1959). *A.I.Ch.E.J.* **5**, 69.

Sternberg, J. C. (1966). *Adv. Chromatogr.* **2**, 205.

Taylor, G. (1953). *Proc. R. Soc. Lond.* **A219**, 186.

Virtanen, R. (1974). *Acta Polytech. Scand.* **123**, 7–67.

Wiktorowicz, J. E., and Colburn, J. C. (1990). *Electrophoresis 1990* **11**, 769–773.

Detection Methods in Capillary Electrophoresis

Teresa M. Olefirowicz and Andrew G. Ewing

I. Introduction

In this chapter, we have focused on some of the more recent and significant developments in detector design for capillary electrophoresis (CE), excluding the commercially available instrumentation. Several reviews containing a substantial amount of information about CE detectors have been published (Ewing *et al.,* 1989; Wallingford and Ewing, 1989a; Kuhr, 1990). The format of the chapter and the consideration given to each topic reflect the general use and applicability of the particular detection schemes. Much work has focused on the detection of analytes of biological origins, whereas a few others have described additional applications with respect to particular detection problems. The rapid advance in detection development for CE is attributable mainly to the relative ease of adaptation of some high-performance liquid chromatography (HPLC) detectors, mainly optical detectors, to CE. With minor modifications, several of these detectors have been adapted to accommodate the use of the fused-silica capillary as the detector flow cell. However, many of these detectors suffer a loss in sensitivity owing to the decreased detection-cell dimensions and the reduced operating volumes of CE.

The diversity of CE detection needs makes it desirable and necessary to use a variety of detection principles. Several performance criteria must be considered when choosing an appropriate detector for a particular analysis. These criteria include sensitivity, selectivity, linear range, and noise. The most useful type of detection system aids in the separation and identification of analytes by its selectivity, where the signal depends on the nature of the analytes. Detector response should produce a known and reproducible relationship with amount or concentration of a solute and have a wide linear-response range. In other cases, it is preferable to use a detector that is "universal" and responds similarly (in terms of both sensitivity and selectivity) to all solutes. Detectors for CE as for other separation techniques should respond independent of the buffer, should not contribute to extra-column broadening, and should be reliable and convenient to use. Unfortunately, no single detector provides all these properties; therefore, an appropriate detector must be chosen based on the particular application.

There are two main types of detectors: bulk-property and specific-property detectors. The bulk-property detector measures the difference in some physical property of the solute relative to that of the buffer alone. These include refractive index, conductivity, and indirect methods. These detectors are generally more universal than specific-property detectors. However, these detectors usually have lower sensitivities and dynamic ranges. This is because the detector signal depends not on the properties of the solutes but on the difference in the properties of the solute and the buffer. Specific-property detectors measure the specific properties of solutes and include ultraviolet (UV) absorbance, fluorescence, mass spectrometry, electrochemical, radiometric, and Raman detectors. These methods limit detection to only those analytes that possess the required specific properties. Use of these selective detectors is very advantageous when the sample matrix is complex and in situations where it is desirable to minimize "background" interferences. This type of detector is normally more sensitive than the bulk-property detector, possesses wider linear ranges, provides more acceptable signal-to-noise ratios, and is used most often in CE.

Detection of the small analyte zone volumes in CE can be accomplished either while they are migrating through the capillary (UV absorbance, fluorescence, conductivity, refractive index, etc.) or as they elute from the capillary (post-column derivatization, electrochemical, mass spectrometry). For on-column detection, the detection cell is part of the electrophoresis capillary; thus, zone broadening due to joints, fittings, and connectors is eliminated. In off-column detection, however, the detector region usually contributes to band broadening and needs to be evaluated by researchers utilizing specific off-column detection techniques.

II. Detector Types Used in Capillary Electrophoresis

A. Ultraviolet-Absorbance Detectors

In CE, UV-absorbance detection is the most widely used. This is mainly because of its ease of operation and almost immediate adaptation from HPLC detectors by use of the capillary as the on-column detection cell. In fact, all commercially available CE instruments employ a ultraviolet-visible (UV-VIS) absorbance detector. High-quality, fused-silica capillaries are usually used for UV detection, since fused silica has a UV cutoff of approximately 170 nm. Additionally, it is believed that at least 65% of all compounds analyzed by liquid chromatography can absorb light at 254 nm (typically that of a fixed-wavelength detector), and more than 90% absorb somewhere in the range of variable-wavelength detectors (Ahuja, 1989). Thus, it should be possible to detect a broad range of compounds using UV absorbance with CE.

Sensitivity of UV detection depends on the path length of the detection cell; hence, detection limits in CE tend to be lower than those of other specific-property detectors. This limits the usefulness of UV absorbance to capillaries having inner diameters of 25 μm or greater. Typical detection limits range from approximately 10^{-13} to 10^{-15} moles detected (Ewing *et al.*, 1989). For UV detection, a trade-off exists between the use of low cell volumes for high performance and the use of larger-diameter capillaries for sensitivity. Increased sample loading can overcome the sensitivity problems, but this can adversely affect the electrophoretic separation process. Also, slit widths for the UV light source should be sufficiently small that only the capillary is illuminated, thereby reducing stray light levels that could result in excessive background noise. Illumination volumes should be smaller than the analyte zones; this is best accomplished by focusing the UV light on the capillary. Finally, many materials absorb light in the UV region, so minimizing background interferences can be difficult.

To be more generally useful, in terms of selectivity, a UV detector with continuously variable wavelength is desirable, even though there will be some loss in sensitivity. With variable-wavelength operation, the monochromator is preferred over filters for wavelength selection since it is more versatile, can be used to scan wavelengths, and produces better resolution. The main disadvantage of instruments utilizing a monochromator is that they are usually less sensitive than systems utilizing filters, which provide more light throughout.

1. Direct Detection

Ultraviolet absorbance detection is useful for a large number of compounds that contain a chromophore. Early work in capillary electrophoresis featured the use of modified HPLC UV detectors. Terabe *et al.* (1984) have carried out on-column UV detection with micellar electrokinetic capillary chromatography (MECC) using a commercial detector. Since no pump is used in CE, this detector has been modified to obtain shorter response times and higher gain than that needed for HPLC detection, where pump pulses can lead to excessive noise. Foret *et al.* (1986) have described the modification of a commercial photometer in combination with optical fibers (0.2-mm light-conducting core), which are in direct contact with the walls of the electrophoresis capillary. The UV radiation from a mercury lamp is filtered using a 254-nm interference filter before entering the optical fiber, and the transmitted light is collected opposite the incident light using another optical fiber. The advantage of using optical fibers is that the capillary inner diameters and the light-conducting core of the fibers can be matched for optimal sensitivity. The disadvantages of using optical fibers include transmission losses in the fibers and reduced efficiency at low wavelengths.

Kobayashi *et al.* (1989), of the Shimadzu Corporation, have discussed the use of a photodiode array detector for CE. This work discusses the application of a full 512-element diode array that can be used in the 200- to 380-nm range. This work shows spectra that were obtained for a mixture of aromatic compounds and water-soluble vitamins that have been separated using MECC. The advantage of this type of detector is that qualitative information about unknown analytes can be obtained from their absorbance spectra.

Sepaniak *et al.* (1989) have discussed some instrumental developments for MECC including a versatile on-column flow cell that employs a unique laser-etched, on-column optical slit. This flow cell has been adapted for photometric detection using a modified UV-absorbance detector and spectrophotometric detection using a photodiode array detector. Small solute bands in CE and MECC are usually detected on-column; however, short optical path lengths and excess stray light can lead to detection difficulties. The flow cell described has been adapted to use with commercially available detectors. External light sources and fiber optics can be used for light transmission. Multidimensional detection has been accomplished by combining the flow cell with a photodiode array detector.

Cohen and Karger (1987) have also used a modified commercial UV detector for the detection of peptides and proteins that are separated using polyacrylamide gel capillary electrophoresis. Detection has been performed by placing the 75-μm inner diameter (i.d.) gel-filled capillary directly in the optical path of the detector. Absorbance detection with gel-filled capillaries is possible at wavelengths of 220 or 280 nm. However, absorbance detection of proteins at 210 nm or less is difficult because of interference from the gel (Karger *et al.,* 1989). This can be avoided by coupling an open piece of capillary to the end of a gel-filled capillary; however, band-broadening effects from coupling should be considered and minimized.

Aguilar *et al.* (1989) have presented work demonstrating the need to account for the electrophoretic mobilities of the solutes in order to detect some species of interest. On-column absorbance detection at 214 nm is performed using a modified commercial UV detector to determine highly negatively charged iron cyanide complexes in liquid samples from the zinc electroplating process. Under the conditions of these experiments, the electrophoretic mobilities of the iron complexes are very large and opposed to the direction of electroosmotic flow. Detection has been carried out at the anodic end of the electrophoresis capillary to circumvent this difficulty. Thus, electrophoretic mobilities of analytes must be considered, since these species would not be detected in the normal CE mode, where detection is usually carried out at the cathodic end of the capillary. Alternatively, the buffer composition and pH could be adjusted to yield a stronger electroosmotic flow that could carry these highly charged anions to the cathode. However, this approach might result in a loss of resolution.

Several researchers have designed UV detectors to eliminate some of the problems associated with modified commercial HPLC detectors. Walbroehl and Jorgenson (1984) have described a fixed-wavelength, on-column UV detector for CE. The key problem with other previously used detectors is that the illumination slit is often too large for use with capillaries smaller than 80 μm. Slit widths that are larger than the capillary result in lower linear ranges and lower signal-to-noise ratios, since most of the light reaching the detector has not passed through the capillary. The detector described by Walbroehl and Jorgenson provides

1. a high-intensity light source;
2. secure positioning of the capillary in the beam; and
3. minimization of the slit width to less than the capillary diameter.

The system design consists of a cadmium "pen-ray" lamp, which can be interchanged with zinc or mercury lamps. The UV radiation passes through a $100\text{-}\mu m$ pinhole onto the capillary, which is held firmly in place with nylon set screws on either side of the pinhole slits. Applications shown include the separation and detection of protein standards and organic bases in non-aqueous media. Detection limits have been determined to be 15 pg (120 fmol) for isoquinoline and 250 pg (18 fmol) for lysozyme, with the detector response being linear over more than three orders of magnitude for each analyte.

Green and Jorgenson (1989) have described the design of a variable-wavelength UV absorption detector for on-column detection (Fig. 1) and have compared its

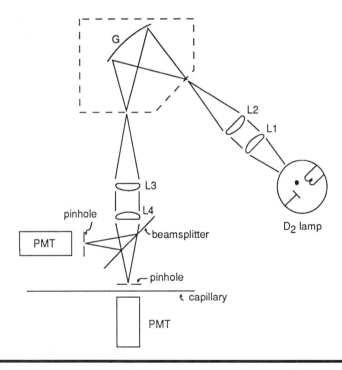

Figure 1   Schematic diagram of the optical layout of the D_2 lamp-based UV-absorption detector. L1 and L2 are fused-silica, plano-convex collimating and collection lenses. G is the monochromator. L3 and L4 are plano-convex collection and collimating lens. Reproduced with permission from Green and Jorgenson (1989).

performance to a fixed-wavelength detector. Design, sensitivity, and noise characteristics of a deuterium lamp-based, variable-wavelength UV detector are discussed, in addition to the use of other fixed-wavelength lamps. The authors point out that, based on the absorbance spectra of proteins, absorption at 193 nm can provide absorbances of 20 times those at 229 nm (Mayer and Miller, 1970), the wavelength commonly used for proteins. In addition, the 193-nm absorbance is approximately proportional to the number of peptide bonds in a protein; thus, approximate molecular weight information becomes obtainable. Because the path length is very small, solvent absorption does not seem to present the problems associated with systems having longer path lengths, such as HPLC. Finally, a deuterium lamp provides a continuous UV source with a peak intensity at approximately 220 nm. An advantage of this design is the relatively simple and inexpensive instrumentation required. The authors investigated the use of several other lamps with this system, including zinc (Zn) (214 nm), cadmium (Cd) (229 nm), and arsenic (As) (200 nm), and have concluded that a zinc lamp is the best choice for fixed-wavelength operation, since it gives a better signal-to-noise ratio and more uniform response factors than the deuterium (at 193 nm), Cd, or As lamps. However, the advantage in using the deuterium lamp over the Zn, Cd, or As lamps is that the deuterium lamp can be operated over the wavelengths of 180 to 400 nm, whereas the other lamps have UV bands only at specific wavelengths, limiting their usefulness.

2. Indirect Detection

If solutes do not contain a chromophore, an absorbance signal can be obtained by indirect detection. Optical systems used for indirect detection are identical to those used for direct UV-absorbance detection. The only difference in this detection mode is that the electrophoretic buffer contains a UV chromophore. Hjertén et al. (1987) have described the use of on-column detection of non-UV-absorbing ions, which are detected when the separation is performed in a UV-absorbing buffer. The buffer system, the sodium salt of diethylbarbituric acid (25 mM at pH 8.6), has been used for the separation and detection of bromide, acetate, and cacodylate anions as well as the four organic acids: formic, acetic, propionic, and n-butyric acids. Indirect detection was carried out at 225 nm.

Foret et al. (1989) have also used indirect UV absorbance for the detection of anions. In their system, the principal component of the background electrolyte is either the light-absorbing benzoic (20 mM) or sorbic acid (7 mM). An example of a separation using benzoic acid as the absorbing buffer component is shown in Fig. 2. Nonabsorbing ionic species are revealed by changes in light absorption due to charge displacement of the absorbing co-ion. The highest sensitivities were achieved for sample ions with effective mobilities similar to those of the absorbing co-ion. The useful dynamic range is limited by the linearity and noise of the detector. The best sensitivities are obtained in low-concentration background electrolytes containing a co-ion with high UV absorption at a given detection wavelength. Detection limits with modified commercial single-beam detectors approach 0.5 pmol.

Grant and Steuer (1990) have described an extended path length UV detector (millimeter range) that is achieved by axial illumination of the capillary using a

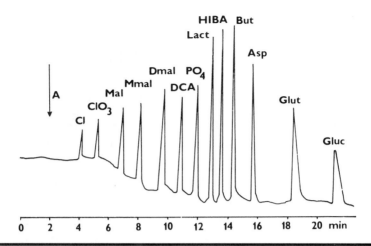

Figure 2 Separation by CZE of thirteen non-UV-absorbing ions in a 100-μm i.d. capillary. Indirect photometric detection was carried out using a Varian 2550 double-beam spectrometric detector (Varian Associates) operating at 254 nm. The background electrolyte was 20 mM benzoic acid-histidine at pH 6.2 with 0.1% Triton X-100 (Sigma Chemical Co.). Mal, malonate; Mmal, methylmalonate; Dmal, dimethylmalonate; DCA, dichloroacetate; Lact, lactate; HIBA, hydroxyisobutyrate; But, butyrate; Asp, aspartate; Glut, glutamate; Gluc, glucuronate. Reproduced with permission from Foret *et al.* (1989).

10 mW 325 nm helium–cadmium (HeCd) laser, as shown in Fig. 3. When UV light passes in this direction through the capillary, it is attenuated as it passes through absorbing zones within the capillary. With a fluorescent species added to the buffer, the absorbance within a given segment of capillary can be indirectly measured. By measuring the fluorescence intensity at a given point in the capillary, the fluorescence signal varies exactly the same as the transmitted light. Thus, the UV-absorbance measurement is taken indirectly. The value of this type of illumination is that the effective path length of the capillary is extended from the micrometer to the millimeter range. This leads to an increase in the absorbance signal. The authors state, however, that for full exploitation of this technique, the background noise must be reduced, possibly by using a more stable light source.

3. Thermooptical Detection

Thermooptical detection involves the use of two intersecting laser beams. A pump laser is focused at a right angle to the electrophoresis capillary, and a second laser beam is used to probe the refractive index of the fluid in the capillary. Absorbance of the pump beam by analytes in the capillary produces a temperature rise. Since the refractive index of most solvents changes with temperature, absorbance of the pump beam produces refractive index changes that are monitored by the probe laser. Thermooptical detection is performed on-column using two lasers, which are focused

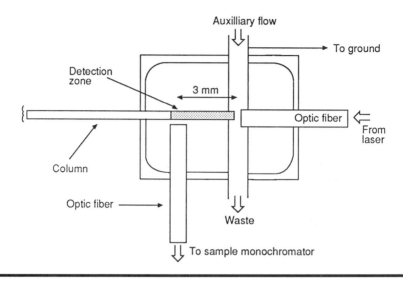

Figure 3 Details of the detection zone for the extended path length UV-absorbance detector. Light from a 325-nm HeCd laser is conveyed through fiberoptics and focused onto the end of the electrophoresis capillary. The fluorescence-collecting fiberoptic is placed perpendicular to the capillary, and light is directed to a photomultiplier tube. Reproduced with permission from Grant and Steuer (1990).

on the capillary. Either a 4-mW HeCd laser (442 nm) (Yu and Dovichi, 1988) or a 130-mW argon ion laser (458 nm) (Yu and Dovichi, 1989) serves as the pump laser, whereas a 1-mW helium–neon (HeNe) (632.8 nm) laser serves as the probe laser.

The absorbance sensitivity for thermooptical detection is proportional to the power of the pump laser, and the signal is independent of the capillary diameter in the range of 50 to 500 μm. Thus, unlike typical absorbance measurements, sensitivity is not lost owing to short detection path lengths. The detection volume is determined by the radius of the tube in combination with the point of interaction of the two laser beams. Yu and Dovichi (1988, 1989) have described the detection of 4-(dimethylamino)-azobenzene-4′-sulfonyl (DABSYL) chloride-derivatized amino acids using a thermooptical absorbance detector. One of the advantages of using the DABSYL chloride derivatives with absorbance detection is the resulting high molar absorptivity; hence, high sensitivity and low detection limits are possible. Also, since excitation light in the visible range is used, the background signal from solvent absorption is significantly reduced (relative to using 254 nm). Finally, the wavelength of maximal absorption of the DABSYL chloride derivatives conveniently matches two commonly used lasers (HeCd and argon ion). Detection limits of 5×10^{-8} M (37 amol) for methionine to 5×10^{-7} M (450 amol) for aspartic acid have been achieved using this detection scheme, representing the best obtained to date for any absorbance detector.

B. Fluorescence Detectors

The second most widely used CE detector is based on fluorescence, using either an arc lamp or a laser as the excitation source. This highly sensitive and selective detector is especially important in biological applications for analytes containing primary amines, such as amino acids, peptides, and proteins, and is important in DNA sequencing (Karger *et al.,* 1989; Drossman *et al.,* 1990). Most fluorescence detectors, as with all other detectors except UV absorbance, have been custom built by individual researchers.

The approaches that provide the best sensitivities in fluorescence detection use a light source that can be tightly focused on the detection volume. The advantage of fluorescence detection is that the signal-to-noise ratio is not a function of detection-cell path length. Fluorescence is an important CE detection scheme mainly because it aids in analyte specificity and can provide extremely low detection limits. Detection limits range from approximately 10^{-15} to 10^{-20} moles, depending on the excitation source, derivatization procedure, and fluorescent tag (Ewing *et al.,* 1989).

1. Direct Detection

One disadvantage of fluorescence detection for samples of biological interest is that many do not fluoresce with sufficient quantum yields at easily evaluated wavelengths. Thus, detection typically involves the use of a fluorescent tag that is covalently attached to the molecules of interest through a derivatization reaction. Phenols, amino acids, amines, carboxylic acids, aldehydes, and ketones can be converted to fluorescent derivatives (Lawrence, 1979; Ohkura and Nohta, 1989).

There are basically two ways to derivatize these types of samples for detection, pre-column and post-column derivatization. Typical derivatization reagents for biological amines include 5-dimethylaminonaphthalene-1-sulfonyl (DANSYL) chloride, fluorescein isothiocyanate, fluorescamine, *o*-phthalaldehyde (OPA), and naphthalene-2,3-dicarboxaldehyde (NDA). In the case of oligonucleotides, fluorescent dyes such as ethidium bromide can be used for the selective detection of double-stranded DNA molecules (Kasper *et al.,* 1988).

Pre-column derivatization requires time for sample derivatization, where reaction times can take several hours, before injection. Also, the reagents used should not be detectable under the same conditions as the derivative. Typical pre-column derivatization reagents include DANSYL chloride, fluorescein isothiocyanate, and NDA. Fluorescamine and OPA are very useful for post-column derivatization, but have also been used for pre-column derivatization, because the derivatization takes place in seconds to minutes under mild conditions, and the reagents are nonfluorescent.

A fluorescence detector based on a mercury–xenon (Hg–Xe) arc lamp as the excitation source combined with filters to isolate the excitation wavelength was first described by Jorgenson and Lukacs (1981). The use of a double monochromator to isolate the excitation wavelength was shown to be superior to use of interference filters alone (Green and Jorgenson, 1986). This system decreased the background signal and thus resulted in better detection limits. Use of an arc lamp and monochromator allows the selection of excitation wavelengths over a 200- to 800-nm range, and filters can be used to isolate the emission wavelength. Detection of DANSYL- and fluorescamine-derivatized amino acids, dipeptides, and proteins as well as an unlabeled tryptophan-containing protein were presented in this work. Detection limits ranged from 4.6 pg for doubly labeled DANSYL-lysine to 91 pg for bovine pancreatic trypsinogen, a naturally fluorescent tryptophan-containing protein, with the linear range spanning 3 to 3.5 orders of magnitude for similar analytes. The advantages in the use of a monochromator are in the broad range of excitation wavelengths available, the precise selection of wavelength possible with a monochromator, and the relative simplicity of this system.

Lasers are very useful for on-column detection in capillaries with small diameters. Their use becomes essential as the capillary diameter is further decreased from 100- or 50-μm i.d. to 10 or even 5-μm i.d. The laser output can be focused to a very small volume within an electrophoresis capillary (thus increasing the excitation power density), leading to increased sensitivity. The difficulty with the use of lasers is the limited number of wavelengths that are conveniently available. Two lasers that are used most frequently are the HeCd (325 nm and 442 nm) and argon ion lasers (330 nm and 488 nm), which have lines in the UV and visible regions of the spectrum. For very sensitive detection, the fluorescence from an analyte must be isolated from the scattered emission of the excitation laser. This is usually done by placing the detector at a 90-degree angle from the laser beam, while tilting the capillary to Brewster's angle (Kuhr and Yeung, 1988a). Further isolation of the fluorescence signal can be carried out with the use of appropriate spatial and optical filters. Analyte fluorescence is typically detected using a photomultiplier tube (PMT).

Laser-induced fluorescence detection with CE has been applied to a variety of separations. Gassmann et al. (1985) have separated racemic mixtures of dansylated amino acids and detected fmol amounts of each using laser–fluorescence detection. Electrophoresis was performed in a 75-cm long, 75-μm i.d. capillary containing a chiral electrolyte, Cu(II)-L-histidine complex. The on-column fluorescence detector described uses a 10 mW (325 nm) HeCd laser as the excitation source. The laser output is directed onto the electrophoresis capillary using an 85-μm fused-silica optical fiber, and the emitted fluorescence is collected at a right angle using a second, 600-μm glass optical fiber. This fluorescence then passes through a monochromator to a PMT for detection. Resolution of dansylated amino acids with another complex, Cu(II)-aspartame has been described by Gozel et al. (1987). The linearity and sensitivity of this system has been tested using DANSYL-arg, -glu, -val, and the detector has been found to be linear over four orders of magnitude (1×10^{-3} to $1 \times 10^{-7} M$).

Roach et al. (1988) have described the determination of methotrexate (a widely used anticancer drug) and 7-hydroxymethotrexate (a metabolite) by laser-induced fluorescence detection. A 100-cm long (75 cm to detector), 75-μm i.d. separation capillary is used with a HeCd laser (17 mW) as the excitation source. A 66-μm optical fiber directs the laser light to the capillary, and the emitted fluorescence is collected with a 1000-μm optical fiber and directed to a PMT after passage through a 450-nm filter. This detection system is shown in Fig. 4. The detection limit for methotrexate is

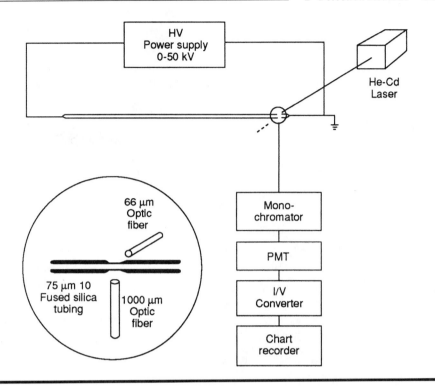

Figure 4 Experimental set-up for capillary zone electrophoresis with laser-induced fluorescence detection. Reproduced with permission from Roach *et al.* (1988).

5×10^{-10} *M* and for 7-hydroxymethotrexate, 2×10^{-9} *M*. The linearity of this detector extends over nearly four orders of magnitude.

Swaile and Sepaniak (1989) have described the use of a fluorometric diode array detector for capillary electrophoresis. By focusing a laser beam to a 10-μm spot inside a 50-μm i.d. capillary, the detection volume becomes just 20 pl. Figure 5 shows a three-dimensional electropherogram (wavelength–intensity–time) of sodium fluorescein and two laser dyes. The two-dimensional results overcome the difficulty in optimizing emission-wavelength selection for increased detection sensitivity and selectivity. By optimizing the spectral bandpass, it is possible to maximize the signal-to-noise ratio. For this experiment, fluorescence excitation is performed perpendicular to the electrophoresis capillary, using a 30-mW (442 nm) HeCd laser. The emission is collected at a 90-degree angle to the laser through collimating lenses onto the entrance slit of a spectrometer that disperses the emission across a 1024-diode array with a spectral resolution of 4 nm per channel. A 1.2-mm entrance slit serves to spatially filter specular scatter from the edges of the capillary. Detection of sodium fluorescein

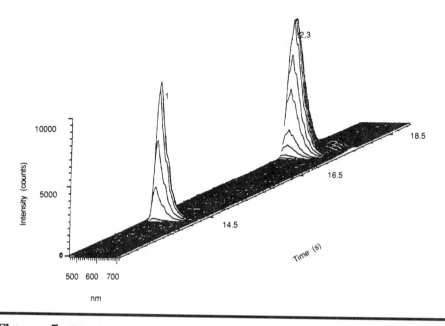

Figure 5   Wavelength–intensity–time electropherogram for (1) sodium fluorescein, (2) Coumarin 102, and (3) 4-(dicyanomethylene)-2-methyl-6-(*p*-dimethylaminostyryl)-4H-pyran (DCM), obtained using a fluorometric photodiode array detector. Reproduced with permission from Swaile and Sepaniak (1989).

has been found to be linear in the range from 1×10^{-7} to 1×10^{-4} *M* with a minimum detectable amount of 60 fg in a volume of 3.4 nl. Maximum sound to noise (S/N) is obtained with an emission wavelength of 514 nm with a 40-nm bandpass. The advantage of this detection system is that detection parameters for complex mixtures can be optimized for the best S/N in a single injection.

Fluorescence detection of NDA-derivatized amino acids has been carried out on a 10-μm i.d. capillary using the 442-nm line (10 mW) of a HeCd laser (Nickerson and Jorgenson, 1988). A 490-nm filter is used as the emission filter, and NDA is used as a pre-column derivatization reagent. The electrophoretic buffer is 25 : 75 MeOH : buffer (10 m*M* boric acid, 20 m*M* KCl, pH 9.5). Detection limits [signal to noise evaluated at three times the root mean square (rms) noise] of 0.42 amol have been reported for a 4×10^{-8} *M* injection of NDA derivatized phenylalanine (Fig. 6).

Sheath flow can be used to minimize the scatter of an excitation laser beam off the walls of the electrophoresis capillary (Cheng and Dovichi, 1988; Wu and Dovichi, 1989). In the sheath-flow method, shown in Fig. 7, detection is carried out in a sheath-flow cuvette placed after the separation capillary. Sheath fluid, the same as the electrophoresis buffer, is introduced into the cuvette at the end of the capillary using an

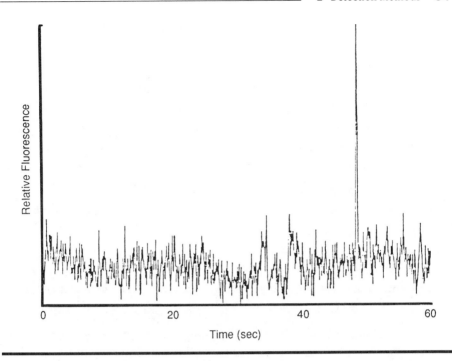

Figure 6 Fluorescence detection of 4×10^{-8} M (2.5 amol injected) phenylalanine labeled with the fluorescent tag, NDA. Reproduced with permission from Nickerson and Jorgenson (1988).

HPLC pump. A 1 W argon ion laser (488 nm) is used as the excitation source. This beam is focused to a 10-μm spot in the cuvette approximately 0.2 nm from the end of the capillary. Laser fluorescence detection of fluorescein isothiocyanate amino acid derivatives have been reported, with detection limits ranging from 5×10^{-12} M to 9×10^{-11} M for alanine and lysine, respectively. This work, although difficult for routine analysis, exhibits the best detection limits obtained for laser-induced fluorescence detection in CE. Of note, however, is that detection limits reported have been based on fluorescein isothiocyanate as the limiting reagent and not the amino acid analytes of interest. Thus, it is not clear whether the reported detection limits can be obtained with "real" samples of unknown concentration.

2. Post-Column Derivatization
The limitations of pre-column derivatization techniques arise from the fact that sufficient sample volumes, typically greater than a few μl, must be available for derivatization. Also, multiple derivatizations for the same molecule are possible, yielding several peaks for one analyte, and migration times are of the derivative, not the analyte. This can make identification and quantitation difficult. Finally, the derivatized samples must be stable throughout the preparation and analysis time, or degradation

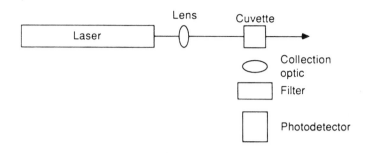

Figure 7 Laser-induced fluorescence detector utilizing the sheath-flow cuvette. Light from a laser is focused with a lens into the sample cuvette. Fluorescence is collected at right angles, filtered, and detected with a photodetector. Reproduced with permission from Wu and Dovichi (1989).

products might also be detected, and quantitative analysis would be inaccurate. Post-column derivatization avoids many of these problems.

It should be noted that when performing extra-column reactions of any kind, extra-column band broadening should be avoided or minimized in order to fully realize the high efficiency that is provided by CE. This becomes critically important when considering capillaries less than 100 μm i.d. Injection, detection, and connective tubing should individually contribute less to band broadening than on-column factors during the separation process. In general, connection devices should be avoided where possible in CE, and the most common detection schemes involve on-column detection (Pentoney *et al.*, 1988). Despite this, several excellent systems for post-column derivatization have been devised to circumvent these difficulties with pre-column derivatization. Post-column derivatization has the advantage that the sample is unaltered during the separation process, and migration times are those of the analyte, not the fluorescent tag. One difficulty is that optimization of the technique is experimentally quite difficult.

Using the arc-lamp fluorescence detector described previously (Green and Jorgenson, 1986), Rose and Jorgenson (1988) have designed a post-capillary reactor for detection, using OPA as a derivatizing reagent. The use of a coaxial capillary reactor, consisting of two concentric fused-silica capillaries, served to mix the OPA with migrating zones. In this post-column reactor (Fig. 8), the reaction capillary is held in a stainless-steel tee with Vespel ferrules, and the detection window is located on the reaction capillary. The electrophoretic capillary (25 μm i.d.) has an outside diameter (o.d.) that is smaller than the i.d. of the reaction capillary and is inserted through the tee and into the reaction capillary, forming a coaxial reactor. Reagent enters the third part of the tee through a 200-μm i.d., 325-μm o.d. capillary that is 70 cm long. Hydrostatic pressure drives the reagent into the tee, where it mixes with migrating zones at the tip of the electrophoresis capillary.

Mixing in the reaction capillary is the result of a combination of diffusion,

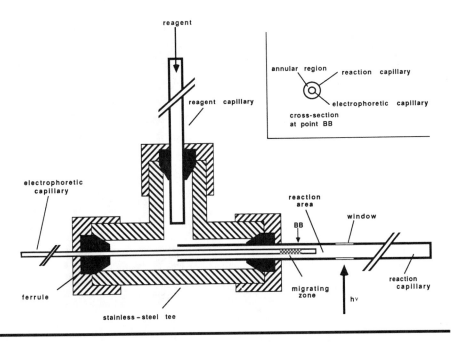

Figure 8 Cross-sectional schematic of post-capillary reactor in stainless-steel mixing tee. Reproduced with permission from Rose and Jorgenson (1988).

convection, and migration effects. Various combinations of electrophoresis capillary o.d. versus reaction capillary i.d. have been tested to determine the largest efficiency : peak area ratio. Rose and Jorgenson (1988) found that the best combination is a 40-μm o.d. electrophoresis capillary and 50-μm i.d. reagent capillary. Reaction products are detected 1–2 cm beyond the electrophoresis capillary by fluorescence. A high-pressure HgXe arc lamp has been used with 312-nm excitation and two 400-nm cut-on highpass filters to isolate the emission wavelengths. Linearity of the detector has been determined by plotting the log of peak area versus log of sample concentration and is found to be linear over three orders of magnitude. The detection limits are in the amol range for amino acids and proteins (Rose and Jorgenson, 1988).

Figure 9 directly compares the UV detector (229-nm fixed wavelength) response to that using the post-capillary fluorescence detector. The improvement with respect to S/N ratio is approximately 100-fold with the post-capillary fluorescence detector. The main advantage of this type of detection scheme is that minimal sample handling and preparation is necessary before analysis.

A post-capillary reactor for laser fluorescence detection, laser-induced fluorescence (LIF) has also been developed (Nickerson and Jorgenson, 1989). This is essentially the same derivatization scheme as described above by Rose and Jorgenson, with

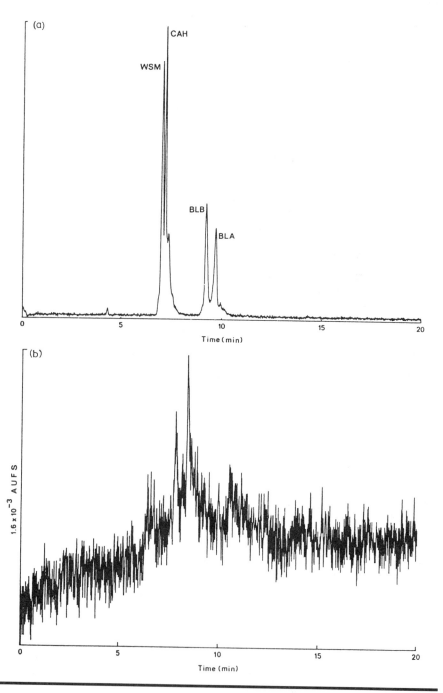

Figure 9 Comparison of post-capillary fluorescence (a) where OPA is the fluorescent tag, and UV-absorbance detection (b) of 100 μg/ml whale skeletal muscle myoglobin (WSM), 100 μg/ml carbonic anhydrase (CAH), 50 μg/ml β-lactoglobulin B (BLB), 50 μg/ml β-lactoglobulin A (BLA). Reproduced with permission from Rose and Jorgenson (1988).

the exception that the arc lamp has been replaced with a HeCd laser as the excitation source. Because the laser is more easily focused, band broadening has been reduced relative to that using an arc lamp. One noted difficulty with this method is that a variation in performance is observed between different post-capillary reactors. However, this type of reactor does avoid problems associated with multiply labeled samples, which can be a major problem in the analysis of proteins. Detection limits for this system were reported to be 1.2×10^{-8} M (44 amol) for horse heart myoglobin.

Pentoney *et al.* (1988) have described an on-line connector applied to derivatization of amino acids. Figure 10A shows the tee connector, which is fabricated from the fused-silica tubing itself. The on-column connector is made by making a hole (approximately 60 μm in diameter) perpendicular to the inner bore of the 75-μm i.d. fused-sil-

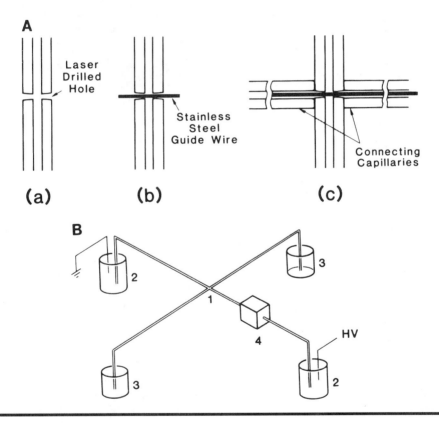

Figure 10 (A) Construction of the on-column capillary connector. In part (c), the guide wire extends beyond one connecting capillary to facilitate removal. (B) Experimental set-up of the laser-induced fluorescence (LIF) detector system for capillary electrophoresis: (1) on-column connector; (2) electrophoresis buffer reservoirs; (3) derivatization reagent reservoirs; (4) LIF detector housing; (HV) high voltage. Reproduced with permission from Pentoney *et al.* (1988).

ica capillary with a 40-W CO_2 laser. A stainless steel wire is inserted through the hole and serves as a guide to align two additional 75-μm i.d. capillaries. The three capillaries are temporarily sealed in place with warm poly(ethylene glycol) and permanently sealed with epoxy. Hydrostatic pressure is used to force the derivatization reagents into the separation capillary (Fig. 10B). The extra zone broadening in this system, due to the on-line connector, is minimized to approximately 10% for CE separations performed in 75-μm i.d. capillaries. Fluorescence detection has been accomplished using the 325-nm line of a HeCd laser (7 mW), which is directed to the capillary by use of a 100-μm fused-silica fiber optic. Emitted light is collected with a 600-μm fiber, through a 450-nm filter for OPA derivatives and through a 520-nm filter for DANSYL derivatives. The response for this detector is linear over more than three orders of magnitude, with detection limits in the sub-fmol range.

Two limitations associated with post-column derivatization include band broadening, and the exact timing that is necessary for the derivatization and detection of derivatized analytes. Since the signal is a function of the reaction rate, post-column reactors must be optimized for maximum signal and carefully calibrated for quantitative work. The main advantage of post-column derivatization is that sample preparation is minimized. Band-broadening considerations may be outweighed by the ease and minimization of "hands-on" time involved in the particular analysis with post-column derivatization.

3. Indirect Detection

The major difficulty of direct fluorescence detection is that a limited number of compounds exhibit fluorescence or have sufficient quantum yields at analytically useful wavelengths. Also, pre- and post-column derivatization can be problematic with respect to multiple derivatization and unstable derivatives or when sample volumes are smaller than needed for derivatization. Indirect fluorescence detection can be used to overcome these problems. Indirect fluorescence-detection methods are typically less sensitive than direct fluorescence detection; however, it is usually more sensitive than direct UV detection. The key advantage of indirect detection is that it is a universal detection scheme for charged analytes. Detection limits are limited by the dynamic reserve of the signal (ratio of background signal to noise on the signal), and similar detection limits can be expected for most charged analytes. The main disadvantage of indirect detection is that its universal nature could become a hindrance with extremely complex samples.

Kuhr and Yeung (1988a) have described indirect laser-induced fluorescence detection for CE. In this work, underivatized, anionic amino acids (pH 9.7) are visualized through the direct displacement of a fluorescent anion (salicylate) in the electrophoretic buffer. The electrophoretic buffer consists of 1×10^{-3} M salicylate as the fluorophore and 2×10^{-4} M sodium carbonate to increase the buffer capacity of the separation medium.

The instrumental set-up for indirect fluorescence is identical to that used for direct detection. The 325-nm line of an HeCd laser (8 mW) is used as the excitation source, and the capillary is mounted at Brewster's angle to minimize the amount of stray light entering the PMT, which is located 90 degrees to the incident laser beam. The PMT is fitted with a spatial filter, and the salicylate fluorescence is isolated using a 405-nm interference filter. The authors reported approximately 200-fmol detection limits for 10 underivatized amino acids, and indicated lower detection limits can be obtained by decreasing the concentration of the buffer and analyte proportionally. These detection limits are about three orders of magnitude higher than those for direct detection.

Figure 11 shows the separation of eight mono- and diphosphate nucleotides and two nucleosides injected at 20 fmol each. Sensitivity and separation with indirect fluorescence detection have been improved by stabilization of the laser power and improving the dynamic reserve by 1000 times (Kuhr and Yeung, 1988b). Additionally, the background laser fluorescence signal can be made more stable by silylation (Kuhr and Yeung, 1988b) of the electrophoresis capillary. Capillary deactivation through silylation is thought to stabilize the indirect signal by minimizing the adsorption/de-

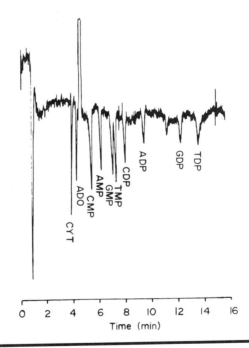

Figure 11 Separation and indirect fluorescence detection of nucleotides and nucleosides in a 25-μm i.d. capillary. The buffer was 0.5 mM salicylate at pH 3.5. The following (20 μm each) were injected: cytidine, CYT; adenosine, ADO; cytidine 5′-monophosphate, CMP; adenosine 5′-monophosphate, AMP; guanosine 5′-monophosphate, GMP; thymidine 5′-monophosphate, TMP; adenosine diphosphate, ADP; cytidine diphosphate, CDP; guanosine diphosphate, GDP; and thymidine diphosphate, TDP. Reproduced with permission from Kuhr and Yeung (1988b).

sorption of ions with the capillary wall, which is more important when using dilute buffer solutions. These modifications improve the detection limits for several anions, nucleotides, and proteins to the 50- to 100-amol range for indirect detection, whereas the detection limits for pre-column derivatized DANSYL-alanine is reported to be 0.5 amol using direct fluorescence detection (Kuhr and Yeung, 1988b).

Gross and Yeung (1989) have discussed the indirect fluorometric detection and quantification of inorganic anions and nucleotides. Using an argon ion laser (330 nm), the typical detection limits for this method are in the 50-amol range, as opposed to the 10-fmol detection limit range for UV absorption in 50-μm i.d. capillaries. The significant aspect of this work is that very low concentrations of sample in very small volumes make this method more sensitive than alternative methods, such as refractive index and conductivity. Concentration detection limits of 1×10^{-7} M were achieved for dihydrogen phosphate.

Hogan and Yeung (1990) have described the indirect fluorometric detection of sub-fmol quantities of a tryptic digest of β-casein, which can be separated and detected in three minutes. An argon ion laser is used here for excitation, and detection limits are at least 180 times better than those reported with UV detection. It is important to note that in order to obtain maximal zone displacement and thus signal, it is necessary to use fluorophores that are of the same charge as the sample components to be analyzed.

Indirect fluorometric detection of cationic analytes has been described by Gross and Yeung (1990). The fluorescent quinine cation of quinine sulfate is used as the fluorophore in this method. Very low quantities of mono- and divalent cations, amines, and oligopeptides have been detected. Since both cationic migration and electroosmotic flow are in the same direction, longer electrophoresis capillaries have been used to compensate for the decreased migration times. An argon ion laser that lases in the UV region on a pair of lines around 350 and 360 nm has been used. These lines are 10 times more intense than the 330-nm line previously used. Ten fmol of calcium, sodium, and magnesium have been detected in less than 6 min using this system.

4. Fluorescence-Detected Circular Dichroism

Christensen and Yeung (1989) have developed a selective, on-column detection method that has been applied to biological molecules that are chiral and exhibit fluorescence. Circular dichroism (CD) provides some additional selectivity in complex biological systems, taking advantage of the natural optical activity that is inherent in chiral molecules. Circular dichroism is the observed elliptical polarization of right and left circularly polarized light produced by an optically active medium. It can be related to the molecular conformation, and since fluorescence-detected circular dichroism (FDCD) can be related to CD, this detection method can be used to provide structural information on-line.

The authors have compared CE and HPLC applications of this detector and have found CE to be simpler, faster, more selective, and yielding better detection limits than HPLC. Since FDCD is a variation of fluorescence, it is readily interfaced with CE. In this work, FDCD (see Fig. 12) is carried out with an argon ion laser (488 nm) as the

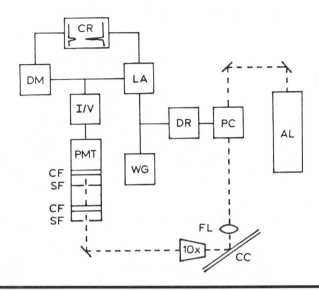

Figure 12 Schematic diagram of a FDCD and FL detection system: AL, argon ion laser; PC, Pockels cell; DR, driver; WG, waveform generator; LA, lock-in amplifier; DM, digital multimeter; CR, dual-pen chart recorder; FL, focusing lens; CC, capillary column; 10×, microscope objective; SF, spatial filters; CF, color filters; PMT, photomultiplier tube; I/V, current-to-voltage converter. Solid lines indicate electrical connections, and dashed lines show the optical path. Reproduced with permission from Christensen and Yeung (1989).

excitation source, and the linearly polarized light is converted to alternating left and right circularly polarized light (1 kHz) by a Pockels cell electrooptic modulator. The modulated light is focused onto an electrophoresis capillary, which is held perpendicular to the incident beam in order to preserve the circular polarization inside the capillary. The fluorescence signal is collected in a manner similar to that previously described (Kuhr and Yeung, 1988a). Since the FDCD signal is derived from the change in molar absorptivity rather than from the direct molar absorptivity, it will typically be less sensitive than direct fluorescence detection. The detection limits for riboflavin using FDCD-CE are reported to be 0.2 fmol in pl volumes, representing an improvement in detection limits of three orders of magnitude for FDCD. These are not exceptional detection limits relative to some other CE detectors; however, FDCD-CE promises some important applications in the biological sciences, where optical activity can be measured in very small volumes. The information content of FDCD is increased since the total fluorescence can be detected simultaneously, and compounds can be identified based on their unique FDCD : fluorescence ratios. It may be possible to determine the enantiomeric purity of eluting solutes using the FDCD : fluorescence measurements. This is especially important in the pharmaceutical industry. Additionally, the detection system described can be used to carry out FDCD, indirect fluorescence, and direct fluorescence detection.

C. Mass Spectrometry Detectors

Electropherograms obtained from the analysis of unknown samples or samples containing contaminants can contain several peaks, making solute identification a major problem. Identification problems can be solved by coupling CE with mass spectrometry (MS). Mass spectrometry is an ideal technique for the structural elucidation of the very small quantities of analytes eluting from an electrophoresis capillary. There is an advantage of coupling CE with MS, relative to HPLC. This advantage lies in the compatibility of low flow rates and small quantities of materials, eluting from the capillary, with the ability of the MS to remove the eluent. The main difficulty in developing CE-MS has been in coupling the two systems and in the removal of the electrolyte system. The difficulty involved is that CE is operated with aqueous buffers, whereas the MS instrument normally operates at high vacuum. It is therefore necessary to use an interface that effectively transfers the analytes from solution to the vapor phase without thermal degradation. Most of the research in combining CE with MS has been in the design and evaluation of reliable interfaces. Two ionization techniques have been successfully coupled to CE. These include electrospray ionization (EI) and fast atom bombardment (FAB).

The sensitivity of MS detection is governed by the mode in which spectra are obtained. Either the entire mass range can be scanned, or selected ions can be monitored as analytes elute from the capillary. Single-ion monitoring (SIM) is faster and provides improved detection limits relative to scanning the entire mass range, but is less useful for the analysis of completely unknown samples. Both techniques have their advantages, and it is up to the analyst to determine which is best for a particular application. Typical detection limits range from approximately 10^{-12} to 10^{-15} moles, with the best detection limits reported to be 10^{-17} moles (Smith and Udseth, 1988) using SIM. Routine CE/MS is a technique that could provide the analytical chemist with a wealth of structural information about unknown samples.

Smith and co-workers (Olivares *et al.,* 1987; Smith and Udseth, 1988; Smith *et al.,* 1988a) have developed a means to couple CE with on-line mass spectrometric detection using an electrospray interface (ESI). This system, as shown in Fig. 13, is unique in that no cathodic buffer reservoir is used. Electrical contact is made directly to the solvent in the capillary through an electrospray needle at the capillary outlet. Ions are produced at the end of the electrophoresis capillary at atmospheric pressure by an electrically induced nebulization process. A countercurrent flow of dry nitrogen encourages evaporation of the highly charged droplets. Continued evaporation results in further droplet breakup, to the point at which direct field-assisted ion desorption can occur from the droplet surface. The result is that ESI desorption is gentle, does not require excessive heating, and can produce both positive and negative ions, depending on the field polarity.

Detection limits have been in the fmol range for a variety of polypeptides and in the amol range for quaternary ammonium salts (Olivares *et al.,* 1987). An example of one of these separations is shown in Fig. 14. Using single-ion detection, detection limits as low as 10 amol can be obtained. Separation efficiencies and detection limits vary from analyte to analyte and are sensitive to buffer composition. Smith *et al.* (1988b)

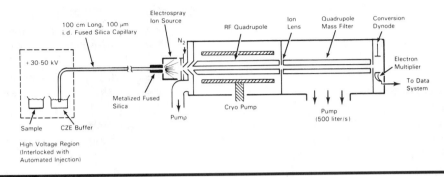

Figure 13 Schematic of the electrospray CZE-MS instrumentation. Reproduced with permission from Smith *et al.* (1988a).

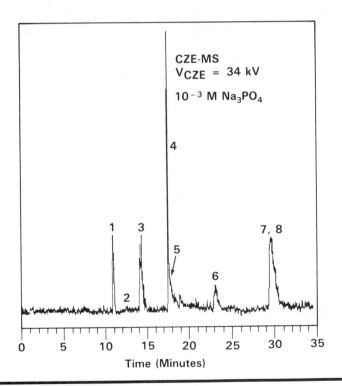

Figure 14 Reconstructed total ion electropherogram of an eight-component mixture obtained using the electrospray ionization interface. The mixture consisted of the following: 1, thiamine; 2, pyridoxamine; 3, pyridoxine; 4, tetrabutylammonium hydroxide; 5, cytidine; 6, adenosine; 7, L-phenylalanyl-L-phenylalanine; 8, L-tryptopyl-L-phenylalanine. Reproduced with permission from Smith *et al.* (1988a).

have subsequently described an improved ESI interface for CE-MS of peptides and proteins (Loo *et al.,* 1989), nucleotide coenzymes, nucleotide mono-, di-, and triphosphates (negative ion electrospray), and oligopeptides and proteins (positive ion electrospray) (Edmonds *et al.,* 1989).

Lee *et al.* (1988) and Muck and Henion (1989) have described the use of on-line capillary zone electrophoresis (CZE)-ion spray tandem MS for the determination of small peptides such as dynorphins, methionine enkephalin, and leucine enkephalin. Low pmol amounts have been determined by full-scan MS and tandem MS. Using selected ion monitoring conditions, low fmol levels can be determined, as shown in Fig. 15.

In methods described by Henion and co-workers, the cathodic end of the electrophoresis capillary is held at 3 kV and connected to the ion spray interface by use of a liquid junction connector. A triple-quadrupole MS with an atmospheric ionization source is used to sample ions formed at the ion spray interface. The ion spray interface produces gaseous ions by a mild form of ion evaporation. The advantage of this ionization technique is that the ions produced are either singly charged or protonated, or they are multiply charged or protonated. Because of this, the MS can be operated in the single-ion monitoring mode, which results in high sensitivity for species that are characteristic of the analyte.

Two basic approaches to coupling flow fast atom bombardment (FAB) mass spectrometry to CE have been reported. The first of these involves the use of a seg-

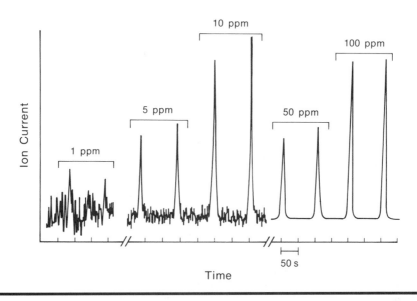

Figure 15 Capillary electrophoresis/ion spray mass spectrometry showing the detection limit (single-ion monitoring mode) for methionine enkephalin. One ppm corresponds to 15 pg (about 30 fmol). Reproduced with permission from Muck and Henion (1989).

mented capillary, in which the first segment is the electrophoresis capillary, and the second segment carries both the CE effluent and a glycerol solution into the mass spectrometer via a flow FAB probe (Minard *et al.*, 1988; Reinhoud *et al.*, 1989). Although this approach is simple and functional, separation efficiency is severely compromised, and detection limits are only in the pmol range.

Caprioli *et al.* (1989) have refined the segmented capillary approach to combining of continuous-flow, FAB mass spectrometry with capillary electrophoresis (Fig. 16). Mixtures of peptides can be analyzed by connecting the electrophoresis capillary to an interface that allows a total flow of 5 μl/min into the mass spectrometer. This interface has been shown to operate at the 50- to 100-fmol level with minimal zone broadening for the analysis of chemically synthesized peptides and proteolytic digests of proteins.

Moseley *et al.* (1989a,b) have described a second type of interface for coupling CE with continuous flow FAB tandem sector mass spectrometry (Fig. 17). The interface can be used to determine fmol levels of nonvolatile or thermally labile compounds. This interface uses two coaxial fused-silica capillaries that deliver capillary eluent and FAB matrix to the probe tip. The coaxial design permits the addition of the viscous FAB matrix very close to the FAB probe tip.

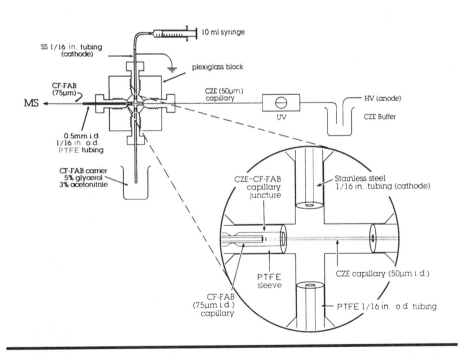

Figure 16 Schematic representation of a coupled CZE-CF-FAB instrument and interface. Reproduced with permission from Caprioli *et al.* (1989).

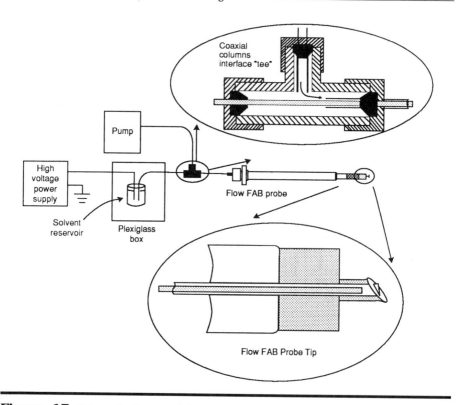

Figure 17   Schematic diagram of the on-line coaxial continuous-flow fast-atom bombardment CZE-FAB-MS system interface with the coaxial columns interface tee and flow FAB probe tip. Reproduced with permission from Mosely *et al.* (1989a).

An important aspect of this interface is that the FAB probe tip, which is held at +8 kV, serves as the cathode of the electrophoretic system. Thus, no transfer line is needed between the electrophoresis capillary and the FAB probe tip. This minimizes the band broadening that would be associated with the connections and parabolic flow that are present in a transfer line.

Figure 18 shows a spectrum obtained for low fmol levels of peptides. The coaxial continuous-flow FAB interface can be used without overwhelming loss of separation efficiency as is observed with some other methods. Efficiencies remain relatively high, in the range of several hundred thousand theoretical plates. These authors found that the choice of electrophoresis buffer is important when using CE in combination FAB-MS. The greatest problem with some buffers is with the formation of buffer–analyte adducts during ionization. Two buffer systems were investigated: potassium phosphate and a more volatile ammonium acetate. It was found that the phosphate buffer formed adducts, whereas the acetate buffer yielded only the protonated molecu-

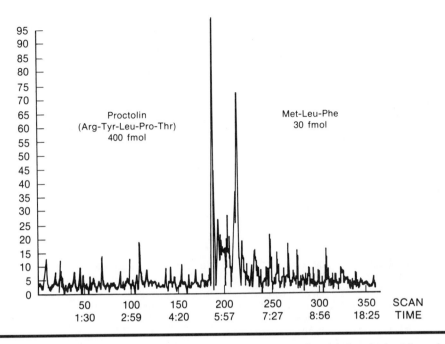

Figure 18 Summed ion electropherogram of the protonated molecules obtained from the on-line coaxial CF-FAB CZE-FAB-MS separation of 30 fmol of Met-Leu-Phe from 400 fmol of proctolin in an ammonium acetate buffer at pH 8.5. Reproduced with permission from Moseley *et al.* (1989b).

lar ions of the analytes. All of their work, therefore, used ammonium acetate buffers. This method, however, has proven to be very difficult to establish, owing to the formation of gas bubbles in the separation capillary and the sensitive nature of the interface. Therefore, most researchers in the area appear to be moving to the use of the electrospray interface.

D. Amperometric Detectors

1. Direct Detection

Electrochemical detection, in the amperometric mode, has been shown to allow the sensitive detection of many biologically important molecules, including catecholamines and indoleamines. The main advantage of electrochemical detection is that it is highly selective and useful for very sensitive detection of many electroactive species without prior sample derivatization. In amperometry, current is measured at a working electrode as analytes undergo electron transfer. In its simplest form, a working electrode is held at one potential, relative to a reference electrode, and current is recorded as a function of time as analytes elute from the electrophoresis capillary. The major difficulty in combining electrochemical detection with CE is in the electrical isolation of the working electrode from the high-potential field applied across the electrophoresis capillary. This difficulty can be overcome by using a segmented capillary. Since this detection mode involves a complex surface reaction that depends on both chemical and physical properties of the electrolyte, judicious choice of operating buffer is necessary. Also, since the working electrode is in direct contact with the buffer, periodic electrode cleaning is necessary for optimal detector performance. Typical detection limits range from approximately 10^{-17} to 10^{-19} moles.

Wallingford and Ewing (1987) introduced the concept of off-column amperometric detection with capillary electrophoresis for easily oxidized analytes. This system, shown in Fig. 19, represents a departure from normal CE, since both ends of the capillary are not submerged in buffer. Isolation of the carbon fiber electrochemical detector from the high voltage of the electrophoresis capillary is accomplished off-column by use of a conductive, low-volume connector. This conductive joint connects two pieces of capillary, a long separation capillary (50–100 cm), and a short detection capillary (1–2 cm), and is constructed at the cathodic end of the capillary. The joint is submerged in the cathodic buffer reservoir, which is held at ground potential, and the

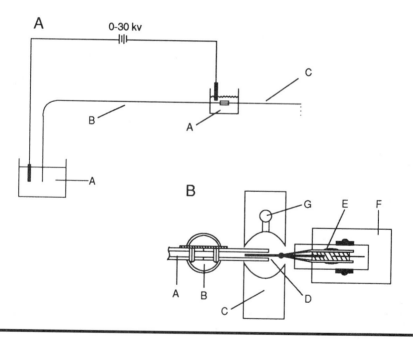

Figure 19   (A) Schematic of CE system with electrically conductive joint: A, buffer reservoirs; B, separation capillary; C, detection capillary. (B) Top view of amperometric detection system: A, capillary column; B, porous glass joint assembly; C, Plexiglas block; D, carbon-fiber working electrode; E, glass slide; F, micromanipulator; G, reference electrode port. Reproduced with permission from Wallingford and Ewing (1988a).

electric field is applied over only the separation capillary. Strong electroosmotic flow forces the buffer and solutes past the conductive joint and to a carbon fiber electrode, which is inserted into the end of the detection capillary. A major concern when coupling capillaries with off-column detectors is band broadening (Wallingford and Ewing, 1987). The length of the second capillary, the detection capillary, plays a significant role in band broadening. This broadening can be kept to a minimum using as short a detection capillary as possible, typically less than 2.5 cm.

The electrochemical detectors used in this method are locally constructed 10-μm diameter cylindrical carbon fiber electrodes. An approximately 200-μm length of the carbon fiber electrode is inserted into the detection capillary, using a micropositioner. Detection is carried out in the amperometric mode using a mercury battery to control the working electrode potential, while a sodium-saturated calomel electrode (SSCE) serves as reference electrode. This configuration was found to be superior to the use of a three-electrode potentiostat in terms of noise minimization.

Micellar solutions have been employed in CE with amperometric detection (Wallingford and Ewing, 1988a) to enhance the separation of nonionic and cationic solutes; however, they also affect the detection. Amperometric detection with micellar solutions has been demonstrated for a mixture of catechols on a 52-μm i.d. capillary. In order to evaluate the effects of sodium dodecyl sulfate (SDS) micelles on the electrochemical response of carbon fiber electrodes, voltammetric experiments have been carried out in various concentrations of SDS, ranging from 3 mM to 80 mM SDS. Higher concentrations of SDS shift the half-wave potentials of various catechols to more negative potentials, thus indicating that the oxidation products are more soluble in the micelles than are the reduced solutes, with the effect being more pronounced for cationic catechols. In addition to the shifts in half-wave potentials, the limiting currents are also affected in the presence of SDS, again with cations being affected to a greater extent. A unique increase in limiting current is observed for neutral catechols only at low SDS concentrations. As a practical consideration, the limiting current dependence on SDS concentration can affect the detector sensitivity and calibration by 60 to 80% for cationic solutes. Therefore, a trade-off between sensitivity and resolution exists when using electrochemical detection with MECC.

Detection limits and linearity have been determined for CE of catechol in 10 mM and 25 mM SDS solutions. The same electrode and capillary have been used to perform comparative experiments. Linearity is verified in each SDS solution with correlation coefficients of 0.997 and 0.999 for separation and detection in 10 mM SDS and 25 mM SDS, respectively. Minimal amounts injected are 19 (S/N = 3.6) and 17 fmol (S/N = 2.4) of catechol for the two respective buffer systems.

Capillary electrophoresis with amperometric detection has also been carried out in 12.7-μm diameter capillaries (Wallingford and Ewing, 1988b). Previously, electrophoresis had been carried out in 25- to 500-μm i.d. capillaries with 50- to 80-μm i.d. being the most common. The use of small-bore capillaries had not been previously explored, owing to difficulties of detection of the extremely small solute zones. Work in small-bore capillaries demonstrates both improved sensitivity and mass detection limits for amperometric detection. As the capillary diameter decreases with a constant electrode area, an enhanced coulometric efficiency is observed, resulting in increased sensitivity. Typically, greater than 50% of the analyte zone undergoes electrochemical reaction at the electrode surface. Detection limits have been reported in the range from 6 to 22 amol (8.5×10^{-9} to $3.4 \times 10^{-8}\ M$) for several electroactive analytes separated in a 12.7-μm i.d. capillary (Fig. 20). Detection of components removed from a single nerve cell has also been presented with this system (Wallingford and Ewing, 1988b). The best detection limit reported to date is 0.7 amol for serotonin separated and detected in a 9-μm i.d. capillary (Wallingford and Ewing, 1989b).

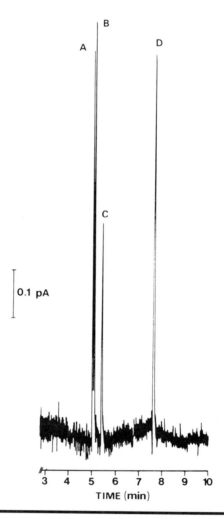

Figure 20   Capillary electrophoretic separation in a 12.7-μm i.d. capillary with amperometric detection: A, serotonin; B, norepinephrine; C, isoproterenol; D, 4-methylcatechol. Reproduced with permission from Wallingford and Ewing (1988b).

The smallest i.d. electrophoresis capillary that can be used with amperometric detection is limited by the size of the detection electrode. Olefirowicz and Ewing (1990a) have described the use of electrochemically etched and methane-flame-etched carbon fiber electrodes for use in 5-μm and 2-μm i.d. capillaries. The oxidation efficiency obtained with a 2-μm i.d. diameter electrode in a 5-μm i.d. capillary has been reported to be as high as 50% for the analytes tested. The linear range of this detector was found to be approximately three orders of magnitude for dopamine, serotonin, 1-dihydroxyphenylalanine, and catechol. Injection volumes as small as 270 fl have been reported with sub-amol detection limits. This method has been used to detect the neutrotransmitters dopamine and serotonin in pl samples of cytoplasm removed from single nerve cells (Olefirowicz and Ewing, 1990a).

2. Indirect Detection

Olefirowicz and Ewing (1990b) have described a system for indirect amperometric detection that is the same as that described for direct electrochemical detection, except that an electroactive component is added to the electrophoretic buffer (Fig. 21).

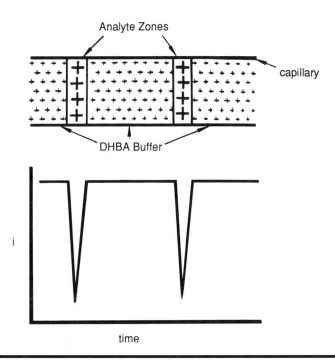

Figure 21 Schematic diagram of cation displacement by migrating zones of cationic solutes in indirect amperometric detection. Reproduced with permission from Olefirowicz and Ewing (1990a).

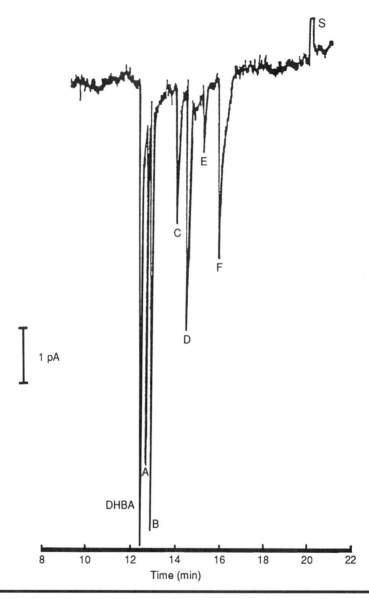

Figure 22 Electropherogram of amino acids and peptides with indirect amperometric detection in a 9-μm i.d. capillary. The electrophoretic buffer consisted of 10 μM 3,4-dihydroxybenzylamine (DHBA) with 25 mM 2-morpholinoethanesulfonic acid (MES). Between 5.8 and 8.5 fmol of the following analytes were injected: A, lysine; B, arginine; C, histidine; D, arginyl-leucine; E, histidyl-glycine; F, histidyl-phenylalanine, and S is the system peak. Reproduced with permission from Olefirowicz and Ewing (1990a).

Detection of nonelectroactive amino acids and dipeptides has been accomplished by adding the cationic electrophore, dihydroxybenzyl amine (DHBA), to the electrophoretic buffer. Continuous oxidation of DHBA at the electrode surface results in a high background oxidation current. As zones of nonelectroactive analyte pass through the detector region, a decrease in the background current is detected as a negative peak. Indirect amperometric detection of several amino acids and dipeptides in a 9-μm i.d. capillary is shown in Fig. 22. The lowest detection limit for this method was 380 amol for the amino acid arginine. One advantage of this method of detection is that both electroactive and nonelectroactive analytes can be detected simultaneously. This technique should prove to be as useful as indirect fluorometric detection and is experimentally much simpler.

E. Conductivity Detectors

Conductivity detection was one of the first detectors used for CE (Mikkers *et al.,* 1979). Conductivity is a universal, relatively simple mode for detection of ionic species in solution; thus, it is ideally suited for CE. In this detection mode, solution flows between two inert indicator electrodes across which a small, constant current is applied. When analytes of differing conductivity than the electrophoretic buffer pass between these electrodes, the voltage between the electrodes is measured. For optimal sensitivity, it is necessary to use very low conductivity buffers, such as the zwitterionic biological buffers. Typical detection limits have been comparable to those of UV-absorbance detection; however, the developments reported by Huang *et al.* (1987) have lowered the detection limits to the 10^{-16} to 10^{-18} range.

Zare's laboratory has introduced a sophisticated on-column conductivity detector that has been optimized for small-diameter capillaries. Huang *et al.* (1987) have described a detector for CE that is constructed using readily available components. Conduction measurements are made on-column by placing electrodes through holes in opposite walls of a 50- or 75-μm i.d. electrophoresis capillary (Fig. 23). These holes are formed using a CO_2 laser, and the platinum wire electrodes are fixed in place with epoxy. The detector end is made the cathode. In this method, an AC conductivity circuit with an oscillation frequency of 3.5 kHz is employed to measure and amplify the voltage change between the two indicator electrodes. The detector measures a constant value for the conductivity of the buffer until a solute zone migrates between the detector electrodes and changes the conductivity. Hence, this is a bulk-property detector, and conductivity is a nearly universal detection scheme. The advantage of using an on-column conductivity detector is that it does not appreciably disturb the flow or cause band broadening due to excessive dead volume. Peak area has been shown to be linear over three orders of magnitude from 0.0025 to 2.0 mM for lithium (Li$^+$) with detection volumes as small as 30 pl (volume between electrodes). Rubidium (Rb$^+$), sodium (Na$^+$), potassium (K$^+$), and Li$^+$ have been separated and detected using this system (Fig. 24), where the migration time of each ion is approximately proportional to the reciprocal of its mobility. The areas of the peaks represent the mobility differences between the ions in the detection zone and histidine, the counterion of the background

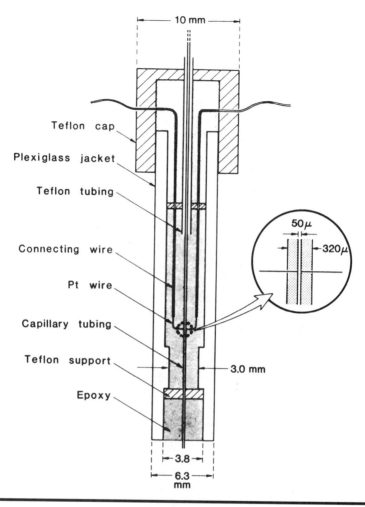

Figure 23   Diagram of the conductivity detector with the detector cell constructed by fixing platinum wires through diametrically opposed holes in a 50-μm or 75-μm i.d. capillary. Reproduced with permission from Huang *et al.* (1987).

electrolyte. The detection limit for Li$^+$ was determined to be about 1×10^{-7} M. Additionally, the separation and detection of tetramethyl ammonium, triethylamine, arginine, and histidine have also been demonstrated (Huang *et al.*, 1988). Conductivity detection has been used for the quantitative determination of Li$^+$, Na$^+$, and K$^+$ in human serum (Huang *et al.*, 1988).

Huang *et al.* (1989a) have used conductivity detection for the quantitative analysis of low-molecular-weight carboxylate anions in standard materials and also in two wine samples and a yogurt sample. In general, these carboxylic acids are not readily

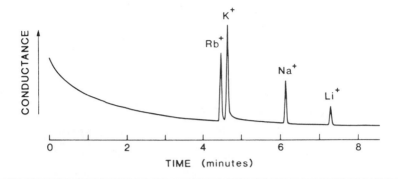

Figure 24 Electropherogram of a mixture of four cations detected with the conductivity detector; Rb^+, K^+, Na^+, and Li^+ (2×10^{-5} M each) were separated in a 75-μm i.d. capillary. Reproduced with permission from Huang *et al.* (1987).

detected by UV absorption. In these experiments, the electrophoretic mobilities of some anions are greater than and opposite in direction to the electroosmotic flow. Thus, in order to allow all carboxylate anions to pass through the detector, it is necessary to add tetradecyltrimethylammonium bromide to control the electroosmotic flow.

In contrast to other CE-detection methods, conductivity detection shows a direct relationship between the ionic mobility and the detector response. Thus, conductivity detection in CE allows accurate determination of absolute concentrations in a mixture by use of an internal standard, without separate calibration of the response of each component. Increased detection sensitivity can be obtained by maintaining a constant ratio of analyte concentration to background electrolyte concentration, while diluting the background electrolyte (Huang *et al.*, 1989b). The increase in sensitivity is primarily caused by a decrease in the background electrolyte conductivity. A fourfold decrease in the electrolyte concentration causes an increase in absolute sensitivity of more than 12 times. Using this method, detection limits have been extended to 1×10^{-6} M for a mixture of carboxylic acids.

F. Radiometric Detectors

Radioisotope detection is an extremely sensitive and highly selective detection mode based on scintillation counting of the CE eluent. As the radionuclei undergo disintegration, excess energy is transferred to a scintillator, which relaxes by emission of a photon. Photon counts are detected using a photomultiplier tube (Vickery and Stevenson, 1983). Since the background radiation is essentially negligible, the detection limits obtained are generally excellent.

Pentoney *et al.* (1989a) have described an on-line semiconductor radioisotope detector for CE, as shown in Fig. 25. The commercially available semiconductor device (CdTe), which is placed near the outlet of the electrophoresis capillary, is sensi-

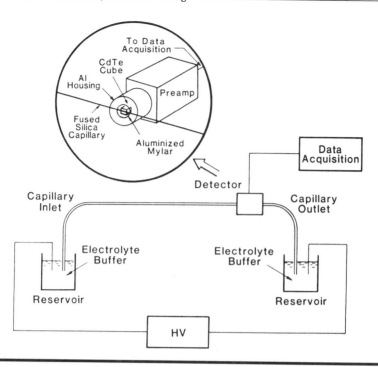

Figure 25 Experimental setup of the capillary electrophoresis/semiconductor radioisotope detector system. The inset shows the positioning of the CdTe probe with respect to the capillary tubing. Reproduced with permission from Pentoney *et al.* (1989b).

tive to gamma or high-energy beta radiation that passes through the walls of the capillary. Detection of ^{32}P-labeled molecules by the detector placed external to the separation capillary is possible because of the large energy (1.7 MeV) associated with the beta decay of ^{32}P. Penetrating distances for emitted beta particles range from approximately 2000 μm for water to 950 μm for fused silica (Knoll, 1979). Capillary dimensions for CE generally fall well within this range, making this detector extremely useful for very sensitive detection of ^{32}P-labeled solutes. Capillaries used in this work had an inner diameter of 100 μm, and the detection volume was approximately 15 nl.

The radioactivity detector has been tested using ^{32}P-labeled solutes, and the detection limit has been determined to be in the amol range for ^{32}P-labeled adenosine triphosphate (ATP) and guanosine triphosphate (GTP). Lower detection limits can be obtained by reducing the rate of flow past the detector. This "flow" programming is carried out manually, since flow programming is difficult to automate, by reducing the separation voltage as solute zones elute. Without flow programming, the detection limits have been found to be in the concentration range of about 1×10^{-9} *M*. These detection limits were reduced to about 1×10^{-10} *M* by use of flow programming. The efficiency of monitoring the radioactive decay for this detector is approximately 26%.

One difficulty with this detector is the positioning of the capillary within the detector cube. However, proper positioning can be accomplished by filling the capillary with radioactive material and monitoring the signal as the detector is translated with respect to the capillary. Maximal signal is obtained when the capillary is positioned ± 1.5 mm from the center of the detector housing.

Pentoney *et al.* (1989b) have also described another radioisotope detector, which uses a commercially available plastic scintillation material that completely surrounds (360 degrees) the electrophoresis capillary. The efficiency of monitoring radioactive decay for this detector is 65%, indicating an improved geometry over the former device. Radiolabeled ATP, cytidine 5′-triphosphate (CTP), and (TTP) have been separated and detected. Detection limits in the subnanocurie range, corresponding to a solute concentration of about 10^{-10} M, have been obtained with flow programming. The extremely low noise levels of the CdTe device compared to the plastic scintillator results in similar sensitivities for the two detection schemes described. The main advantage of these detectors is that they are extremely sensitive; however, applications are limited to high-energy beta and gamma emitters.

G. Raman-Based Detectors

Raman spectroscopy is a useful detection system for CE since qualitative information about analytes can be obtained in the form of Raman spectra. In this technique, the analyte is irradiated with a monochromatic light source, and the intensity and frequency of the scattered light are monitored. The differences in frequency between the incident and scattered radiation generally lie in the midinfrared region (Skoog and West, 1980).

Chen and Morris (1988) have described an on-line Raman spectroscopy/CE system. Figure 26 shows the capillary, which is held in a Lucite cylinder with an optical fiber relay system used to collect the signal. Excitation is achieved by a 40-mW HeCd laser operated at 442 nM, which is focused onto the capillary using a $5\times$ microscope objective. The signal is gathered by an array of 10- to 200-μm quartz optical fibers. This configuration yields the best signal-to-noise ratios. The fibers are arranged at 15 degrees in the plane normal to the plane of polarization of the laser beam and 30 degrees above and below the central fiber. The distal ends of the fibers are held in a vertical line to match the monochromator slit.

Methyl red and methyl orange have been used as test substances because they are water soluble and have well-characterized resonance Raman spectra in both acidic and basic solutions. The intense N=N stretch at about 1410 cm^{-1} has been used in this work with a monochromator slit width of 550 μm. Two different-diameter capillaries have been used in this work, one with 100-μm i.d. and 100-μm wall thickness, and the other with 80-μm i.d. and 160-μm wall thickness. Only the thick-walled capillary has proved useful for Raman detection. Some movement of the capillary is observed when the electrophoresis voltage is higher than 4.5 kV, and it is thought that the mechanically stronger wall (thicker-wall capillary) minimizes capillary vibrations. The concentration dependence of the signal has been measured between 2.5 $\times$ 10^{-6} M and 1 $\times$ 10^{-4} M, and the slope of the log/log plot of signal intensity versus concentration

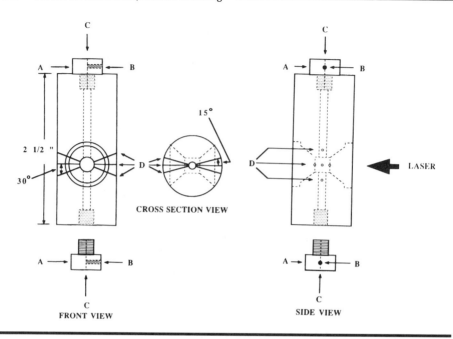

Figure 26 Lucite capillary holder and optical fiber relay system for Raman spectroscopic detection with capillary electrophoresis showing front, cross-sectional, and side views of the holder. A, threaded cap; B, set screw; C, bore for capillary; D, bore for optical fiber. Reproduced with permission from Chen and Morris (1988).

is 0.862. The detection limit reported is $2.5 \times 10^{-6}\ M$, comparing favorably with absorbance detection in CE, while providing structural information.

Further developments (Chen and Morris, 1990) in Raman detection, using a charge-coupled device (CCD), have lowered the detection limits for methyl orange to $1 \times 10^{-7}\ M$.

H. Refractive-Index Detectors

Refractive-index detection is a nonspecific detection method in which the sensitivity depends solely on differences in refractive index (RI) between the buffer and the solute. Although RI detection is generally one of the least sensitive detection modes for either CE or HPLC, it has the advantage of being useful with buffers whose physical properties (such as UV absorption) might otherwise interfere with detection.

Chen *et al.* (1989) have described a detection scheme for CE that utilizes analyte velocity modulation applied to RI detection. A schematic of this system is shown in Figs. 27A and 27B. This detection system is insensitive to thermal effects, laser-beam drift, small capillary-position changes, and other background fluctuations. The voltage

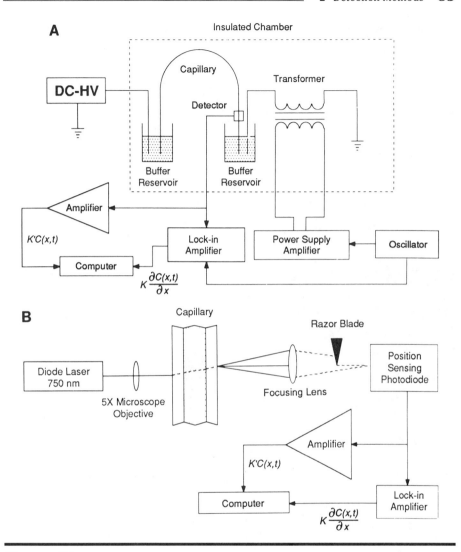

Figure 27 (A) Instrument for capillary electrophoresis with an analyte velocity-modulation system. (B) Laser beam deflection refractive index detector. Reproduced with permission from Chen *et al.* (1989).

across the capillary is modulated, yielding a response that is a derivative of the conventional electropherogram. The advantage of analyte velocity modulation is that the implementation is electrical, not optical, and can produce stable background-derivative responses for a variety of detectors, not just RI. The signal can then be further refined by rejecting signal that does not fall within the bandpass of the demodulation system.

For most electrophoretic runs using the RI detector, the modulation frequency has been about 390 Hz, and the peak AC voltage is about 50% of the DC voltage. This system has been used to detect carbonic anhydrase, with detection limits of 0.0015% in solution. This detection limit is comparable to that of absorbance detection but is not so sensitive as that of fluorimetry.

The key aspect of this work is that the background is more stable with analyte velocity modulation, and it should be useful with any detectors that are shot-noise or thermal-noise limited, such as RI, absorbance, conductivity, and indirect detectors.

References

Aguilar, M., Huang, X., and Zare, R. N. (1989). *J. Chromatogr.* **480,** 427–432.

Ahuja, S. (1989). "Selectivity and Detectability Optimizations in HPLC." Chemical Analysis Series, Volume 104. Wiley, New York.

Caprioli, R. M., Moore, W. T., Martin, M., DaGue, B. B., Wilson, K., and Moring, S. (1989). *J. Chromatogr.* **480,** 247–258.

Chen, C.-Y., and Morris, M. D. (1988). *Appl. Spectroscopy* **42,** 515–518.

Chen, C.-Y., and Morris, M. D. (1991). *J. Chromatogr.* **540,** 355–363.

Chen, C.-Y., Demana, T., Huang, S.-D., and Morris, M. D. (1989). *Anal. Chem.* **61,** 1590–1593.

Cheng, Y-F., and Dovichi, N. J. (1988). *Science* **242,** 562–564.

Christensen, P. L., and Yeung, E. S. (1989). *Anal. Chem.* **61,** 1344–1347.

Cohen, A. S., and Karger, B. L. (1987). *J. Chromatogr.* **397,** 409–417.

Drossman, H., Luckey, J. A., Kostichka, A. J., D'Cunha, J., and Smith, L. M. (1990). *Anal. Chem.* **62,** 900–903.

Edmonds, C. G., Loo, J. A., Barinaga, C. J., Udseth, H. R., and Smith, R. D. (1989). *J. Chromatogr.* **474,** 21–37.

Ewing, A. G., Wallingford, R. A., and Olefirowicz, T. M. (1989). *Anal. Chem.* **61,** 292A–303A.

Foret, F., Deml, M., Kahle, V., and Bocek, P. (1986). *Electrophoresis* **7,** 430–432.

Foret, F., Fanali, S., Ossicini, L., and Bocek, P. (1989). *J. Chromatogr.* **470,** 299–308.

Gassmann, E., Kuo, J. E., and Zare, R. N. (1985). *Science* **230,** 813–814.

Gozel, P., Gassman, E., Michelsen, H., and Zare, R. N. (1987). *Anal. Chem.* **59,** 44–49.

Grant, I. H., and Steuer, W. (1990). *J. Microcol. Sep.* **2,** 74–79.

Green, J. S., and Jorgenson, J. W. (1986). *J. Chromatogr.* **352,** 337–343.

Green, J. S., and Jorgenson, J. W. (1989). *J. Liq. Chromatogr.* **12,** 2527–2561.

Gross, L., and Yeung, E. S. (1989). *J. Chromatogr.* **480,** 169–178.

Gross, L., and Yeung, E. S. (1990). *Anal. Chem.* **62,** 427–431.

Hjertén, S., Elenbring, K., Kilar, F., Llao, J.-L., Chen, A. J. C., Siebert, C. J., and Zhu, M.-D. (1987). *J. Chromatogr.* **403,** 47–61.

Hogan, B. L., and Yeung, E. S. (1990). *J. Chromatogr. Sci.* **28,** 15–18.

Huang, X., Pang, T., Gordon, M. J., and Zare, R. N. (1987). *Anal. Chem.* **59,** 2747–2749.

Huang, X., Gordon, M. J., and Zare, R. N. (1988). *J. Chromatogr.* **425,** 385–390.

Huang, X., Luckey, J. A., Gordon, M. J., and Zare, R. N. (1989a). *Anal. Chem.* **61,** 766–770.

Huang, X., Gordon, M. J., and Zare, R. N. (1989b). *J. Chromatogr.* **480,** 285–288.

Jorgenson, J. W., and Lukacs, K. D. (1981). *Anal. Chem.* **53,** 1298–1302.

Karger, B. L., Cohen, A. S., and Guttman, A. (1989). *J. Chromatogr. Biomed. Appl.* **492,** 585–614.

Kasper, T. J., Melera, M., Gozel, P., and Brownlee, R. G. (1988). *J. Chromatogr.* **458,** 303–312.

Knoll, G. F. (1979). "Radiation Detection and Measurement." Wiley, New York.

Kobayashi, S., Ueda, T., and Kikumoto, M. (1989). *J. Chromatogr.* **480,** 179–184.

Kuhr, W. G. (1990). *Anal. Chem.* **62,** 403R–414R.

Kuhr, W., and Yeung, E. S. (1988a). *Anal. Chem.* **60,** 1832–1834.

Kuhr, W., and Yeung, E. S. (1988b). *Anal. Chem.* **60,** 2642–2646.

Lawrence, J. F. (1979). *J. Chromatogr. Sci.* **17,** 147–151.

Lee, E. D., Muck, W., Henion, J. D., and Covey, T. R. (1988). *J. Chromatogr.* **458,** 313–321.

Loo, J. A., Jones, H. K., Udseth, H. R., and Smith, R. D. (1989). *J. Microcol. Sep.* **1,** 223–229.

Mayer, M. M., and Miller, J. A. (1970). *Anal. Biochem.* **36,** 91–100.

Mikkers, F. E. P., Everaerts, F. M., and Verheggen, Th. P. E. M. (1979). *J. Chromatogr.* **169,** 11–20.

Minard, R. D., Chin-Fatt, D., Curry, P., and Ewing, A. G. (1988). Presented at 36th Annual Conference on Mass Spectrometry and Allied Topics, San Francisco, California, June.

Moseley, M. A., Deterding, L. J., Tomer, K. B., and Jorgenson, J. W. (1989a). *J. Chromatogr.* **480,** 197–209.

Moseley, M. A., Deterding, L. J., Tomer, K. B., and Jorgenson, J. W. (1989b). *Rapid Commun. Mass Spectr.* **3,** 87–93.

Muck, W., and Henion, J. D. (1989). *J. Chromatogr. Biomed. Appl.* **495,** 41–59.

Nickerson, B., and Jorgenson, J. W. (1988). *J. High Res. Chromatogr.* **11,** 533–534.

Nickerson, B., and Jorgenson, J. W. (1989). *J. Chromatogr.* **480,** 157–168.

Ohkura, Y., and Nohta, H. (1989). In "Advances in Chromatography" (J. C. Giddings, E. Grushka, and P. R. Brown, eds.), Vol. 29, pp. 221–258. Marcel Dekker, New York.

Olefirowicz, T. M., and Ewing, A. G. (1990a). *Anal. Chem.* **62,** 1872–1876.

Olefirowicz, T. M., and Ewing, A. G. (1990b). *J. Chromatogr.* **499,** 713–719.

Olivares, J. A., Nguyen, N. T., Yonker, C. R., and Smith, R. D. (1987). *Anal. Chem.* **59,** 1230–1232.

Pentoney, S. L., Huang, X., Burgi, D. S., and Zare, R. N. (1988). *Anal. Chem.* **60,** 2625–2629.

Pentoney, S. L., Jr., Zare, R. N., and Quint, J. F. (1989a). *J. Chromatogr.* **480,** 259–270.

Pentoney, S. L., Jr., Zare, R. N., and Quint, J. F. (1989b). *Anal. Chem.* **61,** 1642–1647.

Reinhoud, N. J., Niessen, W. M. A., Tjaden, U. R., Gramberg, L. G., Verheij, E. R., and van der Greef, J. (1989). *Rapid Commun. Mass Spectr.* **3,** 345–351.

Roach, M. C., Gozel, P., and Zare, R. N. (1988). *J. Chromatogr. Biomed. Appl.* **426,** 129–140.

Rose, D. J., and Jorgenson, J. W. (1988). *J. Chromatogr.* **447,** 117–131.

Sepaniak, M. J., Swaile, D. F., and Powell, A. C. (1989). *J. Chromatogr.* **480,** 185–196.

Skoog, D. S., and West, D. (1980). "Principles of Instrumental Analysis." Saunders, Philadelphia, Pennsylvania.

Smith, R. D., and Udseth, H. R. (1988). *Nature* **331,** 639–640.

Smith, R. D., Olivares, J. A., Nguyen, N. T., and Udseth, H. R. (1988a). *Anal. Chem.* **60,** 436–441.

Smith, R. D., Barinaga, C. J., and Udseth, H. R. (1988b). *Anal. Chem.* **60,** 1948–1952.

Swaile, D. F., and Sepaniak, M. J. (1989). *J. Microcol. Sep.* **1,** 155–158.

Terabe, S., Otsuka, K., Ichikawa, K., Tsuchiya, A., and Ando, T. (1984). *Anal. Chem.* **56,** 111–113.

Vickery, T. M., and Stevenson, R. L. (1983). "Liquid Chromatography Detectors." Marcel Dekker, New York.

Walbroehl, Y., and Jorgenson, J. W. (1984). *J. Chromatogr.* **315,** 135–143.

Wallingford, R. A., and Ewing, A. G. (1987). *Anal. Chem.* **59,** 1762–1766.

Wallingford, R. A., and Ewing, A. G. (1988a). *Anal. Chem.* **60,** 258–263.

Wallingford, R. A., and Ewing, A. G. (1988b). *Anal. Chem.* **60,** 1972–1975.

Wallingford, R. A., and Ewing, A. G. (1989a). *In* "Advances in Chromatography" (J. C. Giddings, E. Grushka, and P. R. Brown, eds.), Vol. 29, pp. 1–76. Marcel Dekker, New York.

Wallingford, R. A., and Ewing, A. G. (1989b). *Anal. Chem.* **61,** 98–100.

Wu, S., and Dovichi, N. J. (1989). *J. Chromatogr.* **480,** 141–155.

Yu, M., and Dovichi, N. J. (1988). *Mikrochim. Acta* **III,** 41–56.

Yu, M., and Dovichi, N. J. (1989). *Anal. Chem.* **61,** 37–40.

CHAPTER 3

Quantitative Aspects of Capillary Electrophoresis Analysis

Stephen E. Moring

I. Introduction

This chapter acquaints the reader with the special problems that pertain to qualitative and quantitative analysis in capillary electrophoresis (CE), and includes a review of quantitative methods in electrophoresis and a discussion of the optimization of modern integration algorithms. Aspects of spectrometric detection in a capillary format, and sampling factors that affect the accuracy and precision of electrophoretic analysis will also be discussed.

II. System Description and Analytical Method

Before a detailed discussion of qualitative and quantitative analysis can begin, a brief overview of a generic CE system is in order. Figure 1 is a block diagram of a CE system. The instrumentation can be broken down into five main units: an injection system; a separation system consisting of a temperature-regulated compartment, the capillary and buffer reservoirs; a detector; a high voltage power supply and controller; and a data-processing system. The injection system generally consists of a mechanism that allows for both hydrodynamic (vacuum or pressure) and electrokinetic (electromigra-

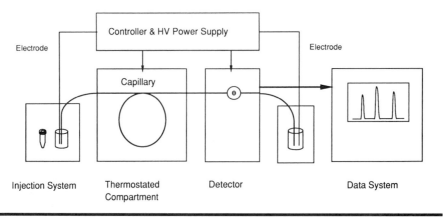

Figure 1   Schematic diagram of a capillary electrophoresis system.

tion) injection of samples from a variety of sample positions in an automated sampling device. The thermostated compartment provides a precisely controlled, temperature regulated environment for the separation capillary and ensures that the Joule heat is efficiently dissipated (see Chapter 1 for a further discussion of Joule heating effects). The capillaries typically used consist of 50- to 150-cm lengths of fused silica tubing with an inner diameter of 25 to 100 μm and an outer diameter of 200 to 400 μm. The detection system used in the majority of commercial CE instruments consists of a ultraviolet (UV)-visible absorbance or fluorescence based detector that has been adapted optically for the very short light path lengths across the capillary (see Chapter 2). The capillary–detector interface consists of a < 1-cm section of capillary where the plastic outer coating has been removed to allow the transmission of light. The high-voltage power supply should have the capability of providing up to 30 kV, with a maximum power output of approximately 10 W. The instrument controllers are microprocessor based and provide the ability to program instrument methods and a variety of operation parameters. The instrument data system can consist of a simple chart recorder, or sophisticated integrator, or computer-based data acquisition system.

The analytical method of capillary electrophoresis is fundamentally the same as the chromatographic techniques of gas chromatography (GC) and high performance liquid chromatography (HPLC) as it relates to data analysis. Chromatographic qualitative and quantitative techniques involve the interpretation of data in two-dimensional form. Qualitative information, e.g., migration times, is represented as the distance the chart paper travels as a function of time. The quantitative information, peak heights and area, is displayed as the deflection of the recorder pen perpendicular to the time axis, as a function of the voltage output from the detector (Fig. 2). The purpose of these methods is to identify the components in the sample mixture and the amounts of

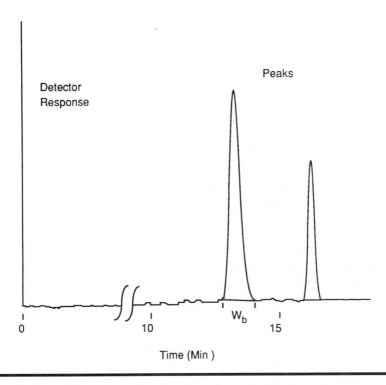

Figure 2 Quantitative presentation of peak area from a typical electropherogram.

the components. Identification can be achieved by matching migration times to known chemical standards. The next step is to compare the relative detector response, e.g., peak area, of the unknown peaks to those of calibration standards.

III. Qualitative Analysis in Capillary Electrophoresis

The terms *retention time* and *migration time* are generally used to describe peak data generated from a chromatographic and electrophoretic system, respectively. In this context the qualitative method involves matching the migration time of a known chemical substance with that of an unknown. Moreover, the accuracy of the measurement must be within the coefficient of variance of the intrinsic precision of the instrumental system. Even when migration times are matched in a particular case, the probability that the migration time for the unknown is also within the same time frame for that of a different molecular species is still significant. To increase the statistical reliability for the identification of an unknown, multiple analyses are necessary under different conditions, i.e., temperature, buffer composition, and pH. If the identification of substances in the sample is limited to only a relatively few potential species (e.g., under 10), then the reliability of the identification of any one of these components is greatly enhanced.

A. Qualitative Precision and Accuracy

The most important elements in good qualitative and quantitative analysis are the precision and accuracy of the measurement device. In this discussion our concern is directed to unique aspects of capillary electrophoresis instrumentation. The terms *precision* and *accuracy* as they pertain to analytical instrumentation are a measure of the random and the systematic errors of the system, respectively. Random error in CE is typically the result of imprecise aspects of the injection system, the temperature-regulation system, and random error due to detector noise, etc. Systematic error in CE is usually the result of improper (or inaccurate) system calibration and integration. The subsequent discussion will deal predominantly with aspects of qualitative precision.

For CE to be a useful analytical technique, migration time reproducibility or electrophoretic mobility reproducibility must be comparable to those of other established techniques, such as GC and HPLC. The optimization of reproducibility can be accomplished by analysis of peak shape over a wide range of sample concentration. A typical value for peak efficiency in CE for small molecules (e.g., less than 2000 daltons) is 250,000 theoretical plates/m. This corresponds to a peak width at the base of 0.085 min for a peak with a migration time of 10 min. If we visualize the peak as a Gaussian distribution of solute concentration versus time in which 2σ represents a window where 66% of the time the same solute should elute under ideal circumstances (Dyson, 1990), then the value 2σ equals 0.043 min (assuming that the width at base = 4σ). This is a relative value of 0.4% of the migration time of the peak. If we were to use this value as the migration-time window (i.e., the same identity as a standard), then a precise instrument should be able to place the elution time of the solute within this window 95% of the time. If an instrument provides migration time reproducibility measured in terms of relative standard deviation [% RSD = standard deviation(σ) of the peak migration time/average migration time $\times$ 100] of less than 0.1%, then by statistical definition (e.g., 4σ represents 95% of the area under a probability distribution), it should fit these criteria. Using this reasoning, CE instrumentation (integrator or data system included) should provide a qualitative measure of less than 0.1% RSD.

In CE the electroosmotic flow in the capillary tube is the predominant factor that affects the variability of migration time (Albin *et al.*, 1991; Moring *et al.*, 1990). If one normalizes for electroosmotic flow variation with the use of an internal standard and calculates an electrophoretic mobility, then the qualitative precision improves roughly by an order of magnitude (Table I). See Chapter 4 for a more detailed discussion of electrophoretic mobility.

Table I **Reproducibility of Mobility and Migration Time**

Solute	Migration time (% RSD)	Mobility (% RSD)
Dynorphin 1–13	0.48	0.030
Dynorphin 3–13	0.57	0.040
β-Lactoglobulin B	0.62	0.037

B. Capillary-Wall Chemistry

The condition of the inside of the capillary tube has a profound effect on the electroosmotic flow in CE and thus the qualitative precision. If the capillary surface is not properly conditioned and the sample and buffer conditions are not properly chosen, then poor migration time precision will result.

Most capillaries used for CE are fused silica, a very pure form of amorphous silicon dioxide. Capillaries composed of this material have many outstanding qualities, which include precise dimensions, a very high dielectric constant, low electrical conductivity, high thermal conductivity, chemical inertness, mechanical strength, and high optical transmission to a wide spectrum of light (190 to 900 nm). The inside surface of the capillary is rich in silanol (—Si—OH) functional groups, which are very acidic [estimated negative log of ionization constant (pK) of approximately 1.5 (Ginn, 1970)]. At most pH values, the surface is composed of predominantly deprotonated silicic acid groups (—Si—O$^-$). The silicic acid groups are responsible for the phenomenon of electroendoosmosis. These groups also exist as abundant sites for coulombic (charge) interaction with oppositely charged solute molecules, and thus can have both an adverse effect on the quality of the separation as well as solute detectability.

In Chapter 1 the relationship between electroendoosmotic flow and buffer pH in fused silica capillaries is discussed. The flow in the capillary is subject to a dramatic change from pH 4.5 to 6.0. Therefore the use of buffers in the pH range of 4.5 to 6.0 may adversely affect qualitative precision. Generally speaking, the greater the ionic strength of a buffer, the better its buffering capacity. Additionally, the effective charge on the wall is decreased with increased ionic strength, and consequently, the electroosmotic flow is reduced (Altria and Simpson, 1986). Using a buffer of moderately high ionic concentration (30 mM) provides satisfactory migration time reproducibility. There is, however, a disadvantage in using a high concentration buffer (>150 mM or a conductivity of 15 mmho/cm), because they can lead to peak broadening and distortion due to high Joule heating inside the capillary (Moring, unpublished data, 1989).

Variations in migration times can be best controlled by the reduction or elimination of electroosmotic flow. An early approach to eliminating electroosmotic flow was to covalently block the silanol sites with a polar organosilane ligand (Hjertén, 1985;

McCormick, 1988) or to utilize wall-coated capillaries, such as those used in gas chromatography. In the former case, the bonded coating has proved to be effective for only limited periods owing to the reversible hydrolysis of the silyl oxygen bond, particularly at higher pH. In the latter case, the wall-coated GC capillaries proved to be very hydrophobic, thus introducing the phenomenon of solute partitioning into the electrophoretic process, with resulting deleterious effects on resolution and efficiency. Recent advances in the chemistry of wall coating promises more stable and longer-lived coatings (Altria and Simpson, 1986; see Chapter 10). However, a more versatile approach to controlling electroosmotic flow and migration time stability, as well as the elimination of coulombic adsorption, is to use a "charge-reversed" polymeric coating. Recent work with high-molecular-weight polyamines has demonstrated that the negative charge on the surface of fused silica, and hence, the electroendoosmosis, can be carefully controlled by the amount of polyamine adsorbed (Wiktorowicz and Colburn, 1990). This treatment has been shown to be very effective in eliminating much of the charge-induced adsorption of multiply charged biopolymers, such as proteins and peptides. Furthermore, the coulombic interactions of strongly basic proteins can be eliminated by the application of thicker polymer films and the formation of a positively charged layer at the surface of the capillary. The result of this approach is a reversal of the electroosmotic flow and the repulsion of cationic solutes from the positively charged surface. The most important benefit that covalent or polymeric coatings provide is the ability to use buffers in the pH range from 4.5 to 8.0 without severe solute–wall interaction.

C. Data Reporting

The reporting of analytical data in CE uses essentially the same format that is common to HPLC and GC. In CE, however, a distinction is made concerning retention time and migration time. Strictly speaking, migration time best describes the position in the time domain for solutes in free solution electrophoresis, and retention time is the best description of separation time in micellar electrokinetic chromatography. Since the reproducibility of migration times, or as retention time in the latter case, are heavily influenced by electroendoosmotic flow variations, the more precise expression of the qualitative data is in terms of electrophoretic mobility or relative retention time, respectively. In order for either of these reporting methods to be useful, the sample must be mixed or coinjected with a "neutral" marker, which provides a response that is unaffected by the electric field or chromatographic partitioning. In Chapter 4, the calculation of mobility is thoroughly discussed. The electrophoretic mobility, μ, is calculated by the expression

$$\mu = \frac{L_d L_t}{V} \left(\frac{1}{t_m} - \frac{1}{t_{eo}} \right) \tag{1}$$

where L_t and L_d represent the capillary total length and the length to the detector respectively, V is the applied voltage, and t_m and t_{eo} represent the migration time of the analyte and the neutral marker, respectively. This equation can be expanded to include the use of a internal mobility standard for conditions in which there is little or no electroosmotic flow. In this case t_{std} is substituted for t_{eo}, and μ_{std} is added to the main term of the equation (Moring *et al.,* 1990). For micellar electrokinetic capillary chromatography (MECC), the relative retention time can be expressed simply as the ratio of the retention time of the solute (t_m) to the internal standard (t_{marker}), where the relative retention time is defined as t_m/t_{marker}.

IV. Quantitative Analysis

The quantitative information from an electropherogram is reported as the area under a peak, Fig. 2. With most detectors that use linearizing electronics, peak area is directly proportional to the sample mass or concentration for a fixed volume of sample. In some cases the peak height can be more accurately correlated to a physical property of a molecular substance. Such is the case with UV-absorbance detection, where concentration can be directly determined from the molar extinction coefficient. The use of peak area, however, is the preferred method for quantitative analysis because it provides a greater linear response with respect to sample amount than peak height (Fig. 9). See Section IV,C for a more detailed discussion of the use of peak height versus area.

A. Sampling Techniques

The manner in which the sample is introduced into the capillary has very important implications for quantitative methodology. Area (or height) reproducibility in CE is predominantly a function of the precision of the injection technique. In general, there are two modes of sample introduction. The first is hydrodynamic injection, which involves the forcing of a small slug of liquid sample into the end of the capillary by applying a pressure difference across the capillary. The second, electrokinetic injection, involves electrophoretically introducing a small volume into the capillary. The amount of sample introduced into the capillary during the electrokinetic injection is both a function of the volume of sample introduced into the end of the capillary owing to electroosmotic flow and the migration rate of sample ions.

A more thorough examination of the differences between hydrodynamic and electrokinetic injection reveals that they both have advantages for different types of applications. Generally speaking, hydrodynamic sampling provides the most precise injection of a sample because it is based strictly on volume loading of the sample. The volume introduced into the end of the capillary, V_a, is defined by the Poiseuille equation and is a function the capillary length (L_t), the inner diameter (d), the sample viscosity (η), and the pressure difference (ΔP) across the ends of the capillary for a given unit of time (t):

$$V_a = \frac{\Delta P \pi d^4 t}{128 \eta L_t} \tag{2}$$

Hydrodynamic injection with various commercial CE instruments utilizes different mechanisms: head-space pressurization injection (Beckman Instruments, Inc., Fullerton, CA); vacuum injection, in which a negative pressure is applied to the opposite end of the capillary (Applied Biosystems, Inc., Foster City, California, and Spectraphysics, Inc., San Jose, CA); or gravimetric injection, in which the sample and the capillary are elevated with respect to its opposite end, and introduction of the sample is achieved by siphoning action (Dionex, Inc., Sunnyvale, CA). Hydrodynamic injection is the preferred injection method for free solution and MECC applications when sample concentration is well within the sensitivity limits of the detector. See Chapter 1 for a discussion of the impact of injection mode on separation efficiency.

Electrokinetic injection has its greatest utility with the use of gel-filled capillaries where volumetric loading of sample is impossible. The technique also has great potential for sample concentration due to electrophoretic stacking. This has particular importance when sample concentrations are below the concentration limit of detection. In cases where the electroosmotic flow is present or impeded due to the presence of a gel ($\mu_{eo} = 0$), the amount of sample material (Q) introduced is a function of the electric field strength (E), the injection time (t), the mobility of sample molecules (μ_{ep}), the electroosmotic mobility (μ_{eo}), conductivity of the sample (λ_s), and the conductivity of the sample buffer electrolyte (λ_b). The equation published by Rose and Jorgenson (1988) can be used to describe the relationship between the parameters affecting this process,

$$Q = \frac{\pi r^2 \, c_s \, (\mu_{ep} + \mu_{eo}) \, Et \, \lambda_b}{\lambda_s} \tag{3}$$

An important consequence of the behavior of molecules with different mobilities is that they will enter the capillary at differing rates, such that solutes with the highest mobility will be loaded in greater amounts. This can be a particular problem when the sample is composed of low mobility solutes that are near the concentration limit of detection.

Sample can be concentrated during injection when the conductivity of the sample solution is lower than that of the electrophoretic buffer solution. Under these conditions the electric field in the sample medium is much greater than that in the capillary. The rapidly moving sample ions migrate until they contact the run buffer zone of higher conductivity (lower field strength). The sample ions "stack" until the solute-zone conductivity approaches the conductivity of the run buffer. In this way the sample zone can be concentrated into a very narrow band, providing very sharp solute peaks. Figure 3 illustrates this phenomenon with comparison of electrokinetic injections with differing sample conductivities. Sample stacking can be achieved using either electrokinetic or hydrodynamic injection modes.

The effectiveness of electrokinetic stacking is a function of the pH of the buffer and the resulting high electroosmotic flow. Generally speaking, the ability to concentrate sample using stacking at buffer pH values above 5.0 is approximately one fifth of that at lower pH values (Moring, unpublished data, 1989). With buffer pH above 5.0, the electroosmotic flow is high, causing sample buffer to be drawn into the capillary. Stacking occurs (the solute molecules slow down) when they encounter the zone of

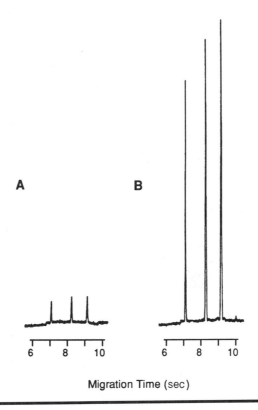

A   B

6   8   10   6   8   10

Migration Time (sec)

Figure 3   The effect of relative buffer conductivity of sample versus electrophoresis buffers on the solute mass injected with electrokinetic sampling. (A) 10-sec electrokinetic injection at 5 kV, sample in electrophoresis run buffer; (B) 10-sec electrokinetic injection at 5 kV, sample in water. Samples: heptapeptide RKRSRKE, dynorphin 3–13 and dynorphin 1–13; electrophoresis run buffer, 20 mM sodium phosphate, pH 2.5 from Moring (unpublished data, 1989).

higher buffer conductivity. If the conductivity of the sample zone is very low, then the electric field becomes the greatest across the sample zone, and significantly diminishes the field along the rest of the capillary. When sample volume exceeds approximately 4% of the capillary volume, this effect results in a significant slowing of migration times of solute molecules, and consequently poor migration-time precision and peak broadening. It is important to note that essentially the same limitation applies when attempting to utilize sample stacking with hydrodynamic injection. In general, the sample zone should not exceed 2% of the total usable capillary volume.

The linearity of electrokinetic injection is also affected by sample conductivity. Figure 4 shows the relationship between the amount of sample detected and injection duration for vacuum or voltage injection. In this example, the vacuum-injection technique provides a linear relationship ($R = 0.9987$ and 0.9945 for two model proteins, curves A and B, respectively). A nonlinear electrokinetic injection ($R = 0.9021$ and

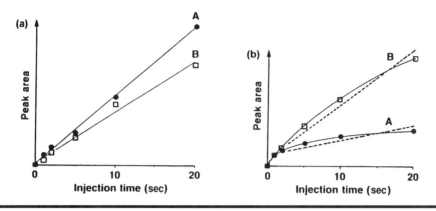

Figure 4 Linearity of (a) vacuum and (b) electrokinetic injection as a function of injection time. (a) Injection, vacuum: 5.0 in of Hg; run buffer: 20 mM bicine-TEA, pH 8.5; run voltage: 25 kV; current 4 μA. (b) Injection voltage: 5 kV; other conditions same as for (a). Curves A, β-lactoglobulin; B, horse heart myoglobin. Samples were dissolved in water (100 μg/ml). Reprinted with permission from Moring *et al.* (1990).

0.9906 for curves A and B, respectively) is observed as a changing rate of sample loading with time. This has been attributed to a continuously changing conductivity in the sample zone (Moring *et al.*, 1990). One can minimize the variation in sample loading, and thus the quantitative precision, by matching the conductivity of the sample and run buffer. This is done, however, at the expense of sample concentration due to stacking.

B. Precision and Accuracy

Vacuum injection has been shown to offer the greatest reproducibility, due to the fact that better control over physical parameters can be maintained. Electrokinetic injection, which might appear at first glance to be under the same degree of control, is influenced by many factors, including sample type, sample solution conductivity, as well as the surface chemistry of the capillary. Table II compares the migration time and area reproducibilities of each type of injection.

As can be seen from the reproducibility data for electrokinetic injection, a higher percentage of relative standard deviations (> 3.0% RSD) can be attributed to the low conductivity of the sample buffer (e.g., water).

C. Quantitative Methods

The task of qualitative and quantitative determination requires instrument calibration with a series of standards having a known concentration range. Proper calibration requires the preparation of standards that are in a solvent or buffer compatible with the separation method and whose concentration is accurately measured. The electrophoretic system must be sufficiently equilibrated or stabilized as to permit repro-

Table II   Quantitative Reproducibility with Hydrodynamic and
Electrokinetic Injection Modes[a]

| Solute | Sample buffer | Area (% RSD[b]) | |
		Electrokinetic	Vacuum
Dynorphin 3–13	20 mM^c	1.8	1.3
	Water	5.4	0.7
Myoglobin	20 mM^d	2.9	3.1
	Water	3.1	2.9
	20mM^d	1.5	0.9
	Water	6.5	2.1
β-Lactoglobulin B	20 mM^d	2.1	1.6
	Water	3.1	2.0

[a] From Moring *et al.* (1990).
[b] Data based on seven consecutive injections.
[c] Sodium citrate buffer, pH 2.5.
[d] Bicine-Triethylamine (TEA), pH 8.5.

ducible analyses. Primarily this requires that the inside surface of the capillary, the buffer ionic concentration, and pH are in a state of equilibrium to allow a consistent electroosmotic flow rate. A constant electrical current is a good indication of equilibrium. After the calibration standard is run and migration time and area data are reported, migration time windows and response factors then can be determined. Modern integrators and data systems provide menu-driven methods to do this.

The calibration of migration time is carried out by establishing a time window in which the calibrated peak is expected to be detected. This is done by choosing a plus/minus time increment on either side of the peak apex. The increment can be based on a percentage of the peak migration time or an absolute time. If a detected peak in a subsequent analysis falls within the window defined by the calibration, then the peak is identified as the "same" chemical type (or molecular substance) as the standard. If the peak falls outside the window, then it is considered an unknown.

Retention (migration) time drift is a phenomenon characteristic of all separation systems and is not exclusive to CE. A means to compensate for this is provided by the use of a reference component or standard. This is defined as one or more components that are always present in the sample to be analyzed. The purpose of the reference standard is to allow adjustment of a proportional shift in the migration times of all components in the separation. For example, if a reference peak drifts from the center of the calibrated window by approximately 0.5%, then the integrator peak identification algorithm shifts the time windows for all the nonreference peaks proportionally (Fig. 5). Since the identification of the reference peak is crucial, the reference recognition windows are typically 50% larger than the nonreference windows. The ability to compensate for migration time drift with the use of a reference peak often is

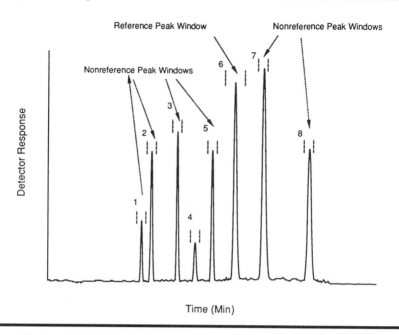

Figure 5   Peak recognition windows used for qualitative identification and calibration methods.

essential for the success of the quantitative method and therefore should always be used.

The calibration of peak response is a straightforward process of determining the ratio of response (area or height) with respect to the sample mass injected or its concentration for a specific injection volume. This amount (mass or concentration)/area (height) ratio is defined as the response factor:

$$RF = Amount/Area \qquad (4)$$

The RF represents the slope of the presumed linear response of the detection system. In the case of nonlinear response, it represents the slope on any given point on the resulting response curve.

Four methods are used for the quantitative determination of a sample component, based on its peak area or height. These are the percent normalization method, the external standard method, the internal standard method, and the standard addition method.

The percent normalization method requires that all the components that result from a given separation must be calibrated, and response factors, generated. This method has meaning only when every possible component in a sample is represented in the system calibration. If an unidentified substance appears as a result of the separation, then the resulting data are of little value.

With this method, the following formula is used in the calculation of the percent normalization of an individual component (i):

$$NORM\% = \frac{AREA_i \times RF_i \times 100}{\sum_{i=1,n} (AREA_i \times RF_i)} \tag{5}$$

This method is used for the analysis of well-characterized mixtures in which a portion or all of the calibrated components will be analyzed.

The external standard method requires calibration of only the sample constituents of interest. If one is interested in the determination of only one or more sample components, then this method should be used. The following formula is applied for the calculation of the external standard method:

$$AMOUNT = AREA_i \times RF_i \tag{6}$$

This method, however, requires precise measurement of the sample-injection volume.

The internal standard method is similar to the external standard method, in that only the sample constituents of interest need be calibrated; however, in this case, the standard is mixed with the sample. The difference with this method is that sample volume is no longer a critical parameter because the use of an internal standard normalizes for the change in response due to variations in the sample volume. The criteria for the selection of an internal standard include the availability of a commercial standard that is chemically similar (e.g., similar extinction coefficient, and solubility) to the components present in the sample and is chemically stable in the buffer conditions used for the separation. An internal standard also should never occur naturally in the sample, must be present in the same concentration range as the sample, and should elute away from the peaks of interest. Finally, the internal standard must be calibrated as any other standard and must be accurately added to each sample. The calculation used for the internal standard determination is

$$AMOUNT = \frac{AREA_i \times RF_i \times AMOUNT\,(ISTD)_i}{AREA\,(ISTD)_i \times RF\,(ISTD)_i} \tag{7}$$

The standard addition method is a variation of the external standard or internal standard method, and is used when the sample contains a relatively large amount of interfering material. Frequently the sample matrix can influence the detector response because of quenching, particularly in the case of fluorescence detection. The use of the method involves the analysis of a given sample, first with the addition or spiking of a known amount of calibration standard, followed by the analysis of the original sample. The response factor for the calibration standard is determined by the calculation of the difference of the areas of the spiked sample and the original sample. The final quantitation is accomplished using the external or internal standard calculations.

The use and description of these methods may vary from one integrator or computer data system manufacturer to another; therefore, it is recommended that beginners in this methodology consult their instrument manuals for specific details.

V. Integration and Data-Reporting Devices

A. Fundamentals

Modern integration data systems range from relatively inexpensive dedicated micro-processor-controlled reporting integrators to minicomputers with analog-to-digital signal conversion devices and versatile software packages. In either case, an analog detector signal from the analytical instrument is converted into a digital signal before any qualitative or quantitative information can be extracted.

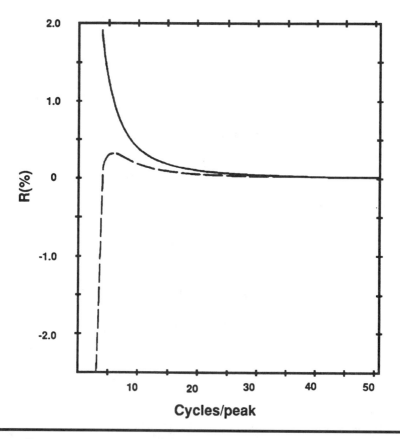

Figure 6   The calculated average sampling error as a function of the number of sample cycles per peak. The solid line represents calculations based on Gaussian peaks, whereas the broken line represents a typical tailing peak. Reprinted with permission from Matthews and Hayes (1976).

The rate at which the analog signal is digitized is entirely dependent on the type of electrophoretic signal to be processed. A general rule used by many manufacturers or software developers of chromatographic data systems is that a peak should be represented by a minimum of 10 data points in order to represent the area within 0.5% accuracy (Fig. 6). For a typical HPLC peak having a width at the base of between 10 and 60 sec, the minimum data sampling rate is on the order of 1 to 0.2 data points per sec (or 1 to 0.2 Hz). High efficiency CE separations, in which the peak widths range between 0.2 and 5 sec require a data sampling rate of 2 to 50 points per sec. The computer algorithms process this data by the analysis of the change in response above a certain threshold value or as a rate of change in the detector response. The onset of the chromatographic peak is determined once a given threshold or slope sensitivity is reached (Fig. 7). In a typical example of an integration algorithm, the beginning of the peak is determined when the signal rises more than two data points beyond a preset slope threshold. The actual beginning of the peak (first baseline point) is then extrapolated back three data points. This is done to ensure that the resulting baseline generated between the beginning and end boundary point closely approximates a tangent to the baseline. The algorithm continues from the baseline point by summing data slices as the signal reaches an apex and declines to within the same or a user-determined height or slope threshold. The peak apex is determined either as the time increment with the largest response value or by interpolation of a true apex by using a Gaussian curve fit or with a moments analysis of the area under the curve (Dyson, 1990). The

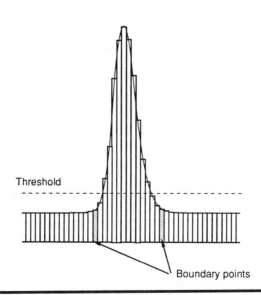

Figure 7 A digital representation of an electrophoretic peak.

area under the peak is calculated from the difference of the total summed response increments from the start to the end peak times and the summed response below a base line intersecting the beginning and ending points of the peak (Fig. 8).

As is shown in Fig. 9, peak area provides better quantitative information than peak height. The poor linearity exhibited by the use of peak height is attributable to solute overload that results in peak broadening and the concomitant plateauing of peak height. The use of peak area compensates for this response behavior. If all the components in a given sample are completely resolved, then it is relatively straightforward to do a quantitative extrapolation from a standard curve. However, when peaks are merged or exist on top of an unstable or changing baseline, then accurate determination of the area or height can be very difficult. The various methods that are used to assign baselines in these cases are discussed thoroughly by Dyson (1990) and Katz (1987).

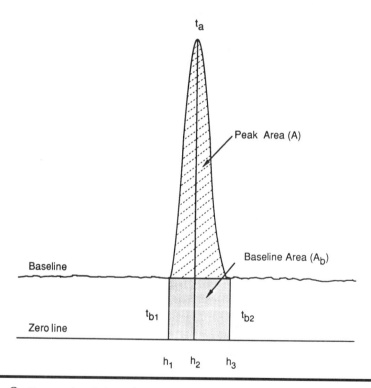

Figure 8 Processed peak data used for the calculation of the true area under a peak. Values t_{b1}, t_{b2}, t_a, h_1, h_2, h_3, A_b, and A are the cardinal points. The begin-peak time increment is t_{b1}, h_1 is the begin-peak height increment, h_2 is the peak height apex, t_{b2} is the end-peak time increment, h_3 is the end-peak height increment, and t_a is the peak apex time increment.

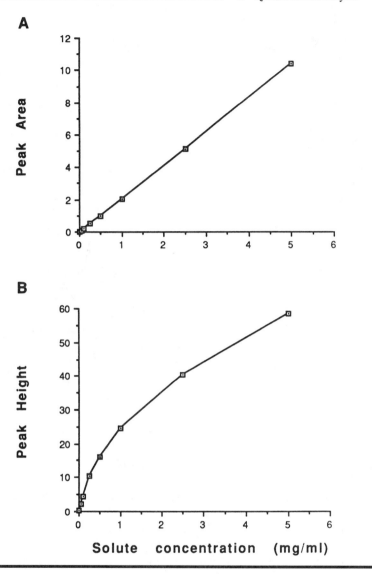

Figure 9 Comparison of UV-absorbance detector response for dynorphin 1–13 at 200 nm with measurement of peak area (A) and peak height (B).

B. Optimization of Integration

The optimization of signal integration in capillary electrophoresis involves overcoming the relatively high noise (compared to HPLC) that is characteristic of using absorbance detection. The increased noise is a consequence of the short path lengths that

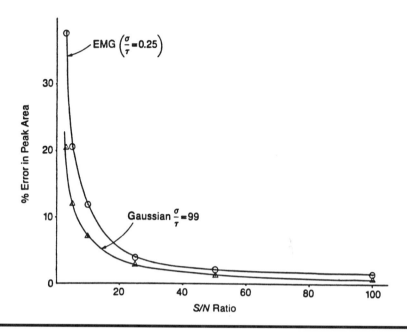

Figure 10 Accuracy of area measurement as a function of signal-to-noise ratio. Peaks are 10 sec at base with a sampling interval of 1 sec. Reproduced with permission from Rossi (1988).

are dictated by the 50- to 100-μm internal diameters typical of capillaries used in CE. Proper suppression of noise can do much to maximize the accuracy of area integration. Figure 10 illustrates the relationship between integration accuracy and noise level. This effect is most often observed when area reproducibility is measured with repetitive analyses. For sample concentrations that provide large signal-to-noise ratios, little ambiguity is experienced with the assignment of the baselines during integration. As the signal-to-noise decreases, the relative height of the changing baseline compared to the peak height increases, resulting in different baseline assignments from analysis to analysis. It has been suggested that the use of peak height provides better accuracy and precision for cases of low S : N values, in which response problems due to sample overload are negligible (Dyson, 1990).

The precision of a given analysis can be optimized with the proper setting of detector signal filtering (time constant or rise time) and bunching factors (peak width parameter). Detector signal filtering provides a means of dampening detector noise by slowing the rate at which the signal changes with manipulation of an electronic time constant. If too high a filtering rate is used, the sensitivity (signal-to-noise) is significantly improved, but at the expense of efficiency and resolution, due to peak distortion (Fig. 11). Quantitative sampling error due to excessive noise can be best dealt with by the proper use of filtering of the detector signal. Most absorbance and fluorescence detectors are equipped with electronic filtering capabilities, such as the rise

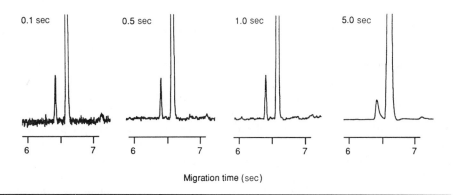

Figure 11 Effect of rise time on the baseline noise, peak shape, and resolution for different rise-time values.

time, or a time constant. In the former case, the rise time is defined as the minimum time in which the detector responds to a change in signal from 10% of the signal maximum to 90% of its maximum. Typical rise times range from 0.02 sec to 5 sec. If a value of 0.1 sec or less is used, then the detector is able to respond very rapidly to changes in sample zone concentration. This rise time would be appropriate for sharp electrophoretic peaks with an elution time on the average of 0.6 sec from baseline inflection to baseline inflection. Peak elutions of this speed typically are the result of very high efficiency separations on the order of one million or more theoretical plates or separations that take place on short capillaries (< 50 cm) and at high voltages (> 20 kV). Under these conditions, the resultant noise levels observed will be high, and thus there is a trade-off between sensitivity and efficiency (Table III). In CE, typical efficiencies are less than 500,000 plates/m with peak width (at base) ranging from 2 to 5 or more sec. In the case of a 5-sec peak width, the optimal rise time (RT_{opt} = $PW_{base}/6$) would be 1 sec. The selection of this value in effect reduces the detector noise by a factor of three to four times that when a 0.1 rise time value is used

Table III Comparisons of Figures of Merit for Various Detector Rise Times

Rise time (sec)	Signal noise	Peak width (sec)	Resolution	Theoretical plates	Area	Height
0.1	5.6	4.40	3.76	194,862	20,958	21.0
0.5	8.4	4.32	3.73	203,569	17,621	18.5
1.0	13.3	4.56	3.47	182,818	20,188	20.0
2.0	15.0	6.36	2.64	94,095	20,608	15.1
5.0	18.0	10.56	1.49	30,626	21,115	9.0

(Fig. 11). The use of rise times exceeding one half the peak width typically results in peak distortion, loss of resolution, and a loss of analytical precision.

The integration bunching factor is a function that averages digitized data in order to suppress baseline noise. Similarly, the consequence of excessive bunching is to distort the peak shape, which results in a decrease in resolution. As a general rule, the bunching factor used should not increase the peak width more than 10% from that which results from a data sampling rate corresponding to 10 points across the base of a peak. Once the detector signal filtering and integration-bunching factors are optimized, then accuracy will be dependent on the quality of the commercial integrator or data system algorithm used. Generally speaking, the quality of the integrator or data system will be a function of its ability to sample rapidly for very narrow peaks and the sophistication of the algorithm used, with regard to its ability to accurately determine peak cardinal points.

VI. Detector Response Characteristics of Common Capillary Electrophoresis Detectors

The types of detectors that have been used in capillary electrophoresis include UV-visible absorbance, direct and indirect fluorescence, electrolytic conductivity, refractive index, electrochemical, radioactivity, and mass spectrometry detectors. To date only absorbance and fluorescence detection are available in commercial instrumentation. These detection systems have their own special response characteristics, and some discussion of advantages as well as limitations of the latter two are in order. See Chapter 2 for a complete discussion of detection in CE.

The UV-absorbance detection mode, which has been the most versatile and easy to use in HPLC, has all the same advantages in capillary electrophoresis. Since the flow or separation channel is typically a 25- to 100-μm i.d. capillary tube, the use of conventional LC flow cells with the small sample volumes characteristic of CE (10 to 50 nl) would introduce excessive sample peak dilution with disastrous losses in resolution as well as sensitivity. The fused silica capillary tube is a good optical medium for direct absorbance detection when the protective plastic coating has been removed. The UV cut-off of fused silica is approximately 190 nm. In CE the capillary has been used as the optical flow cell in two ways: the first, by passing light through the capillary radially, and second, by focusing the light along a section of the capillary axially. The first approach, which has been exploited by all of the current commercial CE instruments, will be discussed here. It involves utilizing the internal diameter of the capillary as the flow cell, and therefore limits the path length to a fraction of the internal diameter of the capillary. The area illuminated in this case is very small (e.g., a length of 50 to 2000 μm i.d. of the capillary), and therefore light energy passing through to a photometer is an order of magnitude less than that with conventional HPLC detectors. The consequence of this is a greater operating noise level than is seen in HPLC.

With all detectors, the cell path length and noise level defines the limit of detection (LOD) in any optical detection system. The LOD is typically expressed as the signal level two times that of the noise (Katz, 1987; Snyder and Kirkland, 1979). The linear dynamic range of a detector can be defined as linear range of detector response from the LOD to the point where the response begins to plateau due to detector saturation or solute mass overload of the system. In CE this is observed as peak perturbation or fronting. The dynamic range with absorbance detection is unique for CE, due to the fact that mass overloading occurs in the capillary before the absorbance response begins to saturate (Fig. 12A). With most CE absorbance detectors, the linear dynamic range is typically three or four orders of magnitude at best (Albin *et al.*, 1991; Moring *et al.*, 1990).

With fluorescence detection there are some advantages to be realized as a result of the short optical path length. These include the diminished effect of self-quenching phenomena and the operation of photometers near their dark current noise limits (Albin *et al.*, 1991; Guibault, 1990) (Fig. 12B). In fluorescence detection, the detector

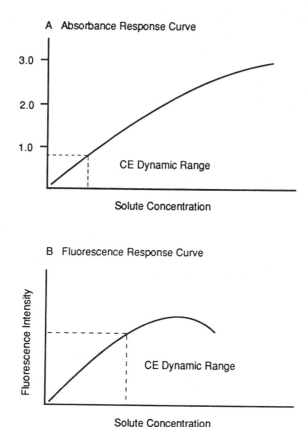

Figure 12   Typical response curves for photometric detectors.

noise behaves differently from that in absorbance detection. The noise level on the top of the signal increases as a function of the intensity of the signal. As a general rule, the S : N increases as the square root of the incident light intensity. In this case the concept of detector dynamic reserve is more appropriate than signal to noise. It is defined as the ratio of the noise generated on top of the detector signal and the signal level itself.

References

Albin, M., Weinberger, R., Sapp, E., and Moring, S. (1991). *Anal. Chem.* **63**, 417–422.

Altria, K. D., and Simpson, C. F. (1986). *Anal. Proc.* **23** (December), 453–454.

Dyson, N. (1990). *In* "Chromatographic Integration Methods" (Roger M. Smith, ed.), pp. 20–25. RSC Chromatography Monographs, Royal Society of Chemistry, Cambridge, England.

Ginn, M. E. (1970). *In* "Cationic Surfactants" (E. Jungermann, ed.), pp. 343–348. Surfactant Science Series Vol. 4, Marcel Dekker, New York.

Guibault, G. G. (1990). *In* "Practical Fluorescence" (G. Guibault, ed.), pp. 17–32. Marcel Dekker, New York.

Hjertén, S. (1985). *J. Chromatogr.* **347**, 191–198.

Katz, E. (1987). *In* "Quantitative Analysis Using Chromatographic Techniques" (R. Scott, C. Simpson, and K. Ogan, eds.), Wiley, New York.

Matthews, D. E., and Hayes, J. M. (1976). *Anal. Chem.* **48**(9), 1375–1382.

McCormick, R. M. (1988). *Anal. Chem.* **60**, 2322–2328.

Moring, S. E., Colburn, J. C., Grossman, P. D., and Laurer, H. H. (1990). *LC/GC* **8**(1), 34–46.

Rossi, D. T. (1988). *J. Chromatogr. Sci.* **26**, 100–105.

Rose, D. J., and Jorgenson, J. W. (1988). *Anal. Chem.* **60**, 642.

Snyder, L. R., and Kirkland, J. J. (1979). *In* "Introduction to Modern Liquid Chromatography." 2nd ed., pp. 127–130. Wiley, New York.

Wiktorowicz, J. E., and Colburn, J. C. (1990). *Electrophoresis* **11**, 769–773.

MODES

OF

CAPILLARY

ELECTROPHORESIS

CHAPTER

4

Free-Solution Capillary Electrophoresis

Paul D. Grossman

I. Introduction

As is the case for all electrophoretic separations, in free-solution capillary electrophoresis (FSCE), separations are achieved as a result of the unequal rate of migration of different solutes under the influence of an externally applied electric field.
However, FSCE differs from the other modes discussed in this book in one important
respect — there is no secondary influence, other than the structure of the analyte and
the solvent, contributing to the selectivity of the separation. There is no polymer network, superimposed pH gradient, or secondary phase.

Because the analyte is typically not chemically altered before the analysis, FSCE
is a direct analog to native gel or paper electrophoresis. However, because there is no
solid support, in the case of FSCE there are no *matrix* effects; thus, FSCE is well
suited to detect subtle changes in the structure of native macromolecules.

In this chapter, we review the factors that determine the electrophoretic mobility
of a solute as measured by FSCE and briefly discuss the dispersive phenomena that,
along with differences in the electrophoretic mobility, dictate the resolving power of
the technique. In addition, we discuss practical aspects of FSCE measurements, and
some experimental studies that use model systems to explore the various factors determining selectivity in FSCE.

II. Electrophoretic Mobility

A. Definition of the Electrophoretic Mobility

When a charged particle is placed in an external electric field, E, it experiences a force, F_e, which is equal to the product of its net charge q, and the electrical field strength, E,

$$F_e = q \cdot E \tag{1}$$

In addition to the electrical force acting on the particle, once it begins to move, the particle experiences a drag force in the direction opposite to its direction of motion. This drag force, F_d, is proportional to the particle velocity,

$$F_d = f \cdot v \tag{2}$$

where the proportionality constant f is called the translational friction coefficient. For example, for a spherical particle, f is given by Stokes' law as

$$f = 6\pi\eta R \tag{3}$$

where η is the viscosity of the surrounding medium, and R is the apparent hydrodynamic radius of the particle. Thus, the equation describing the translational motion of a particle under the influence of an electric field is

$$m\left(\frac{dv}{dt}\right) = qE - fv \tag{4}$$

where m is the mass of the particle.

Except for a brief transient when the electrical force is first applied, the electrostatic force is exactly counterbalanced by the drag force, resulting in a steady state velocity, v_{ss}. Typically, for macromolecules, this transient is very short, on the order of 10^{-11} sec. By equating Eqs. (1) and (2), the resulting steady-state velocity can be related to the charge and frictional properties of the particle by the simple relationship

$$v_{ss} = \frac{qE}{f} \tag{5}$$

Furthermore, the electrophoretic mobility, μ, of a particle is defined as the steady-state velocity per unit field strength, or, from Eq. (5),

$$\mu = \frac{q}{f} \tag{6}$$

Clearly, differences in the electrophoretic mobility of molecules can arise as a result of differences in frictional properties, i.e., size or shape, or as a result of differences in the net charge on the molecule. These differences in the properties of molecules form the basis for the selectivity in FSCE separations. We shall discuss the relationship between the molecular structure and the electrophoretic mobility of an analyte in the following subsections.

B. Relationship between Mobility and Molecular Size

The way in which the frictional coefficient, f, is related to the size of a molecule depends on what model is used to describe the conformation of the molecule and on the nature of the solvent. In this section we discuss the case of a solute migrating through free solution.

If, in free solution, the migrating molecule is modeled as a solid sphere, the relationship between the translational frictional coefficient and molecular size would be straightforward. From Stokes' law we know that

$$f = 6\pi\eta R \tag{7}$$

Furthermore, for a solid sphere, the radius and the mass, m, are related by

$$m = \rho_p \left(\frac{4}{3}\pi R^3 \right) \tag{8}$$

where ρ_p is the density of the particle. Thus,

$$R \sim m^{1/3} \tag{9}$$

where the "~" symbol indicates proportionality. Therefore, from Eqs. (6) and (9)

$$f \sim m^{1/3} \tag{10}$$

If, rather than being a solid sphere, the solute behaves as a loose coil or a rod, we can no longer use Eq. (8) to directly relate the molecular mass to an apparent hydrodynamic radius and thus to a value for f. However, the relationship between mass and the translational friction coefficient has been established for a number of practically important molecular conformations. Some of these are given in Table I. Table I clearly shows that the way in which electrophoretic mobility in free solution is related to molecular size is strongly dependent on the model chosen to represent the conformation of the solute.

Table I   **Proportionality Relationship between Frictional Coefficient and Molecular Weight**[a]

Molecular model	Proportionality relationship
Solid sphere	$f \sim (MW)^{1/3}$
Random coil — unperturbed chain	$f \sim (MW)^{0.5 \text{ to } 0.6}$
Long rod	$f \sim (MW)^{0.8}$
Wide thin disk	$f \sim (MW)^{2/3}$
Free-draining coil	$f \sim (MW)^{1.0}$

[a] Reproduced with permission from Cantor and Schimmel (1980).

As it turns out, for many practically important applications, DNA and sodium dodecyl sulfate (SDS)-treated protein applications in particular, separations based solely on differences in free-solution electrophoretic mobilities are not possible. This is a direct result of the relationship between f and the molecular size characteristic of these solutes. This can be demonstrated through a simple argument, using DNA as an example.

The structure of the DNA molecule is such that the total net charge on the molecule is directly proportional to its size; i.e., approximately two charges per base pair. Thus,

$$q \sim N \tag{11}$$

where N is the number of units (base pairs) in the DNA chain. In addition, since DNA exists in solution as a free-draining coil, it can be seen from Table I that

$$f \sim N \tag{12}$$

(A free-draining coil is one in which each of the units of the chain contributes equally to the overall drag of the chain.) Finally, by combining Eqs. (11) and (12) with the definition for electrophoretic mobility, Eq. (6), it can be seen that μ is no longer a function of molecular size, i.e.,

$$\mu = \frac{q}{f} \sim \frac{N}{N} = N^0 \tag{13}$$

where N^0 indicates that μ is constant with respect to changes in N. Thus, because both the charge and the frictional drag are proportional to molecular size, free-solution electrophoretic separations of DNA are impossible. This argument also applies to the case of SDS proteins in which charge is also directly proportional to N, and the polymer behaves as a free-draining coil. Therefore, in order to perform a separation of DNA fragments, or of any molecule with a constant ratio of net charge to translational friction coefficient, one must exploit an alternate separation mechanism. If, instead of allowing the DNA molecule to migrate in free solution, one forces it to travel through a porous polymer network, one can impart a size dependence to the electrophoretic mobility. The migration of chainlike molecules through a polymer network will be discussed in Chapter 8.

C. Effect of Buffer Ions on the Effective Charge and Mobility

1. Charge Screening

In an electrolyte solution, when a charged solute is moving, the effective charge of the solute, q, is less than the total charge, owing to the screening influence of the counterions in solution. This shielding is attributable to the presence of counterions, located within a stagnant layer immediately adjacent to the surface of the solute (see Fig. 1). The stagnant layer is caused by the viscous force acting between the solvent and the surface of the solute. The interface between the stagnant layer and the surrounding solution is called the surface of shear. The exact position of the surface of shear is not

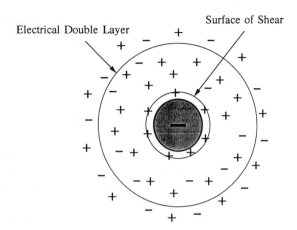

Figure 1 Schematic illustration of the ion atmosphere surrounding a negatively charged spherical particle.

known, but typically it is assumed that it has a thickness on the order of a few molecular diameters.

As might be expected, the degree of counterion shielding depends on the nature and concentration of the counterions as well as properties of the solvent and the analyte. The complete general description of this phenomenon is complex and beyond the scope of this chapter. However, in order to gain an appreciation of the key features of this effect, we shall explore a simplified case — that of a "small" spherical particle in a dilute electrolyte solution, where the thickness of the counterion atmosphere is large compared to the radius of the solute particle.

From elementary physics, we know that the electrostatic potential, ψ, at a distance r from the surface of a uniformly charged sphere of radius R and total charge q in a medium having a permittivity of ε is (Atkins, 1978)

$$\psi(r) = \frac{q}{4\pi\varepsilon r}; \qquad r > R \tag{14}$$

To find an expression for the effective charge of the same charged sphere in an electrolyte solution, we will first determine the effect of the ions on $\psi(r)$ and then compare the resulting expression with that of Eq. (14). This discussion will follow closely that of the electrical double layer presented in Chapter 1. However, in this case, spherical rather than planar coordinates will be used.

To determine the electrostatic potential in the vicinity of a charged spherical solute in an electrolyte solution, we must solve Poisson's equation in cylindrical coordinates,

$$\frac{1}{r^2}\frac{d}{dr}\left(r^2\frac{d\psi}{dr}\right) = \frac{-\rho^*}{\varepsilon} \tag{15}$$

where ρ^* is the charge density in the surrounding medium due to the ions in solution. In Eq. (15) it has been assumed that ρ^* is only a function of the r direction and not a function of the θ or ϕ directions. The charge density is related to the properties of the electrolyte solution and the electrostatic potential using the Debye-Hückel approximation,

$$\rho^*(r) = \sum_i \frac{z_i^2 e^2 n_{i0}}{kT} \psi \qquad (16)$$

where z_i is the valence of the ions, e is the electronic charge, n_{i0} is the number concentration of ions in the bulk solution, k is the Boltzmann constant, T is the absolute temperature, and the sum is taken over all ions in the solution. Because of the assumptions made in its derivation, Eq. (16) is valid only for potentials less than 25 mV. This constraint is easily satisfied for most macromolecular solutes in typical buffer systems. A derivation of Eq. (16) is provided in Chapter 1. Combining Eqs. (15) and (16) results in the expression

$$\frac{1}{r^2} \frac{d}{dr} \left(r^2 \frac{d\psi}{dr} \right) = -\kappa^2 \psi \qquad (17)$$

where the constant κ^2 is defined as

$$\kappa^2 = \frac{e^2}{\epsilon kT} \sum_i z_i^2 n_{i0} \qquad (18)$$

κ^2 is an important parameter related to the charge-shielding properties of an electrolyte solution. The "thickness" of the electrical double layer adjacent to a charged surface is given by κ^{-1} (see Chapter 1). Eq. (17) can be solved by recognizing it as a form of Bessel's equation and applying the boundary condition that as $r \rightarrow \infty$, $\psi \rightarrow 0$, resulting in the expression

$$\psi(r) = B r^{-1/2} J_{-1/2}(\kappa r) \qquad (19)$$

where B is a constant yet to be determined and $J_{-1/2}$ is the Bessel function of the first kind. However, when the particle radius is smaller than the thickness of the electrical double layer, i.e., for $\kappa r < 1$, it can be shown that

$$r^{-1/2} J_{-1/2}(\kappa r) \approx \frac{1}{r} \exp(-\kappa r) \qquad (20)$$

Thus, for $\kappa r < 1$, Eq. (19) can be expressed in terms of exponentials,

$$\psi(r) = \frac{B \exp(-\kappa r)}{r} \qquad (21)$$

Next, to determine the value of B, we must apply the boundary condition that at $r = R$, $\psi = \zeta$, where ζ is the zeta potential, the potential at the surface of shear. Note that this boundary condition implies that the surface of shear coincides with the surface of the particle. This is a result of the assumption that the thickness of the double layer is large compared to the size of the particle. Applying this boundary condition results in the expression (Rice and Nagasawa, 1961)

$$\psi = \frac{\zeta R}{r} \exp[\kappa(R - r)] \tag{22}$$

To express Eq. (22) in terms of charge rather than the zeta potential, we can recognize that the total charge on the solute, q, must equal the charge of the counterions in the surrounding solution. Thus,

$$q = \int_R^\infty \rho^*(r) \, dV \tag{23}$$

where dV indicates the integral is taken over the surrounding volume. Substituting Eq. (15) for ρ^* and recognizing that $dV = 4\pi r^2 dr$ results in the integral,

$$q = -4\pi\varepsilon \int_R^\infty \frac{1}{r^2} \frac{d}{dr} \left(r^2 \frac{d\psi}{dr} \right) r^2 \, dr \tag{24}$$

Performing the indicated integration results in an expression relating q to ζ, where

$$q = 4\pi\zeta\varepsilon R \, (\kappa R + 1) \tag{25}$$

Solving Eq. (25) for ζ and substituting into Eq. (22) gives,

$$\psi = \frac{q}{4\pi\varepsilon R} \frac{1}{(\kappa R + 1)} \tag{26}$$

Comparing Eq. (14) and Eq. (26) evaluated at $r = R$, it can be seen that

$$q_{\text{eff}} = q \left(\frac{1}{1 + \kappa R} \right) \tag{27}$$

Thus, as the value of κ increases, the effective charge is more effectively shielded. From Eq. (18) we can see that as either the ionic concentration or the permitivity of the solvent is increased, the value of κ will increase, resulting in more effective charge shielding. It must always be remembered that Eq. (27) is valid only for situations in which $\kappa R < 1$.

2. The Relaxation Effect

When an external electrical field is applied to a solution of ions, each with its atmosphere of opposite charge, the situation is dynamic, and we have to think of the counterion atmosphere as a structure with motion. Because the counterion atmosphere does not establish itself infinitely quickly, as an ion is moving through the solution, the atmosphere is incompletely formed in front of it, and has not fully decayed in its wake (see Fig. 2). The overall effect is to displace the center of charge of the counterion

Electrical Double Layer

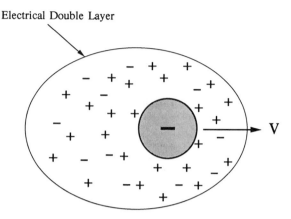

Figure 2 Schematic illustration showing the distortion of the ion atmosphere resulting from the translational motion of the charged particle. This distortion leads to the relaxation effect.

atmosphere a short distance behind the moving ion. Because the charge of the ion and the atmosphere are opposite, the result of this charge separation is to retard the motion of the moving ion. This phenomenon is called the relaxation effect, because the formation and decay of the atmosphere is a kind of relaxation process of the ions into an equilibrium distribution. As might be expected, as the velocity of the ion increases, the retardation effect becomes more pronounced. Thus, as the net charge of an ion increases, a nonlinear increase in μ would be expected.

D. Measurement of Mobility

In order to measure the electrophoretic mobility (μ) of a solute by FSCE, one must first take into account the influence of electroosmosis on the measurements, where electroosmosis is the bulk flow of liquid due to the influence of the electric field on the layer of counterions adjacent to the negatively charged capillary wall. A detailed discussion of both theoretical and practical aspects of electroosmosis is provided in Chapter 1.

What is directly measured in an FSCE experiment is an apparent electrophoretic mobility, μ_{app}. The apparent mobility contains both an electrophoretic component, μ, and an electroosmotic component, μ_{os}. The relationship between μ_{app} and the actual mobility of the solute, μ, is given by

$$\mu_{app} = \mu \pm \mu_{os} \tag{28}$$

where μ_{os} is defined as the electroosmotic velocity per unit field strength (see Fig. 3). Note that if the electrophoresis and the electroosmosis both act in the same direction, μ_{app} is given by the sum of μ and μ_{os}, whereas if μ and μ_{os} act in opposite directions, μ

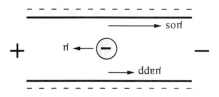

Figure 3 Diagram showing the relative magnitude and direction of the electroosmotic mobility (μ_{os}), the apparent electrophoretic mobility (μ_{app}), and the electrophoretic mobility (μ). In this diagram, it is assumed that the capillary wall and the analyte are negatively charged and that $\mu_{os} > \mu$.

is given by the difference between the two terms. Thus, in terms of easily measurable quantities, μ_{app} can be expressed as

$$\mu_{app} = \frac{v_{net}}{E} = \frac{LL_{tot}}{tV} \tag{29}$$

and μ_{os} can be expressed as

$$\mu_{os} = \frac{v_{os}}{E} = \frac{LL_{tot}}{t_{os}V} \tag{30}$$

where v_{net} is the net (measured) migration velocity of the analyte, E is the electric field strength, L is the length from the point of sample injection to the detection region, L_{tot} is the total length of the capillary between the electrode reservoirs, t is the measured migration time of the analyte, V is the voltage applied across the capillary, v_{os} is the electroosmotic velocity, and t_{os} is the time required for an uncharged solute to be swept past the detector by the electroosmotic flow. By combining Eqs. (28) through (30), it can be easily shown that

$$\mu = \frac{LL_{tot}}{V}\left(\frac{1}{t_{os}} - \frac{1}{t}\right) \tag{31}$$

where it has been assumed that the direction of the electrophoretic migration is opposite to the direction of the electroosmotic flow. This equation gives the relationship between the migration time of a neutral "marker" molecule, t_{os}, the migration time of the analyte, t, and the electrophoretic mobility of the analyte, μ. Rather than using a neutral molecule as a marker to measure electroosmotic flow, one can use a marker that has a finite electrophoretic mobility, μ_s. This is useful when the electroosmotic flow is very slow. In this case,

$$\mu = \frac{LL_{tot}}{V}\left(\frac{1}{t_s} - \frac{1}{t}\right) + \mu_s \tag{32}$$

where t_s is the migration time of the marker. Again, it has been assumed that the direction of the electrophoretic migration (for both the marker and the analyte) is opposite to the direction of the electroosmotic flow. Note that Eq. (32) reduces to Eq. (31) when the marker is neutral; i.e., $t_s = t_{os}$ and $\mu_s = 0$.

It is important to report FSCE data in terms of electrophoretic mobility rather than migration time. Whereas it is tempting to carry on the customs from chromatography, in the case of CE, it is not at all appropriate. First, because μ_{os} can be difficult to control, electrophoretic mobility values will be more reproducible than migration times. Because μ_{os} is dependent on the condition of the surface of the capillary wall, it can vary greatly from run to run or column to column. By reporting mobility values, this variability is eliminated. Second, unless one measures μ, it is impossible to tell whether a change in operating conditions has affected μ, μ_{os}, or both. This could be particularly important information if one is trying to discover the cause of the change in electrophoretic mobility. Third, it is important to remember that μ is the only parameter measured by FSCE that has any physical significance with regard to solute structure. Thus, if one wants to correlate the electrophoretic behavior of a solute with its structure, one must use the electrophoretic mobility. Finally, because values of μ are independent of field strength, capillary dimensions, and time, it is much more convenient to compare values of μ from different experiments than values of the migration time.

III. Resolution — Effect of Dispersion on Separation Performance

Thus far we have discussed FSCE separations only in terms of how the structural features of the solute and the nature of the buffer affect the average electrophoretic mobility of a solute. Implicitly we have assumed that each molecular species migrates as an infinitely thin zone, traveling at a velocity equal to $\mu_{app} \cdot E$. This picture is clearly a vast over-simplification. In order to give a more realistic picture of the electrophoretic separation process, we must take into account the influence of diffusion and other dispersive phenomena. These dispersive effects work against the separation, serving to remix the separated bands and decrease the resolving power of the process.

The parameter that combines the influence of mobility differences and dispersive phenomena on the ultimate separation performance is the resolution, Res, where

$$\text{Res} = \frac{1}{4} \frac{\Delta\mu}{\mu_{avg}} N^{1/2} \tag{33}$$

where $\Delta\mu$ is the difference in mobilities, μ_{avg} is the average mobility, and N is the number of theoretical plates. Equation (33) can be viewed as the product of a selectivity term, $\Delta\mu/\mu_{avg}$, and an efficiency or dispersive term, N. Equation (33) is derived in Chapter 1.

In Chapter 1, an in-depth discussion of the various causes of band dispersion is presented. This discussion includes analytical expressions that can be used to estimate the magnitude of the contribution of various effects to band dispersion as a function of experimental conditions and to minimize their impact on separation performance. This discussion will not be repeated here, but the reader is encouraged to examine the presentation in Chapter 1. In the following section we focus on the factors that determine the value of the selectivity term in FSCE.

IV. Selectivity

In this section we discuss how the structural character of an analyte molecule, and the surrounding environmental conditions, influence its electrophoretic behavior, as measured by FSCE. This will allow us to elucidate the factors that control the unique selectivity of FSCE.

As will be demonstrated, FSCE separations are very sensitive to subtle differences in the native structure of the analyte. Because of the complexity of the native structure of biopolymers, it is impossible to predict a priori how structural features will influence electrophoretic mobility. For this reason, it is necessary to rely on empirical studies to elucidate the relationship between structure, solvent conditions, and selectivity. This is in contrast to studies of separation efficiency, in which much of the behavior can be successfully modeled analytically.

Many of these studies utilize small synthetic peptides as test solutes. Peptides were chosen as model solutes because they are well-characterized species that have defined charge characteristics and minimal secondary or tertiary structure. Moreover, they incorporate the charge character of proteins and peptides and allow the flexibility to explore the effects of various structural and environmental factors.

A. Effect of Buffer pH

For solutes whose charged moieties are ionizable weak acids or bases, it would be expected that changes in buffer pH would greatly affect the overall charge of the solute and thus its electrophoretic mobility. Moreover, if different solutes are composed of ionizable groups with different pK values, changes in pH should affect each solute differently, thus affecting selectivity. Studies examining these effects using synthetic peptides have been performed (Grossman *et al.*, 1988). The major findings of these studies follow.

In order to eliminate the effect of size and to focus on the influence of charge on the electrophoretic mobility, all the peptides used in this study had the same size (seven amino acids). The structure of the model peptides along with the calculated charge on each peptide at three pH values and the calculated isoelectric points are given in Table II.

The separation of peptides 1 through 6 (under acidic conditions at pH 2.5) is shown in Fig. 4. It can be seen that all six peptides are resolved, and that the resolution is such that a calculated charge difference of as little as 0.04 charge units (peaks 4 and

Table II Sequences, Calculated Charges, and Isoelectric Points of Model Peptides[a]

Peptide	Sequence	Calculated charge			Calculated isoelectric point
		pH 2.5	pH 4.0	pH 11.0	
1.	A F A A I N G	0.41	0.02	−0.95	6.0
2.	A F D A I N G	0.37	−0.54	−1.95	3.2
3.	A F D D I N G	0.33	−1.09	−2.95	3.0
4.	A F K A I N G	1.41	1.02	−0.71	10.1
5.	A F K K I N G	2.41	2.02	−0.47	10.5
6.	A F K A D N G	1.37	0.46	−1.71	6.7
7.	A F K K A N G				
8.	A F K A N K G	2.41[b]	2.02	−0.47	10.5
9.	A F K I K N G				
10.	A F K A K N G				

[a] Reprinted with permission from Grossman et al. (1988).
[b] Because peptides 7–10 all contain the same charged amino acid side chains and terminal amino acids, according to the model used in this study, they all have the same charges and isoelectric points.

6) can be partially resolved. Furthermore, the peaks elute in the order predicted by the calculations; i.e., the most positively charged species elute first, with the highest measured electrophoretic mobility. (Note that at pH 2.5, the electroosmotic flow is essentially eliminated.)

Next, the effect of changing the buffer pH was investigated. In addition to separations at pH 2.5, separations using buffers having pH values of 4.0 and 11.0 were studied. As can be seen in Fig. 5, at pH 4.0, peptides 4 and 6 are more fully resolved, which is reflected in the fact that the selectivity ($\Delta\mu/\mu_{avg}$) between peptides 4 and 6 goes from 0.018 at pH 2.5 to 0.620 at pH 4.0, an increase of over 34 times. Furthermore, as expected based on the charge values in Table II, peptides 2 and 3 disappear completely from the electropherogram, presumably because at pH 4.0, these peptides are negatively charged and thus would no longer migrate toward the negative electrode. From Fig. 5 it can also be seen that at pH 11.0, the migration order of peptides 1 and 6 is reversed. (At pH 11.0, because the electrophoretic migration is "upstream" against the electroosmotic flow, the peptides with the largest electrophoretic mobility elute more slowly.)

These experiments highlight the powerful influence buffer pH can have on the resolution of peptide, and by analogy, protein samples. A change of only 1.5 pH units can dramatically change the selectivity of a separation. Furthermore, by changing the buffer pH, one is able to explore different aspects of the charge character of the peptides. At high pH, e.g., pH 11, the contribution of the negatively charged acidic residues is emphasized, whereas at low pH, e.g., pH 2.5, the influence of the positively charged basic residues is accentuated. This is in contrast to isoelectric focusing, in which only an average charge at the isoelectric point can be measured.

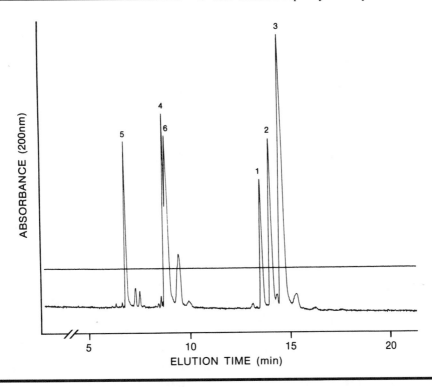

Figure 4 Electropherogram of model peptides 1 through 6 (see Table II). Conditions: field, 277 V/cm; current, 24 μA; buffer, citric acid, 20 mM, pH 2.5; capillary length, 65 cm (45 cm to the detector). The horizontal line across the electropherogram is a measure of the electrical current through the capillary. Reprinted with permission from Grossman *et al.* (1988).

B. Effect of Size and Charge

It is clear from Eq. (6) that the primary factors controlling the electrophoretic mobility of a solute are its charge and, through its translational friction coefficient, size. However, whereas Eq. (6) shows how μ depends on q and f, it says nothing about how q and f relate to the composition and size of a solute.

Various attempts have been made to develop expressions that predict the electrophoretic mobility of biopolymers in aqueous solution, based on molecular structure. In this section we discuss a number of these studies.

1. Peptides

An early attempt to relate the structure of a biopolymer to its free-solution electrophoretic mobility was made by Offord (1966). This study used paper electrophoresis to measure the mobility of natural peptide molecules. Offord found that at acidic pH (pH 1.9), the electrophoretic mobility was proportional to the molecular weight raised to a power intermediate between one third and two thirds. Assuming that the

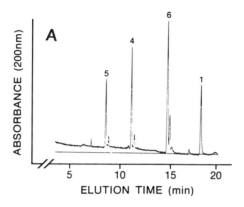

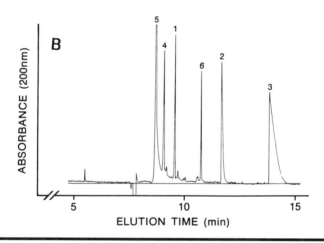

Figure 5   Electropherogram of model peptides 1 through 6 (see Table II). Conditions: (A) field, 277 V/cm; current, 12 μA; buffer, citric acid, 20 mM, pH 4.0; capillary length, 65 cm (45 cm to the detector); (B) field, 250 V/cm; current, 8 μA; buffer, [cyclohexylamino]propanesulfonic acid (CAPS), 20 mM, pH 11.0; capillary length, 120 cm (100 cm to the detector). Reprinted with permission from Grossman *et al.* (1988).

peptide behaves as a freely rotating coil, according to Eq. (6) and Table I, one would expect that $\mu \sim MW^{-0.5 \text{ to } -0.6}$. Thus, at low pH, Offord's results appear to agree with theoretical predictions. However, it was found that at neutral pH (pH 6.5), $\mu \sim MW^{-2/3}$. Offord explains the $-2/3$ exponential term by proposing that the size dependence of μ at pH 6.5 is dominated by interactions between the peptides and the paper matrix. Possibly at the low pH values, both the peptides and the support are positively charged, thus eliminating any coulombic interaction, whereas at neutral pH, significant interaction occurs.

There are a number of shortcomings to the Offord treatment that limit its utility to application to FSCE. First, only *relative* mobilities are considered, because it is impossible to accurately define the electric field strength within the pore structure of the paper matrix, owing to the tortuous path the solute must travel. Furthermore, it is difficult to accurately quantify the electroosmotic flow. Also, no attempt was made to combine the effects of size and charge into a single expression. Peptides having different charges were investigated, but no expression to relate the behavior of peptides having different charges was attempted. Finally, only a narrow range of charge was studied, from $+3$ to -3, where only integral charge values were considered.

Using FSCE, a semiempirical expression was developed that combines the effect of both size and charge on the electrophoretic mobility of peptides in free solution (Grossman *et al.*, 1989). In this study a series of 40 synthetic peptides was used, ranging in size from 3 to 39 amino acids and ranging in charge from 0.33 to over 14 charge units. The expression derived from these data is

$$\mu = 5.23 \cdot 10^{-4} \frac{\ln(q + 1)}{n^{0.43}}$$

(34)

where q is the calculated charge (expressed in terms of the number of positive or negative charges) and n is the number of amino acids (see Fig. 6). Note that in the original publication, a constant term of $2.47 \cdot 10^{-5}$ cm^2/V $\cdot$ sec was included in the above expression. Subsequent to publication, it was found that this nonzero intercept was an experimental artifact resulting from an inaccurate value of the mobility of the peptide used as an "internal standard" to measure electroosmotic flow.)

A number of observations should be made regarding Eq. (34). First, it is important to recognize that this expression is in terms of absolute rather than relative mobilities. Second, it is significant that fractional charges are taken into account. Third, it is important to recognize that the nonlinear charge dependence is included — this effect is particularly important for highly charged peptides. However, the constants in Eq. (34) hold only for the experimental conditions used in these studies.

It is interesting to compare the mobility values predicted by Eq. (34) with those that would be expected based on theory. Equation (34) says that if $n = 1$ and $q = 1$, then $\mu = 3.63 \cdot 10^{-4}$ cm^2/V $\cdot$ sec. If we assume that a single residue can be considered spherical and that its radius is approximately 2.5 Å, then using Eq. (7), $f = 4.71 \cdot 10^{-12}$ N $\cdot$ sec/m. In terms of coulombs, $q = 1 = 1.60 \cdot 10^{-19}$ C. Thus, according to Eq. (6), $\mu = 3.4 \cdot 10^{-8}$ m^2/V $\cdot$ sec or $3.4 \cdot 10^{-4}$ cm^2/V $\cdot$ sec. This close agreement between the theoretical value of μ and that predicted from Eq. (34) lends some credence to the correlation.

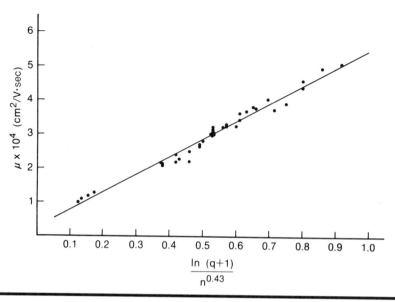

Figure 6   Plot of the measured electrophoretic mobility versus $\ln(q + 1)/n^{0.43}$ for 40 different peptides ranging in size from 3 to 39 amino acids and ranging in charge from 0.33 to 14. Conditions: same as in Fig. 4. The slope of the solid line is $5.23 \cdot 10^{-4}$ cm²/V sec and $r = 0.989$. Reprinted with permission from Grossman *et al.* (1989).

The above correlation should be useful for the identification of peptide species as well as for the detection of structural modifications that affect the overall charge.

2. Synthetic Oligonucleotides

It was stated earlier in this chapter that the electrophoretic mobility of DNA molecules in free solution is independent of the molecular size [see Eq. (13)]. However, this is not necessarily true for short synthetic oligonucleotides. The size dependence of the electrophoretic mobility of single-stranded oligonucleotides as measured by FSCE has been investigated (Menchen and Grossman, 1990).

In this study, it was assumed that the structure of the oligonucleotide was BpBpBpB, where p represents phosphate groups and B the nucleotide bases. It is important to note that for the synthetic oligonucleotides used in this study, the 5′ end of the strand was not phosphorylated. Based on this structure, it can be seen that

$$q \sim N_p = N_B - 1 \tag{35}$$

and

$$f \sim MW \sim N_B + N_p = 2N_B - 1 \tag{36}$$

where N_p is the number of phosphates and N_B is the number of bases in the oligonucleotide. Thus, from Eq. (6) it can be seen that

$$\mu = \frac{q}{f} = K \frac{N_B - 1}{2N_B - 1} \tag{37}$$

where K is a constant of proportionality. Figure 7 shows the measured electrophoretic mobility of seven poly-dT synthetic oligonucleotides ranging in size from 3 to 60 bases in length. The solid curve in the figure is a plot of Eq. (37), where the value of K is $8.29 \cdot 10^{-4}$ cm^2/V $\cdot$ sec and was obtained by fitting the data.

From Fig. 7 it is clear that for small synthetic oligonucleotides having fewer than approximately 10 bases, μ is in fact a strong function of molecular size. However, for fragments longer than 10 bases, the expected size-independent behavior is recovered.

It is important to recognize that the size dependence of μ described by Eq. (37) is not valid if the 5′ end of the oligonucleotide is phosphorylated. In this case,

$$\mu \sim \frac{N_B}{2N_B} \sim N_B^0 \tag{38}$$

where the term N_B^0 indicates that μ is independent of N_B. This is the case for natural oligonucleotides. Thus, these findings are likely to be of little practical importance, but

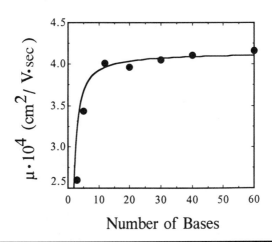

Figure 7 Plot of the electrophoretic mobility of poly-dT synthetic oligonucleotides as a function of oligonucleotide size. The solid curve is a plot of Eq. (36) where $K = 8.29 \cdot 10^{-4}$ cm^2/V sec. Conditions: field, 303 V/cm; current, 60 μA; buffer, TBE [tris (hydroxymethyl)aminomethane, 89 mM boric acid, and 5 mM ethylenediaminetetraacetic acid (EDTA)] with added NaCl 70 mM, pH 8.5; capillary length, 66.1 cm (50 cm to the detector). Reprinted with permission from Menchen and Grossman (1990).

rather they are simply an illustration of the way in which molecular structure can affect the electrophoretic mobility of different solutes, and thus the selectivity of FSCE separations.

C. Effect of Hydrophobicity

It has been observed that even uncharged amino acids can affect the electrophoretic mobility of peptides (Grossman *et al.*, 1988). An example is shown in Fig. 8. It was postulated that this effect was caused by the influence of neighboring neutral amino acids on the dissociation of charged residues.

    To test this hypothesis, in a subsequent report (Grossman *et al.*, 1989), the electrophoretic behavior of a series of peptides, each containing seven amino acids, but

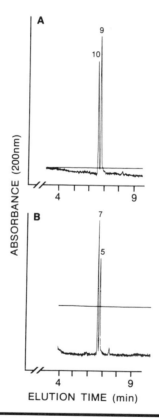

Figure 8 Electropherogram of peptide mixtures in which two peptides differ by only a single uncharged amino acid. Conditions: (A and B) field, 277 V/cm; current, 24 μA; buffer, citric acid, 20 mM, pH 2.5; capillary length, 65 cm (45 cm to the detector). See Table II for the sequence of peptides 5, 7, 9, and 10. Reprinted with permission from Grossman *et al.* (1988).

differing in a single neutral amino acid, was investigated. In each of the peptides, the neutral amino acid was located between two lysine residues. The mobility of each of these peptides was measured and then plotted against a measure of the hydrophobicity (Guo *et al.,* 1986) of the substituted neutral amino acid. As can be seen in Fig. 9, the mobility of the peptides does in fact appear to be related to the hydrophobicity of the substituted neutral amino acid, with the mobility decreasing as the hydrophobicity increases. This is the trend that would be expected for a basic dissociation, such as that of lysine, as the local dielectric decreases.

This effect would be difficult to include in a general relation like Eq. (34), because the position of the neutral residue relative to the charged side-chain, as well as the total hydrophobic content, must be known. This is demonstrated in Fig. 10, which shows the separation of two peptides, each having the same amino acid composition, but differing in amino acid sequence. However, the shallow slope of the curve in Fig. 10 indicates that this effect is minor compared to the effects of size and charge, and results in only a weak perturbation of Eq. (34).

D. Effect of Molecular Orientation

It is well known that molecules having a permanent or induced dipole moment can be oriented by a strong electric field. For a rod-shaped particle, this orientation can affect the value of its translational friction coefficient, f. This in turn, as can be seen from Eq. (6), should reflect itself by a change in the electrophoretic mobility of the particle. In traditional native electrophoresis, this effect is not important, because of the low electric fields typically used. However, in the case of FSCE, the fields are strong

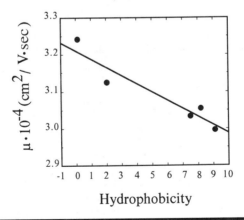

Figure 9   Effect of the hydrophobicity of substituted neutral amino acids on the electrophoretic mobility of five model seven-amino acid peptides. Reprinted with permission from Grossman *et al.* (1989).

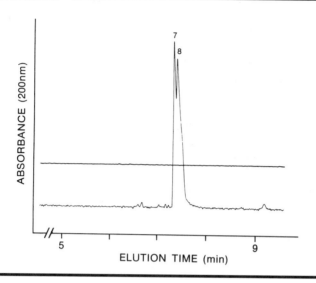

Figure 10   Electropherogram of a mixture of two peptides that have the same amino acid composition, but differ in their amino acid sequence. Conditions: Same as Fig. 4. See Table II for the sequence of peptides 7 and 8. Reprinted with permission from Grossman *et al.* (1988).

enough to make orientational effects worth considering. Studies focusing on the influence of molecular orientation on the electrophoretic mobility of rod-shaped particles have been undertaken, using the tobacco mosaic virus (TMV) as the sample solute (Grossman and Soane, 1990).

   As can be seen from Eq. (7), the translational friction coefficient for a sphere is only a function of the particle radius. However, for a rod-shaped particle, f is also a function of its aspect ratio and its orientation with respect to the direction of motion. The functional form of this relationship for a rigid, rod-shaped particle is given by

$$f = \eta \, (K_1^{-1} \sin^2 \Theta + K_3^{-1} \cos^2 \Theta)^{-1} \qquad (39)$$

where

$$K_1 = \left(\frac{8\pi\phi}{\ln(2\phi) + 0.5} \right) R \qquad (40)$$

and

$$K_3 = \left(\frac{4\pi\phi}{\ln(2\phi) - 0.5} \right) R \qquad (41)$$

and R, L, and ϕ are the radius, length, and aspect ratio ($L/2R$) of the cylindrical particle respectively, Θ is the angle of orientation with respect to the direction of motion, and η is the solution viscosity. As one might expect, the translational friction coefficient assumes a maximum value when the rod is oriented perpendicular to the direction of motion, i.e., when $\Theta = 90$, and approaches a minimum value when the rod is totally aligned with the direction of motion, i.e., when $\Theta = 0$. Thus, as the TMV particle becomes more aligned with its direction of motion, we would expect its electrophoretic mobility to increase, owing to the reduced translational friction coefficient. The expression that relates the orientation angle, Θ, to the electrical field strength and the electrical polarizability of the particle is (Grossman and Soane, 1990)

$$\langle \Theta \rangle = \arccos \left(\frac{[2(\alpha_1 - \alpha_2) E^2]/15kT + 1}{3} \right)^{0.5} \tag{42}$$

where k is the Boltzmann constant, T is the absolute temperature, and $\alpha_1 - \alpha_2$ is the electrical polarizability of the particle, where the subscripts 1 and 2 refer to axial and transverse properties respectively.

To explore the influence of molecular orientation on the electrophoretic mobility of TMV, the influence of electric field strength on the electrophoretic mobility was considered. According to Eqs. (39) and (42), as the field strength is increased, the TMV particle will become more aligned with the direction of motion, and its translational friction coefficient will decrease. The net result should then be an increase in the electrophoretic mobility of the TMV particle. Figure 11 shows the results of such an experiment. Note that the electrophoretic mobility of TMV does indeed increase with electric field strength. In fact, the mobility of TMV increases by over 6% when the

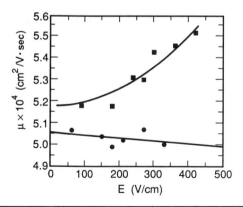

Figure 11   Electrophoretic mobility of TMV virus (■) and 0.364 μm latex spheres (●) as a function of electrical field strength. Conditions: field, 273 V/cm; current, 0.11 μA; buffer, potassium borate, 2 mM; capillary length, 66 cm (50 cm to the detector). Reprinted with permission from Grossman and Soane (1990).

electric field is varied from 60 to 400 V/cm. The solid curve in Fig. 11 is generated using Eqs. (6), (39), and (42) where $\alpha_1 - \alpha_2 = 7.0 \cdot 10^{-14}$ cm^3 (O'Konski *et al.*, 1959).

A control experiment to confirm that the observed increase in μ is attributable to an orientational effect was performed on a spherical latex particle. The experimental conditions were identical to those used for the TMV experiments. The latex particle was chosen to have comparable dimensions to TMV. This spherical-particle system was subject to the same effects of temperature, electroosmotic flow, and ion atmosphere distortion; however, its spherical symmetry precludes it from exhibiting any orientational effects. Thus, if an increase in the buffer temperature or variations in the electroosmotic flow velocity were responsible for the observed increase in the electrophoretic mobility of TMV, this would also have been reflected in the mobility of the latex particle. On the other hand, if the mobility increase of TMV is attributable only to orientational effects, the mobility of the latex particle should not change with increasing field strength. As Fig. 11 shows, the mobility of the spherical particle does not appear to increase with increasing field strength; in fact, there appears to be a slight decrease in μ as E increases. This decrease in μ with increasing field strength is most likely owing to the relaxation effect. The fact that the electrophoretic mobility of the spherical particle does not increase with increasing electrical field strength provides strong evidence that the observed increase in μ for TMV is in fact a result of molecular orientation.

References

Atkins, P. W. (1978). "Physical Chemistry." W. H. Freeman, San Francisco, California.

Cantor, C. R., and Schimmel, P. R. (1980). "Biophysical Chemistry." W. H. Freeman, New York.

Grossman, P. D., and Soane, D. S. (1990). *Anal. Chem.* **62**, 15, 1592–1596.

Grossman, P. D., Wilson, K. J., Petrie, G., and Lauer, H. H. (1988). *Anal. Biochem.* **173**, 265.

Grossman, P. D., Colburn, J. C., and Lauer, H. H. (1989). *Anal. Biochem.* **179**, 28–33.

Guo, D., Mant, C., Taneja, A. K., Parker, J. M. R., and Hodges, R. S. (1986). *J. Chromatogr.* **359**, 499–517.

Menchen, S., and Grossman, P. D. (1990). Unpublished work.

Offord, R. E. (1966). *Nature* **211**, 591–593.

O'Konski, C. T., Yoshioka, K., and Orttung, W. H. (1959). *J. Phys. Chem.* **63**, 1558–1565.

Rice, S. A., and Nagasawa, M. (1961). "Polyelectrolyte Solutions." Academic Press, New York.

Capillary Gel Electrophoresis

Robert S. Dubrow

I. The Role of Capillary Gel Electrophoresis in Separation Science

The first spectacular electropherograms, published in the late 1980s using capillary gel electrophoresis, demonstrated an opportunity for significant advances in the practice of separation science (Cohen and Karger, 1987; Cohen *et al.*, 1987). The promise of combining the known ability of hydrophilic polymer gels to separate biomolecules with the speed, quantitative, and microsample capabilities associated with capillary electrophoresis has resulted in a great deal of scientific interest in this new field.

As techniques have improved for generating gels in capillaries, the promise of a powerful new separation tool has turned into reality. Single-base separations of poly-oligonucleotides have been achieved in minutes (Karger *et al.*, 1989), and DNA sequencing runs resolving 350 bases have been completed in 1 hr (Drossman *et al.*, 1990). Accurate molecular-weight sizing of proteins has been achieved through gel-filled capillaries containing sodium dodecyl sulfate (SDS) (Cohen and Karger, 1987), and chiral compounds have been separated when β-cyclodextrin was included (Guttman *et al.*, 1988). All of these separations were performed using nanogram sample sizes.

With the commercial introduction of gel-filled capillaries (Applied Biosystems, Inc., 1990) and the availability of automated capillary electrophoresis instruments, this technique promises to become an important part of the modern analytical laboratory.

II. Theoretical and Practical Considerations in Capillary Gel Electrophoresis

Capillary gel electrophoresis, a hybrid of traditional slab-gel and free-solution capillary open-tube electrophoresis technology, shares a great deal of commonality with these techniques, but also has a number of unique features. This section will define and examine the variables affecting these features, as well as expose some potential benefits and drawbacks of capillary gel electrophoresis.

A. Application of Hydrogels to Electrophoresis

Early investigators carried out electrophoresis in a variety of separation media, designed to provide anticonvection and sieving properties for structurally similar macromolecules. Materials including paper, glass powder, silica, Sephadex, and polyurethane foam were all tried (Janick, 1986). Hydrophilic polymer gels have proven to be the most useful and are used almost exclusively today in slab-gel electrophoresis.

Gels can be defined as cross-linked networks swollen with a fluid component or sol. The average distance between cross-links, distribution of cross-links, and fluid content generally dictate the size and size distribution of the pores in the gel. Cross-links can be produced by covalent or ionic bonds, Van der Waals interactions, or by physical entanglements of polymer chains. The nature of the pores in the gel dictates its ability to separate molecules of various sizes that are driven through it electrophoretically.

Traditionally for slab-gel electrophoresis, porous gels have been made *in situ* by the polymerization of polyacrylamide and a suitable cross-linker dissolved in an aqueous buffer, or by the gelation of a hot solution of agarose dissolved in an aqueous buffer. These gels have been typically formed into slabs of between 0.2 and 5 mm in thickness or tubes from 0.5 to 10 mm in diameter. By varying the concentration of the agarose or acrylamide and cross-linker, the pore size can be varied in the gel to tailor it for a particular separation. During electrophoresis the molecules separate into bands on the slab or tube gel, with the smallest molecules moving the greatest distance. Once electrophoresis is completed, the bands can be visualized by staining. A variety of specialized stains have been developed for proteins and nucleic acids (Chrambach and Rodbard, 1971).

B. Parameters Affecting Capillary Gel Electrophoresis

Capillary diameter, capillary length, gel concentration, and field strength all affect separation efficiency and time. This section will examine these parameters.

1. Capillary Diameter

Generally the inside diameter of capillaries used in capillary gel electrophoresis is less than 200 μm, with a practical lower limit of about 25 μm. Most gel-filled capillaries today have diameters between 50 and 100 μm, which represents a compromise between minimizing thermal gradients across the capillary, maximizing the detection path length, and the practicalities involved in gel fabrication (e.g., Section III). Commercially available capillaries made of fused silica are quite uniform, typically varying less than 1 μm in inside diameter per meter, and have been the tubing of choice to date.

2. Capillary Length

Capillary gel electrophoresis has been performed in capillaries having effective separation lengths (distance from sample injection to detector) as short as 7 cm (Guttman *et al.,* 1991) and as long as 1 m (Smith, 1989). Resolution can often be improved by increasing the capillary length, although a penalty is paid in the run time, and a higher voltage is required to achieve a given field strength. For this reason, most workers use gel-filled capillaries with 15 to 40 cm effective separation lengths, and vary gel concentration to tune their separations. Separation distances in this range are comparable to those used in slab-gel formats. Figure 1 demonstrates the improvement in resolution achieved in the separation of an oligonucleotide ladder by increasing the effective separation length of the gel-filled capillary.

3. Gel Concentration

As with capillary length, varying polymer concentration is an exercise in optimization. Increasing the gel concentration often improves peak separation and allows the use of shorter capillaries, but it generally decreases the effective separation range and complicates gel fabrication. Figure 2 illustrates resolution improvements achieved in the separation of an oligonucleotide standard by increasing the gel concentration from 6 to 10%. Maintaining separation range is a particularly acute problem, since unlike slab gels, where the distance a molecule travels is related to its size, all of the analytes must travel the same distance through the capillary to reach the detector.

4. Field Strength

A major advantage of capillary gel electrophoresis is that, owing to efficient heat dissipation (see Chapter 1), resolution is maintained with increasing field strength. Figure 3 is a series of electropherograms showing an oligonucleotide ladder in which the field strength was varied between 100 and 500 V/cm. The chart speed was increased to match the increase in field strength and, as can be seen in the figure, resolution is maintained, and the three electropherograms are superimposable. Operation at high field strengths requires special precautions. In the case of these experiments, the buffer conductivity was high enough (2000 μmhos/cm), that above 500 V/cm, the amperage began to climb nonlinearly with respect to voltage, indicating capillary heating. Once this state was reached, the capillary gel formed voids and lost electrical continuity.

The field-strength insensitivity with respect to resolution opens the door to relatively high speed separations compared to traditional slab gels using capillary gel electrophoresis. This represents a tremendous opportunity for capillary gel electrophoresis as well as a technical challenge to develop higher performance gels.

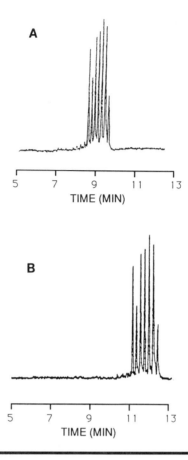

Figure 1   Comparative electropherograms of a p(dA)$_{12-18}$ oligonucleotide ladder run on a gel-filled capillary with an effective separation length of (A) 20 cm, (B) 30 cm. Field strength for each run was 300 V/cm; temperature, 30°C; detection wavelength, 260 nm; 75 mM pH 7.6 Tris-phosphate buffer was used with a MICRO-GEL$_{100}$ gel-filled capillary. Reprinted with permission from Applied Biosystems, Foster City, CA.

C. Sample Injection

Introduction of the sample into the gel-filled capillary is typically done electrokinetically. Because the capillary is essentially plugged with gel, the various hydrodynamic methods of introducing volumes of sample used in free-solution capillary techniques must be abandoned. Electrokinetic injection has various advantages and pitfalls. The main advantage is that it is a relatively simple technique requiring only that an electric field be applied in a timed manner. Typically injection times between 1 and 20 sec are

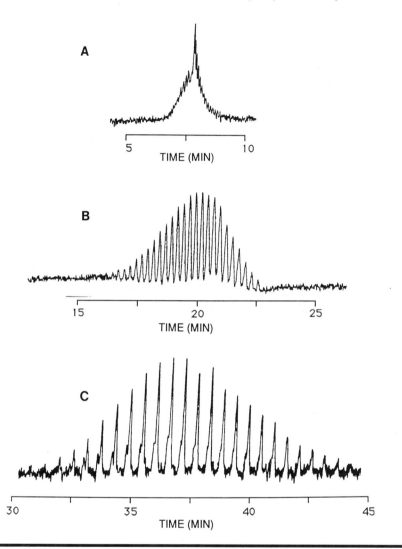

Figure 2   Comparative electropherograms of a p(dA)$_{12-18}$ oligonucleotide ladder run on a gel-filled capillary with (A) 6% total polymer concentration, (B) 8% total polymer concentration and, (C) 10% total polymer concentration. Field strength was 250 V/cm; capillary temperature, 30°C; 30 cm effective separation length; detection wavelength 260 nm; and a 75 mM pH 8.3 Tris-borate buffer was used.

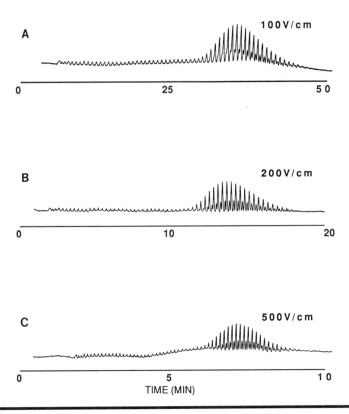

Figure 3 Comparative electropherograms of p(dA)$_{40-60}$ oligonucleotide ladder run on a gel-filled capillary at field strengths of (A) 100 V/cm and a chart recorder speed of 0.5 cm/min; (B) 200 V/cm and a chart recorder speed of 1 cm/min; (C) 500 V/cm and a chart recorder speed of 2 cm/min. Capillary temperature was 30°C; 30 cm effective separation length; detection wavelength, 260 nm; 75 mM pH 7.6 Tris-phosphate buffer was used with a MICRO-GEL$_{100}$ gel-filled capillary.

used at field strengths between 100 and 400 V/cm. At very low sample concentrations, it may be necessary to use the top end of these ranges, and when sample is plentiful, lower injection times and voltages can be used. By preparing the sample in deionized water the phenomenon of "stacking" can be used to load dilute samples more efficiently (see Chapter 3). Actual sample requirements vary with the complexity of the sample and the sensitivity of the detector (Section IV, A, B, C, D). Excessive sample concentration will overload the column and result in peak broadening and loss of resolution.

The primary disadvantages of this injection method are that it is difficult to de-

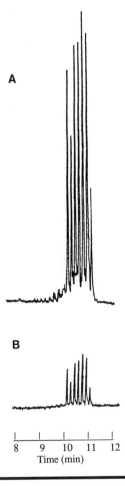

Figure 4 Comparative electropherograms of a p(dA)$_{12-18}$ oligonucleotide ladder run on a gel-filled capillary with (A) no electrolytes added to the sample solution; (B) 1 mM Tris-phosphate added to the sample solution. Field strength for each run was 300 V/cm; temperature, 30°C; 30 cm effective separation length; detection wavelength, 260 nm; a 75 mM pH 7.6 Tris-phosphate buffer was used with a MICRO-GEL$_{100}$ gel-filled capillary.

termine the mass loading on the column and that the conductivity of the sample solution can dramatically alter the loading, as shown in Fig. 4. In general, the lower the salt concentration in the sample solution, the greater the loading on the gel will be. This points to the use of deionized water or nonconducting denaturing solutions such as 7 M urea or formamide for dissolving samples.

D. Reproducibility

Gel networks are tenuous structures, subject to morphological changes caused by temperature, pH, and other environmental fluctuations. These networks are made up of polymers that are classified by their average molecular weight and, when cross-linked, by their average distance between cross-links. In short, most polymers (biopolymers being an exception), are fairly imprecise entities. When this is added to the complexity of forming polymers into a porous network and maintaining the network in a steady state, it is a difficult task to obtain quantitatively consistent results on capillary gels.

Cohen and Karger (1987) reported the importance of cooling the capillary for reproducibility. Slight differences in pH, ionic strength, and gel concentration have been found to have a dramatic impact on migration times and resolution (Dubrow, 1990). If these factors are carefully controlled, the reproducibility of well-made, gel-filled capillaries appears to be equivalent to that of most chromatographic methods. When an internal marker is used, migration times and peak areas from run to run and column to column (Dubrow and Harrington, 1990; Heiger *et al.*, 1990) resulting in relative standard deviations of less than 1% have been reported. As would be expected, because of the complexity of gel matrices, obtaining reproducible absolute migration times is difficult to achieve from run to run as well as from gel to gel.

In summary, carefully made columns and well-controlled run conditions lead to acceptable reproducibility for most purposes. Although improvements will certainly be made as gel technology improves, it would be reasonable to expect that internal standards will be necessary for quantitative work in the near future.

E. Column Breakdown

To date, state-of-the-art capillaries can be run in excess of 20 hr at 300 V/cm without electrical breakdown or loss in resolution. For many applications, this allows upward of 50 typical separations to be run on a single capillary. Gel-filled capillaries usually fail owing to void formation during operation. Small bubbles begin to form and grow, until they interrupt the flow of current. Most investigators have found that the bubbles usually form at the sample-injection end of the capillary. Speculation as to the cause of this phenomenon varies from sample-induced effects to electroosmotic flow at the capillary tip, dragging out gel, but so far no widely accepted theory has been postulated. It has been observed that if the capillary tip is not cleaved evenly and cleanly, the gel will break down prematurely. The capillary ends are also subject to dehydration if left out of buffer for several minutes.

Another difficulty associated with using gel-filled capillaries is the difference in thermal expansion coefficient between the gel and fused-silica capillary that contains it. Fused silica has a very small thermal expansion coefficient ($t = 5.5 \times 10^{-7}$ cm $-$ cm^{-1} °C^{-1}) and a typical 10% hydrogel a high one ($t = 2.1 \times 10^{-4}$ cm $-$ cm^{-1} °C^{-1}). This translates into almost a 1% volume expansion of the gel relative to the capillary over a 40°C temperature change. High stresses and over a centimeter of movement can occur under these conditions, potentially destroying the integrity of the gel. These factors can limit the operating temperature range of gel-filled capillaries.

III. Current Capillary Gel Fabrication Methods

A. Cross-Linked Polyacrylamide-Filled Capillaries

Most workers attempting to build gel-filled capillaries have focused on acrylamide monomers as their building blocks. Considering that this is the standard electrophoresis gel material (Chrambach and Rodbard, 1971), early investigators rightly assumed that the successful polymerization of a polyacrylamide gel in a capillary would lead to a powerful new separation tool.

Although this section will focus on producing polyacrylamide capillary gels, it should be noted that these gels have several inherent weaknesses. Polyacrylamide suffers from hydrolytic instability, and is therefore limited to a narrow pH range. In addition, this instability severely limits its shelf life, even at mild pH. For this reason it is expected that researchers will eventually move toward more stable hydrophilic polymers.

Groups at Northeastern University in Boston and Hewlett-Packard in Palo Alto focused on the challenge of polymerizing acrylamide in capillaries in the mid-1980s. Both groups found two major obstacles in their path. The first was that, owing to shrinkage during polymerization, voids were formed in the gel. The voids led to severe band dispersion and electrical discontinuity when electrophoresis was attempted. The other issue concerned electroosmotic flow in the capillary. Because of the high electric fields used in capillary electrophoresis, the electroosmotic flow can be sufficient to extrude the gel from the capillary. Agarose was generally discarded as a candidate material because it contains charged groups that would cause it also to be propelled out of the capillary when subjected to high field strengths. In addition, its low melting temperature restricts its versatility in the capillary format.

1. Karger and Cohen Technique

Two patents were issued to Northeastern University (Karger and Cohen, 1989a,b) covering capillary gel electrophoresis columns. The first claims a method for preparing a gel-containing microcapillary column for electrophoresis, and the second covers the addition of a hydrophilic polymer to improve the stability of the columns under high electric fields. In these patents, a detailed procedure is presented for polymerizing a gel in microcapillary. The use of a silane coupling agent to covalently link the gel to the capillary wall to prevent gel migration during electrophoresis is heavily stressed. Void formation is surprisingly never mentioned in the patent, although the authors have discussed this problem at length in oral presentations (Karger, 1990). A summary of this process is presented in Appendix 1 for a gel containing 10%

acrylamide monomer (10% T) by weight of the total gel and 3.3% cross-linker (3.3% C) by weight of the acrylamide monomer (0.33% of the total gel).

Several variations of this basic technique have been developed by other researchers, but they still utilize the Karger approach of polymerization and wall binding *in situ*. Two of these will be discussed in the following section.

2. High-Pressure Polymerization

In a patent issued to Bente and Myerson (1988) of the Hewlett-Packard Corporation, high pressures are applied to the acrylamide solution in the capillary as it gels to prevent the formation of voids. The inventors calculate that an acrylamide gel containing 90% buffer will increase in density by 2.2% during polymerization. If the polymerization takes place in a capillary containing a wall-binding agent, the shrinkage associated with densification will cause fractures and voids. Bente and Myerson showed that by applying pressures sufficient to compress the polymerizing solution to approximately its cured density under ambient conditions, void formation could be eliminated. As a general rule, at least half of the pressure required to compress the monomer to its cured density had to be used to eliminate voiding. For a 5% monomer solution, 1700 psi was the minimal pressure required to eliminate voiding, whereas 3400 psi was needed to completely eliminate shrinkage. For a 10% monomer concentration, 8,200 psi was the minimal pressure, with 10,000 psi required to achieve zero shrinkage.

In the patent, the capillaries are placed in a pressure vessel after they are filled with the monomer solution, and then a gas is used to apply the required compressive force (Fig. 5). This technique has also been used successfully to prepare capillary DNA-sequencing gels where water was substituted for gas as the compressive medium (Drossman *et al.*, 1990).

A modified version of this technique was used in which the capillary wall was first covalently coated with a layer of polyacrylamide to eliminate electroosmotic flow, and then a polyacrylamide gel was polymerized in the capillary under more moderate pressures (Lux *et al.*, 1990). The authors claim that wall binding is unnecessary because the coating eliminates electroosmotic flow, and this allows the use of lower pressures (300 psi) during polymerization to prevent voiding.

3. Radiation Polymerization

Cobalt 60 radiation has been used to initiate the polymerization of acrylamide in a capillary (Lux *et al.*, 1990). The authors claim that the use of cobalt 60 to initiate the gel polymerization eliminates the need for chemical initiators, which degrade the performance and lifetime of the capillaries. In addition, they state that the polymerization conditions are mild enough that bubbles do not form in the gel. Radiation doses between 20 and 400 krad were used in the experiments.

All of the previously mentioned techniques for *in situ* polymerization have a major drawback. When a gel is created from nonomers, it is a fairly random event, resulting in a wide distribution of pore sizes as well as significant amounts of free monomer and uncross-linked polymer. Producing gels of higher resolution will no doubt require

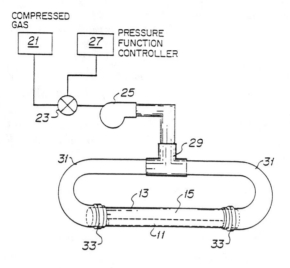

11-Filled tube
13-Pressure chamber
15-Water
21-Gas cylinder
23-Regulator valve
25-Hydraulic amplification pump
27-Pressure function controller
29-T connector
31-Delivery arms
33-Swage-lock fittings

Figure 5 Compressed gas apparatus used in Bente and Myerson (1989) capillary gel-formation method.

cross-linking strategies that can achieve greater control over the production of the polymer network, including the average pore size and pore-size distribution.

B. Uncross-Linked Polymers/Entangled Polymer Networks

Several workers have published separations using capillaries filled with polymer-modified solutions (see Chapter 8; Chin and Colburn, 1989; Hansen *et al.*, 1990; Heiger *et al.*, 1990). Unlike traditional gels, these materials are not chemically cross-linked, but provide molecular sieving, owing to physical entanglements of the polymer chains. Tietz *et al.* (1986) described the behavior of an entangled polyacrylamide polymer network in a slab gel format for the separation of proteins. The authors pointed out that electrophoresis in uncross-linked polyacrylamide is qualitatively the same as that in cross-linked gels. However, Tietz *et al.* found that at concentrations below 10%, polymer thermal convection occurs, degrading resolution. They predicted that use of a capillary would eliminate the convection problem because thermal gradients are minimized.

Uncross-linked polymer solutions have been polymerized by the methods described in Section III,A as well as introduced into capillaries under vacuum and pressure. Several polymers have been used including polyacrylamide (Heiger *et al.*, 1990), polyethylene oxide (Chen *et al.*, 1989), and hydroxylated cellulose derivatives (Chen *et al.*, 1989; Chin and Colburn, 1989). When solutions containing low concentrations of polymer are used, a vacuum or syringe can deliver sufficient force to drive the solution into the capillary. These types of solutions have been used to separate large DNA fragments (Chen *et al.*, 1989). Entangled polymer networks have also been incorporated into capillaries by *in situ* polymerization (Heiger *et al.*, 1990).

Many of the same generalizations can be made for entangled polymers as for cross-linked gels, as to the effects of varying capillary length, concentration, and capillary diameter. An excellent overview of these parameters with respect to entangled polymer gels is given by Heiger *et al.* (1990). Unlike resolution of gel-filled capillaries, resolution of DNA fragments may be degraded by field strengths above 200 V/cm in some of the more dilute polymer solutions (Dubrow, 1991).

C. New Approaches to Capillary Gels

The use of entangled polymer networks in capillary electrophoresis points out that fundamental advances can be expected in separation owing to the advantageous geometry of the capillary. Viscous, viscoelastic, thixotropic, and liquid crystalline materials, although known to have porous network structures, could not be conveniently or effectively handled in traditional electrophoresis formats. The high aspect ratio and minimal thermal gradients in the capillary now make these materials viable candidates as electrophoresis separation media.

IV. Separations Using Gel-Filled Capillaries

A wide variety of compounds have been separated with gel-filled capillaries. These range from stereoisomers to large biopolymers.

A. Proteins and Peptides

Size separations of proteins with gel-filled capillaries were first shown by Hjertén (1983) and later by Cohen and Karger (1987). They adopted the approach to molecular weight determination that is employed in slab-gel electrophoresis. In this strategy SDS is bound to the protein molecules, resulting in a constant charge-to-mass ratio. In addition, β-mercaptoethanol is used to reduce any disulfide bonds. In this state the proteins can be sieved by the gel, independent of charge and secondary structure. An effective procedure to accomplish the denaturation is detailed in Appendix 2.

Once denatured, the proteins can be injected electrokinetically onto the capillary

(Section II,B). Protein concentrations around 0.5 mg/ml are a good starting point, although they may have to be varied somewhat, depending on detection wavelength and sensitivity and sample complexity.

Ideally proteins should be detected at a wavelength of 200 nm, which is close to their maximum extinction coefficient. However, most synthetic water-soluble polymers absorb strongly below 230 nm. For this reason, most researchers have used 230 or 280 nm as a detection wavelength. Of these wavelengths, 230 nm is preferred because a greater number of amino acids absorb strongly at this wavelength, improving both detectability and peak-area analysis accuracy.

Along with the sample buffer, the running buffer and gel buffer should also contain SDS. The standard convention is to use 0.1% throughout the system. Various buffers, including tris(hydroxymethyl)aminomethane (Tris)-phosphate (Cohen and Karger, 1987) and Tris-glycine have been used successfully in gel-filled capillaries. As with slab gels, various gel concentrations are required to separate different protein size ranges. Proteins between 10 and 40 kDa have been effectively sized by acrylamide gels ranging between 10% total monomer (%T) and 3.3% cross-linker, based on the weight of total monomer (%C) to gels composed of 5% T and 3.3% C. Large proteins (100 to 200 kDa) have been separated by entangled solution techniques containing 5% polyethyleneglycol (Chen *et al.,* 1989).

An example of a capillary gel protein separation is shown in Fig. 6. Four proteins were sized in this electropherogram; lactalbumin (MW, 14,200), β-lactoglobulin (MW, 17,500), trypsin inhibitor (MW, 20,100), and carbonic anhydrase (MW, 29,000). As can be seen in Fig. 6, the four proteins are baseline separated and pass by the detector in less than 25 min. When the log of their molecular weights is plotted against migration time, a linear relationship is established (Fig. 7).

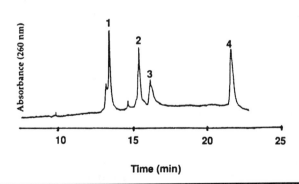

Figure 6   Separations of proteins with molecular weights between 14,200 and 29,000: peak (1) lactalbumin (14,200); peak (2) β-lactoglobulin (17,500); peak (3) trypsin inhibitor (20,100); peak (4) carbonic anhydrase (29,000). Field strength was 300 V/cm; temperature, 30°C; 30 cm effective separation length; detection wavelength, 260 nm; 75 mM pH 7.6 Tris-phosphate buffer with 0.1% SDS was used, and the total polymer concentration was 10.0%.

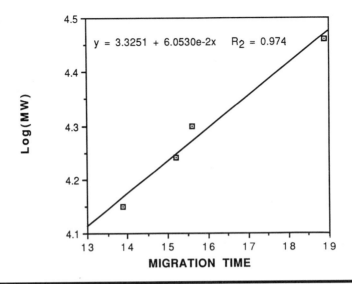

Figure 7 The log of the molecular weight is plotted versus migration time for the electropherogram in Fig. 6.

SDS capillary gel electrophoresis shows potential as a useful analytical tool for the molecular weight determination of proteins. There are, however, several hurdles that must be overcome to make this technique truly viable. The first is that gels must be devised that do not absorb in the 200-nm range, or off-column detection methods must be developed so that sensitivity and quantitative accuracy can be improved. In addition, SDS interacts with most water-soluble polymers, causing instability in the gel matrix and limiting field strengths. Consequently, analysis times are increased, thus dropping sample throughput below that of slab-gel electrophoresis. Resolution using capillary gels appears to be roughly equivalent to that in slab-gel techniques, reducing the incentive to change formats. Improvements in these three areas will make capillary gel electrophoresis a powerful protein analysis tool.

B. Oligonucleotide Analysis

Capillary gel electrophoresis is an efficient technique for directly analyzing the purity of synthetic oligonucleotides (Dubrow, 1991). Methods have been devised in the various disciplines of chemical synthesis to determine the purity of reagents. It would be almost unheard of today to buy a specialty chemical without knowing its purity. For the millions of synthetic oligonucleotides produced annually, purity is not generally determined due to the need to use labor intensive classical techniques such as slab gel electrophoresis or the inability of high-pressure liquid chromatography to deliver adequate resolution.

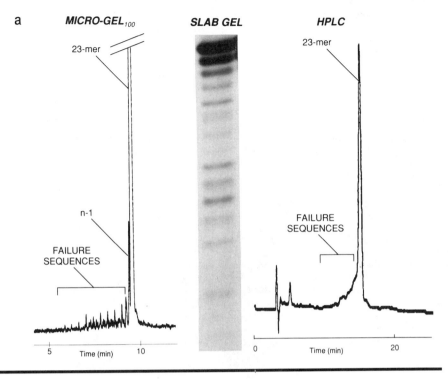

Figure 8 Random sequence oligonucleotides 23 (a), 40 (b), and 120 (c) bases in length were analyzed by capillary gel electrophoresis, slab-gel electrophoresis and HPLC (owing to poor resolution, the 120-mer analyzed on HPLC is not shown). Conditions for capillary gel runs were field strength, 300 V/cm; temperature, 30°C; 30 cm effective separation length; detection wavelength, 260 nm; 75 mM pH 7.6 Tris-phosphate buffer was used with a MICRO-GEL$_{100}$ gel-filled capillary. Slab gels were 20 × 40 × 0.4 cm, 20% polyacrylamide, contained 90 mM Tris-borate and 7 M urea, and were run at 50 mA constant current. The HPLC runs were made on a reverse-phase column with a triethylacetate versus acetonitrile gradient. Detection was performed at 260 nm. *(Figure continues.)*

Figure 8 compares separations of three synthetic oligonucleotides, a 23-, 40-, and 120-mer, by capillary gel electrophoresis, ^{32}P slab-gel electrophoresis, and reverse-phase HPLC. Attempts to analyze the 120-mer by HPLC were unsuccessful, owing to poor resolution (data not shown).

Comparison of the three methods yields the following conclusions:

1. Gel-filled capillaries can provide high resolution up to and beyond 120 bases. This range covers essentially all synthetic oligonucleotides currently produced.
2. One gel concentration can effectively cover the entire range of synthetic oligonucleotides.

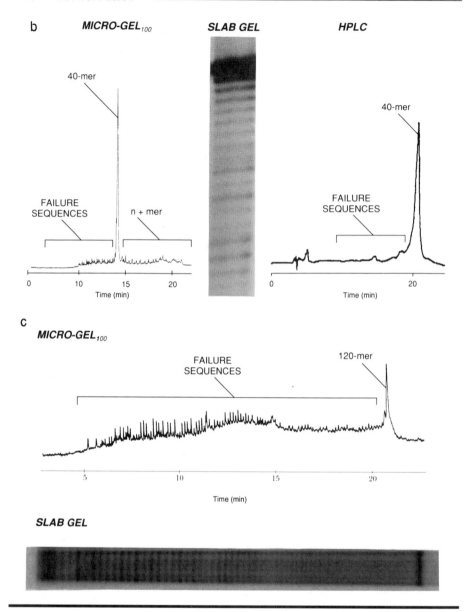

Figure 8   *(Continued)*

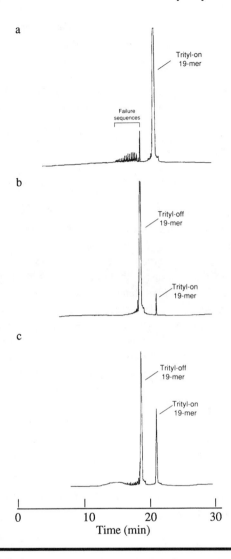

Figure 9 Electropherograms of a 19-base oligonucleotide in its (a) crude trityl-on state, (b) purified trityl-off state, and (c) as a mixture of (a) and (b). Field strength for each run was 300 V/cm; temperature, 30°C; 30 cm effective separation length; detection wavelength, 260 nm; 75 mM pH 7.6 Tris-phosphate buffer was used with a MICRO-GEL$_{100}$ gel-filled capillary.

3. Both capillary gel and slab-gel electrophoresis provide higher resolution than HPLC. For HPLC analysis, resolution is limited to 40 bases or less (Baba *et al.*, 1991). Below 40 bases resolution is significantly poorer than with either gel technique.

4. Detection and data collection with capillary gel electrophoresis is quantitative and done in real time, providing rapid results. Slab-gel detection methods are indirect, requiring staining or radioactive labeling and developing.

5. Capillary gel electrophoresis can analyze samples in minutes, for rapid throughput of critical samples.

Capillary gel electrophoresis can also be used to evaluate the removal of dimethoxytrityl (DMT)-protecting groups from the ends of synthetic oligonucleotides. Complete detritylation is required for the successful use in cloning or ligation experiments and for the use in experiments that require labeling of the oligonucleotide. The trityl-containing species runs significantly slower than the deblocked oligonucleotide. This can be clearly seen in Fig. 9, where electropherograms of a trityl both on and off electropherogram are presented, along with a mixture of the two species.

Oligonucleotides with phosphorylated or dephosphorylated 5′ ends can also be easily detected by capillary gel electrophoresis (Fig. 10). The dephosphorylated species runs slower than the more highly charged phosphorylated oligonucleotide. Clearly, capillary gel electrophoresis can provide a simple and rapid assay for phosphorylation, which often determines biological activity of an oligonucleotide.

Similar to that with slab-gel electrophoresis (Efcavitch, 1991), relatively short oligonucleotides will exhibit a base composition dependence with respect to migration time. Figure 11 shows the electropherogram of a mixture of 12-base-long poly-oligodeoxyadenylic acid (pdA_{12}), polyoligodeoxythymidylic acid (pdT_{12}), and poly-oligodeoxycytidylic acid (pdC_{12}). The order of migration is

$$A > T > C$$

Polyoligodeoxyguanylicacid (pdG) runs considerably slower than pdC, as would be predicted from traditional slab-gel results.

If an internal reference standard of known concentration is included when a crude oligonucleotide sample is prepared for analysis, the synthesis yield can be calculated directly from the electropherogram by determining the ratio of the reference peak to the product peak. This promises to be much more informative than the widely used method of measuring the absorbance of the entire crude product, including failure sequences and impurities, against a reference on a UV spectrophotometer (Fig. 12).

Gel concentrations ranging from 10% T/0% C (Heiger *et al.*, 1990) to 5% T/3% C (Lux *et al.*, 1990) have been used effectively to separate oligonucleotides. Generally Tris-borate buffers containing 7 M urea have been used, although Tris-phosphate has also been used (Applied Biosystems, 1990). Detection is performed using UV absorption at 260 nm.

Sample concentrations between 4 μg/ml and 0.6 μg/ml in water or denaturant solution are typically used. Dilutions from crude synthesis mixtures containing ammo-

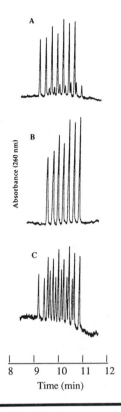

Figure 10 Comparative electropherograms of a $(dA)_{12-18}$ oligonucleotide ladders run on a gel-filled capillary with (A) the oligonucleotides phosphorylated at the 5′ hydroxy end, (B) the oligonucleotides dephosphorylated, and (C) a mixture of the phosphorylated and dephosphorylated ladders. Field strength for each run was 300 V/cm; temperature, 30°C; 30 cm effective separation length; detection wavelength, 260 nm; 75 mM pH 7.6 Tris-phosphate buffer was used with a MICRO-GEL$_{100}$ gel-filled capillary.

nia in deionized water can be injected directly onto the capillary without any intermediate steps. Capillary gel electrophoresis appears to be an efficient method for the analysis of synthetic oligonucleotides. Its combination of ease of use, speed, resolution, and quantitation should make this technique a fixture in synthetic oligonucleotide laboratories.

C. DNA Sequencing

It has been shown that fluorescent-labeled DNA fragments generated in an enzymatic sequencing reaction can be rapidly separated by capillary gel electrophoresis and detected at attomole levels within the capillary (Drossman *et al.,* 1990; Swerdlow and

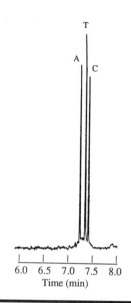

Figure 11 A mixture of four oligonucleotide homopolymers, dA_{20}, dT_{20}, and dC_{20} of equal length were analyzed by capillary gel electrophoresis. Field strength for each run was 400 V/cm; temperature, 30°C; 30 cm effective separation length; detection wavelength, 260 nm; 75 mM pH 7.6 Tris-phosphate buffer was used with a MICRO-GEL$_{100}$ gel-filled capillary.

Gesteland, 1990). A great deal of interest has been focused on this technique in the hope that it might be the cornerstone of a high-speed genomic sequencer.

Current capillary sequencers utilize an argon laser source that is focused onto a capillary, causing excitation of the fluorescent-tagged DNA. The emission is sent through a bandpass filter to select the wavelength region of interest and then detected with a photomultiplier tube, electronically filtered, digitized, and stored on a computer for analysis. Multiple bandpass filters can be used, along with a beam splitter, to detect different labels simultaneously (Drossman *et al.*, 1990). Off-column detection by a sheath flow method has also been used effectively (Swerdlow and Gesteland, 1990).

Polyacrylamide gel concentrations of about 3% T and 5% C seem to be most widely used. Higher concentrations appear to blur longer fragments. Linear acrylamide gels have also been used by Heiger *et al.* (1990) for sequencing. Tris-borate buffers containing 7 M urea have been used exclusively to date for sequencing.

With capillary diameters of typically 50 μm compared to those of slab gels of 300 μm, the time to determine the sequence of a DNA fragment can be reduced by a factor of five to ten. However, unless a relatively simple and cost-effective method for running multiple capillary separations simultaneously is devised, the time per sequenced base on an automated slab gel sequencer is a factor of three faster, since up to 24 lanes can be run simultaneously. The feasibility of ultra-thin (50-μm) slab-gel sequencers is being investigated and may prove to be the optimal format using tradi-

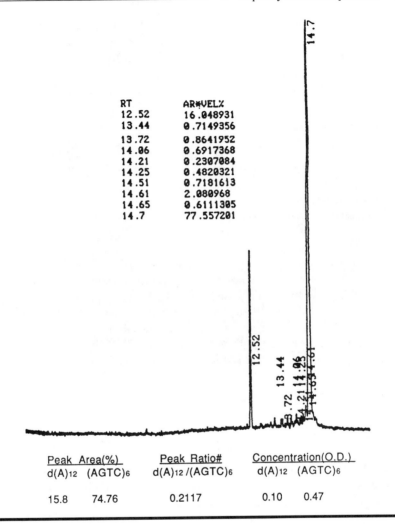

RT	AR*VEL%
12.52	16.048931
13.44	0.7149356
13.72	0.8641952
14.06	0.6917368
14.21	0.2307084
14.25	0.4820321
14.51	0.7181613
14.61	2.080968
14.65	0.6111305
14.7	77.557201

Peak Area(%)		Peak Ratio#	Concentration(O.D.)	
$d(A)_{12}$	$(AGTC)_6$	$d(A)_{12} / (AGTC)_6$	$d(A)_{12}$	$(AGTC)_6$
15.8	74.76	0.2117	0.10	0.47

Figure 12 An electropherogram of a mixture of a crude synthetic (5′-AGCTAG CTAGCT AGCTAG CTAGCT-3′) oligonucleotide and an oligonucleotide marker (pdA_{12}) of known concentration. By determining the area ratio of the two peaks, the concentration of $(AGCT)_6$ is determined. Field strength for each run was 300 V/cm; temperature, 30°C; 30 cm effective separation length; detection wavelength, 260 nm; 75 mM pH 7.6 Tris-phosphate buffer was used with a MICRO-GEL$_{100}$ gel-filled capillary.

tional gels, however, capillary sequencing may show significant advantages in the area of resolution in the future. Current research has utilized conventional acrylamide gels, and resolution has been comparable to that with traditional sequencing results. Moving into novel gel materials made possible by the capillary format could have profound scientific implications, if larger fragments can be sequenced.

D. Double-Stranded DNA

Excellent separations of double-stranded DNA have been performed with gel-filled capillaries (Fig. 13). The most impressive work to date has been the use of a capillary filled with 3% T, 0.5% C acrylamide to base-line separate a DNA ladder from 71 to 12,000 bases (Heiger *et al.*, 1990). With the combination of speed, high resolution, and wide separation range, gel-filled capillary electrophoresis should find widespread use in evaluating the growing number of polymerase chain reaction (PCR) products for size and concentration.

As was mentioned in Section III,D, electrophoretic separations of double-stranded DNA have been accomplished in dilute polymer solutions. Hydroxylated cellulose derivatives have been widely used, generally in the concentration range of 0.1% to 1.0% (Chin and Colburn, 1989). With this technique the polymer solution can be drawn into the capillary under vacuum and subsequently rinsed out and replaced if necessary. This eliminates the problems associated with void formation and subsequent electrical failures in gel-filled capillaries. This technology is covered in depth in Chapter 8.

E. Chiral Molecules

Complexing agents have been incorporated into gel-filled capillaries to carry out stereospecific separations. The addition of β-cyclodextrin to polyacrylamide gel-filled capillaries has been shown to be effective in accomplishing electrophoretic separations of dansylated amino acids (Guttman *et al.*, 1988; Smith, 1990). Figure 14 shows an example of a separation of dansylated amino acids.

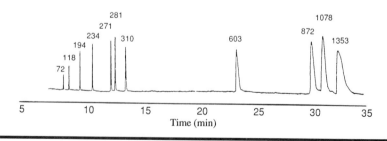

Figure 13   An electropherogram of oX174/*Hae* III restriction fragments separated on a gel-filled capillary is shown. Field strength for each run was 300 V/cm; temperature, 30°C; 30 cm effective separation length; detection wavelength, 260 nm; 75 mM pH 7.6 Tris-phosphate buffer was used with a MICRO-GEL$_{100}$ gel-filled capillary.

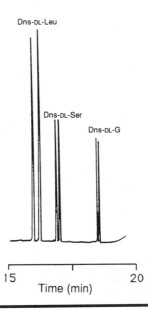

Figure 14 Separation of dansyl DL-isomers of leucine, serine, and glutamine on a gel-filled capillary containing β-cyclodextrin. Field strength was 300 V/cm; temperature, 30°C; 40 cm effective separation length; detection wavelength, 254 nm; Tris-borate pH 8.3 buffer was used, containing 7 M urea. A polyacrylamide gel was used in the capillary. From Smith (1990). Permission of Applied Biosystems.

Initially, chiral separations by capillary electrophoresis are being used to confirm HPLC separations. As research continues in this field, unique capabilities should be developed. Functionalization of the gel matrix with chiral agents is a particularly promising area.

V. Future Directions

The intensity of research and growing list of applications of capillary gel electrophoresis point to the potential of this technique to solve separations problems. The first wave of research has primarily focused on the challenge of reproducibly filling capillaries with polyacrylamide gels and optimizing their separations properties.

Capillaries based on this initial technology have reduced sample requirements and analysis times over existing separations methods. However, it is going to take a second wave of research that focuses on improved resolution and drives analysis times down to several minutes to realize the potential of capillary gel electrophoresis. This potential should be realized because of the geometrical advantage of the capillary, which opens the door to a variety of advanced separations materials.

Appendix 1: Karger and Cohen Gel-Fabrication Technique

1. A 45-cm length of 75 i.d. fused-silica capillary tubing is cut and 1 cm of its polyimide coating is burned off near one end to serve as a detector window.
2. The tubing is heated to 120°C in air, then flushed at room temperature with ammonia gas for 2 hr. It is then filled with a 50% solution of 3-methacryloxypropyltrimethoxysilane and allowed to sit overnight. Successive flushings of methanol and water are done to rinse out any unreacted coupling agent.
3. Buffer solution is made by dissolving 1.1 g Tris and 0.01 g ethylenediaminetetraacetic acid (EDTA) in 100 ml 7 M urea solution and then titrating to a pH of 8.6 with sodium dihydrogen phosphate.
4. A concentrated monomer solution containing 29 g acrylamide and 1 g N,N-methylenebisacrylamide in 100 ml buffer solution is then prepared.
5. Next a catalyst solution is made by dissolving 0.2 g ammonium persulfate in 2.0 ml buffer solution.
6. The final gel solution is made by adding 10 ml of the monomer concentrate to 30 ml buffer solution, then adding 4.0 μl of the persulfate solution and 2.5 μl of N,N,N',N'-tetramethylethylenediamine. This mixture is designed to polymerize in about 45 min.
7. In excess of 50 μl of this solution is forced through the capillary until bubbles no longer emerge. The gel is then allowed to polymerize in the capillary for several hours with the ends dipped in buffer.

Appendix 2: Protein Denaturation Procedure

1. Dilute desired proteins to 10 mg/ml in deionized water.
2. Prepare a denaturing solution composed of 2% SDS and 2% β-mercaptoethanol in deionized water.
3. Combine equal volumes of protein and denaturing solutions, about 100 μl total volume is sufficient, in a 0.5-ml centrifuge tube.
4. Heat this solution to 90°C for 15 min to denature the proteins.
5. Remove an aliquot of the denatured protein solution and dilute to 1 part protein solution to 9 parts deionized water.
6. This final solution contains 0.5 mg/ml protein, 0.1% SDS and 0.1% β-mercaptoethanol.

References

Baba, Y., Enomoto, S., Chin, A. M., and Dubrow, R. S. (1991). *J. High Res. Chromatogr.* **14**, 204–206.
Bente, P. F., and Myerson, J. (1989). U. S. Patent 4,810,456.
Chen, A. J. C., Zhu, M., Hansen, D., and Burd, S. (1989). "Bulletin 1479." Bio-Rad Laboratories, Richmond, California.
Chin., A. M., and Colburn, J. (1989). *Am. Biotech. Lab.* **7**, 10A.

Chrambach, A., and Rodbard, D. (1971). *Science* **172**, 440–451.

Cohen, A. S., and Karger, B. L. (1987). *J. Chromatogr.* **397**, 409–417.

Cohen, A. S., and Paulus, A., and Karger, B. L. (1987). *Chromatographia* **24**, 15–24.

Drossman, H., Luckey, J. A., Kostichka, A., D'Cunha, J., and Smith, L. M. (1990). *Anal. Chem.* **62(9)**, 900–903.

Dubrow, R. S. (1990). *Appl. Biosys. Res. News* **2**, 11–13.

Dubrow, R. S. (1991). *Am. Lab.* March, 64–67.

Dubrow, R. S., and Harrington, S. (1990). 2nd Conference on High Performance Capillary Electrophoresis. Poster 205. February, San Francisco, CA.

Efcavitch, W. (1991). *In* "Gel Electrophoresis of Nucleic Acids," (D. Rickwood and B. D. Hanes, eds.) pp. 125–149. IRL Press, Oxford.

Guttman, A., Paulus, A., Cohen, A. S., Grinberg, N., and Karger, B. L. (1988). *J. Chromatogr.* **448**, 41–53.

Guttman, A., Ohms, J., and Cooke, N. (1991). 3rd Conference on Capillary Electrophoresis. Poster PT-10. February, San Diego, CA.

Hansen, D., Zhu, M., Rodriguez, R., Wehr, T., and Benoy, C. (1990). 2nd Conference on High-Performance Capillary Electrophoresis. Poster 245. February, San Francisco, CA.

Heiger, D. N., Cohen, A. S., and Karger, B. L. (1990). *J. Chromatogr.* **516**, 33–48.

Hjertén, S. (1983). *J. Chromatogr.* **270**, 1–6.

Janick, B. (1986). *In* "Encyclopedia of Polymer Science and Technology," pp. 772–792. Wiley Interscience, New York.

Karger, B. L. (1990). 2nd Annual Conference on High Performance Capillary Electrophoresis. February, San Francisco, CA.

Karger, B. L., and Cohen, A. S. (1989a). U. S. Patent 4,865,707.

Karger, B. L., and Cohen, A. S. (1989b). U. S. Patent 4,865,706.

Karger, B. L., Cohen, A. S., and Guttman, A. (1989). *J. Chromatogr.* **492**, 585–614.

Lux, J. A., Yin, H. F., and Schomburg, G. (1990). *J. High Res. Chromatogr.* **13**, 436–437.

Smith, N. (1990). *App. Biosys. Report.* **8**, 1–6.

Swerdlow, H., and Gesteland, R. (1990). *Nucleic Acids Res.* **18**, 1415–1419.

Tietz, D., Gottlieb, M. H., Fawcett, J. S., and Chrambach, A. (1986). *Electrophoresis 1986* **7**, 217–220.

Fundamentals of Micellar Electrokinetic Capillary Chromatography

Michael J. Sepaniak, A. Craig Powell, David F. Swaile, and Roderic O. Cole

I. Introduction

The commercial availability of narrow-bore (25- to 75-μm i.d.), fused-silica capillaries and their intrinsic ability to dissipate electrophoretically generated heat have engendered the development of capillary zone electrophoresis (CZE), an extremely efficient technique for separations of charged solutes (Jorgenson and Lukacs, 1981). The technique can be used to separate solutes ranging in size from small inorganic ions to large proteins and oligonucleotides. The results of fundamental studies of factors that influence separation performance, instrumental improvements, and numerous applications have been reported during the last decade (Ewing *et al.*, 1989). In addition to the advantage of high efficiency, the technique requires only minute amounts of sample and reagents. In many instances these characteristics represent distinct advantages of CZE over conventional electrophoretic or chromatographic techniques.

The primary limitation of CZE is its inability to separate neutral compounds. During the past decade, efforts to develop open capillary liquid chromatography (OCLC) for the separation of neutrals have met with only marginal success (Novotny, 1988). Similar to the CZE technique, OCLC has the potential for high efficiency, a consequence of high permeability that permits the use of long capillaries. The technique also requires only minute amounts of reagents. Unfortunately, in order to minimize band dispersion caused by poor mass transfer in the mobile phase, extremely narrow-bore (<10 μm i.d.) capillaries must be used in OCLC (Knox and Gilbert, 1979; Knox, 1980; Guiochon, 1981). This places stringent volume requirements on injection and detection procedures. Moreover, the preparation of stable, high-capacity stationary phases, by chemical attachment to the inside surfaces of these narrow-bore capillaries, is not easily accomplished. Column occlusion is commonly encountered. In 1984 Terabe and co-workers introduced a modified version of CZE (Terabe *et al.*, 1984) in which surfactant-formed micelles are included in the running buffer to provide a two-phase chromatographic system for separating neutral compounds. The running buffer represents the primary phase, which is electroosmotically transported through the capillary, and the micelles represent the secondary phase, which is transported through the capillary by a combination of electroosmotic flow and micelle electrophoretic migration (see Section II,A). The buffer/surfactant systems studied to date exhibit a micelle electrophoretic migration that opposes electroosmotic flow, but is of a smaller magnitude; thus, both phases are transported at different velocities toward a common end of the capillary. It is, however, possible to reduce the electroosmotic flow by reducing the pH of the running buffer (below approximately 5) or by modifying the capillary surface (see Section II,A), such that the running buffer and micelles counterpropagate. This condition is less versatile for actual separations since certain solutes may not be transported toward the detection end of the capillary (Otsuka and Terabe, 1989). Partitioning of neutral (or ionic) solutes between these differentially migrating phases imparts a chromatographic mode of separation to CZE. We have dubbed this technique micellar electrokinetic capillary chromatography or MECC (Sepaniak *et al.*, 1987).

The MECC technique offers several advantages over the competing separation techniques discussed. Whereas the instrumentation is identical to that employed in CZE, MECC is more versatile in that differences in electrophoretic mobility can be exploited to separate ionic solutes and, simultaneously, differences in phase distribution can be exploited to separate neutral solutes. We illustrated this in the separation of vitamin B_6 and several of its metabolites, a mixture containing compounds with both acidic and basic functionalities (Burton *et al.*, 1986a). When compared with reverse-phase ion-pair chromatography (a competing, conventional technique), much higher efficiency was obtained using only minute amounts of sample (Swaile *et al.*, 1988). The "pluglike" profile of electroosmotic flow and the very uniform distribution of the colloidal-sized micelles provide for less dispersion due to mass transfer problems, relative to the OCLC technique (see Section II,B). High efficiency is obtained with moderate-sized capillaries, thereby minimizing occlusion as well as difficulties with injection and detection. Moreover, it is not necessary to form a bonded stationary phase. In fact, the micellar phase is continually regenerated and can be easily changed to improve the separation. Despite these advantages, the MECC technique is no panacea in the separation of minute amounts of neutral compounds. Theoretical and practical aspects of the technique are both complicated and limited by the fact that the secondary (micellar) phase has a net velocity and an effective volume that is critically dependent on separation conditions. Thus, reproducible retention requires very strict attention to experimental details. A finite elution range (or "window"), which limits peak capacity, is generally observed (see Section II,A), rendering MECC inappropriate for the separation of extremely complex mixtures. Although retention characteristics usually resemble reverse-phase liquid chromatography (LC), it is not possible to separate hydrophobic compounds, as micelles are not particularly stable in the mobile phases typically used for the separation of these compounds by reverse-phase LC. We elaborate on these and other considerations throughout this chapter. The remainder of this section is devoted to a cursory description of surfactant–micelle systems and their properties as they relate to MECC. More detailed information on micelles can be found elsewhere (Attwood and Florence, 1983; Hinze, 1987).

Surfactants are amphiphilic species, comprising both hydrophobic and hydrophilic regions. They can be anions, cations, zwitterions, or even polar neutral molecules. In most cases the surfactant contains a polar and/or ionic head group attached to a nonpolar hydrocarbon tail. In the appropriate solvent, surfactants can aggregate to form colloidal-sized assemblies such as vesicles and micelles (Hinze, 1987). Although there is some controversy concerning the exact structure and shape of micelles (Hinze, 1987), they are generally depicted as spherical aggregates. The nature of the micelle and the micellization process is dependent on experimental conditions such as temperature, solvent type, the presence of additives, etc. Normal micelles (hydrophobic tails pointing inward, polar head group defining the surface of the micelle) are formed rapidly when the surfactant concentration is above the critical micelle concentration (CMC) via the equilibrium:

$$\text{monomer} \leftrightharpoons N\text{-mer}$$

where N is referred to as the aggregation number. Table I is a listing of CMCs and aggregation numbers for some common surfactants.

Most micelles have N values in the 40–120 range (Hinze, 1987). Solutions containing mixtures of surfactants can form "mixed" micelles (Balchunas and Sepaniak, 1988). At the onset of micelle formation (i.e., at concentrations very near the CMC), it is likely that "small" micelles are present. As the concentration of surfactant is increased appreciably beyond the CMC, N-mer micelle formation dominates, and the free-monomer concentration is maintained at the CMC value. Surfactants and conditions that yield high CMCs are undesirable in MECC, since the resulting large concentration of free charged monomer results in a significant thermal load that must be dissipated by the capillary (see Section II,B). It should also be understood that a reported aggregation number represents an average, and that normal micelles are dynamic entities that are constantly exchanging with monomer to form new micelles with sizes ranging around that of the N-mer micelle. This is a critical point with regard to obtaining high efficiency in MECC. This *polydispersity* of micelles can result in a range of micelle migration velocities, producing a source of band dispersion that is not experienced in other separation techniques, even micellar LC.

Table I Critical Micelle Concentration and Aggregation Number for Some Common Surfactants Used in MECC

Surfactant (abbreviation)	Structure	CMC $(mM)^a$	N
Anionic surfactants			
Sodium dodecyl sulfate (SDS)	$C_{12}H_{25}OSO_3^-Na^+$	8.1	62
Sodium octyl sulfate (SOS)	$C_8H_{17}OSO_3^-Na^+$	136.0	20
Cationic surfactants			
Hexadecyltrimethylammonium chloride (CTAC)	$C_{16}H_{33}N^+(CH_3)_3Cl^-$	1.3	78
Dodecyltrimethylammonium bromide (DTAB)	$C_{12}H_{25}N^+(CH_3)_3Br^-$	15.0	50
Zwitterionic surfactant			
N-dodecylsultaine (SB-12)	$C_{12}H_{25}(CH_3)_2N^+CH_2CH_2CH_2SO_3^-$	1.2	
Nonionic surfactant			
Polyoxyethylene-t-octylphenol (Triton X-100)	$(CH_3)_3CCH_2C(CH_3)_2C_6H_4(OCH_2CH_2)_{9.5}OH$	0.2	143
Bile salt			
Sodium deoxycholate (NaDC)		6.4	14

^{a} Critical micelle concentration (CMC) and aggregation number (N) data reproduced with permission from Hinze (1987).

Changes in experimental conditions can have dramatic effects on micelle properties. Whereas pressure is generally not a factor, temperature changes, whether owing to ambient fluctuations or electrophoretically generated heat, influence the CMC. For ionic surfactants, CMCs generally exhibit a minimum value in the 20 to 30°C range (Hinze, 1987). If high-ionic-strength running buffers are employed, capillary temperatures in MECC can be considerably greater than this and, furthermore, transverse temperature gradients can be created in the capillary (see Section II,B). The addition of electrolytes to aqueous micelle systems usually results in an increase in aggregation number and a reduction in CMC. For example, the CMC of a commonly used surfactant, sodium dodecyl sulfate (SDS), is approximately 8 m*M* in water. However, we show evidence of micelle formation at approximately 5 m*M* using the phosphate/borate running buffers we commonly employ in MECC (Sepaniak and Cole, 1987). Organic solvents also affect micelle formation. Small alcohols enhance micelle formation at very low v/v %, but inhibit micelle formation at higher concentrations (e.g., SDS will not form micelles in mixed organic-aqueous solvents containing more than about 35% v/v methanol). Solvents that hydrogen bond with water, such as acetonitrile, also increase CMC, even at moderate v/v % values (Hinze, 1987). This organic solvent effect is an important consideration in MECC, since these solvents are commonly used as organic modifiers to alter retention in reverse-phase LC. Moreover, reductions in micelle concentration due to the addition of these solvents can result in serious reductions in efficiency (see Section II,B).

Differential association of injected solutes with the micellar phase is an obvious requisite for effective separation by the MECC technique. Moreover, the overall magnitude of the association (as reflected in capacity factors, k's) is more critical in MECC than in conventional chromatography owing to a distinct k' optimum (Terabe et al., 1985a). Solutes can associate with normal micelles by a variety of mechanisms (Hinze, 1987). Depending on the nature of the solute/surfactant system, association can involve different regions of the micelle. Nonpolar to moderately polar solutes interact with the hydrophobic core of the micelle based on dispersive and/or induced dipole interactions. The nature of the interaction and the observed elution order is similar to that experienced in reverse-phase LC. The phase ratio ($V_{micelle}/V_{mobile}$) for a 0.05 M SDS solution is greater than 10^{-2}. This relatively large phase ratio, and the instability of micelle systems in mobile phases with large organic content (see previous discussion), often result in prohibitively large capacity factors for nonpolar solutes in MECC (i.e., components of nonpolar mixtures all completely associate with the micelle and are not separated). The use of nontraditional surfactants (Cole et al., 1990) and unusual methods of imparting organic character to the mobile phase (Terabe et al., 1990) have improved this situation somewhat. Polar solutes can associate with the micelles, near the polar (or charged) surfactant head groups, via dipole–dipole interactions. Amphiphilic solutes (e.g., organic compounds containing acid or base functionalities) tend to align themselves to interact dispersively with the hydrophobic core and electrostatically with the head groups of the micelles. For such solutes, it is quite possible that acid/base strengths can be influenced by the unique environment within and near the micelle. Strong coulombic interactions (attractive or repulsive, depending on solute charge) can be involved when charged solutes associate with charged head groups at the surface of the micelle. Thus, depending on the nature of the solute, many different mechanisms can influence micelle–solute association.

A variety of surfactants (including all of those listed in Table I) have been used in MECC. In some instances, mixed micelles have been employed to either impart a unique selectivity (Dobashi et al., 1989; Dobashi et al., 1989b) or alter micelle velocity (Balchunas and Sepaniak, 1988). Chiral surfactants such as the bile salts (Cole et al., 1990; Cole et al., 1991) and modified amino acids (Dobashi et al., 1989a,b) have been used for enantiomeric separations. More information can be found in Section IV and Table V. Because of the ease with which the composition of the solution in the capillary can be changed, and the existence of many untried surfactants, it is likely that new micelle systems will be employed in MECC in the future.

II. Theory

In MECC, the micelles have a charged surface and possess an electrophoretic velocity, $v_{e,m}$, that opposes the electroosmotic velocity, v_{eo} (see Fig. 1 and discussion in Section I). The net velocity of the micelles, $v_M = v_{eo} + v_{e,m}$, and the velocity of a solute that partitions between the phases, v_s, are depicted in Fig. 1. Differential solute–micelle association (see Section I), provides a retention mechanism by which solutes can be separated. In contrast to conventional chromatography, the secondary phase (micelles) elutes from the column. Consequently, even solutes that totally associate with the micellar phase are eluted, resulting in a limited elution range for MECC. This elution range is often expressed as the parameter, t_0/t_m, which is the ratio of the elution (or migration) time of a totally unretained component to the elution time of the micellar phase. For the most commonly used MECC system (SDS in phosphate/borate buffer) this ratio typically has a value between 0.2 and 0.5. For conventional LC, such a ratio would have a value of zero, since the secondary phase is stationary and t_m is infinitely large. This limited elution range in MECC can result in poor peak capacity, relative to conventional LC, and places constraints on optimal retention characteristics (see discussion below).

A. Capacity Factor, Selectivity

A limited elution range has a number of ramifications that alter some of the fundamental equations of conventional LC. For example, in conventional LC the capacity factor

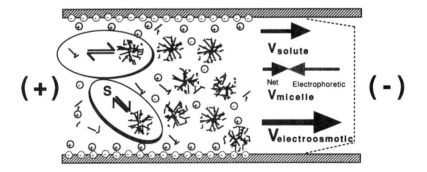

Figure 1   Depiction of flow dynamics in MECC showing velocity vectors for electroosmotic flow ($V_{electroosmotic}$ or v_{eo}), micellar velocity [both its electrophoretic ($v_{e,m}$) and net (v_M) vectors], and solute velocity (V_{solute}). Also depicted are solute–micelle association and surfactant monomer–*N*-mer dynamics.

of a solute (the ratio of the amount of solute present in the stationary phase to that in the mobile phase), k', is expressed as (Johnson and Stevenson, 1978)

$$k' = \frac{t_R - t_0}{t_0} \tag{1}$$

where t_R is the retention time of the solute. For MECC, the k' expression becomes (Terabe et al., 1984)

$$k' = \frac{t_R - t_0}{t_0(1 - t_R/t_m)} \tag{2}$$

Note that as t_m approaches infinity, the parenthetical expression approaches unity, and the expression becomes equivalent to the conventional LC equation.

Likewise, the limited elution range of MECC affects the expression for resolution, R_S (Terabe et al., 1985a):

$$R_S = \frac{N^{1/2}}{4}\left(\frac{\alpha - 1}{\alpha}\right)\left(\frac{k_2'}{1 + k_2'}\right)\left(\frac{1 - t_0/t_m}{1 + (t_0/t_m)k_1'}\right) \tag{3}$$

where N is the number of theoretical plates, k_1' and k_2' are the capacity factors for two solutes, and α is the selectivity factor (the ratio k_2'/k_1'). If t_m has a value of infinity, the resolution equation becomes the same as in the conventional case.

As mentioned previously, the limited elution range influences the peak capacity, n, of the technique (Willard et al., 1981):

$$n = 1 + \frac{N^{1/2}}{4}\ln\frac{t_n}{t_1} \tag{4}$$

where t_n is the retention time of the nth component ($t_n = t_m$ in MECC), and t_1 is the retention time of the earliest-eluting component ($t_1 = t_0$ in MECC). If no effort is made to extend the elution range, $t_0/t_m \sim 0.3$ (Balchunas and Sepaniak, 1987). Assuming $N = 10^5$, this corresponds to a peak capacity of less than 100. Whereas this is reasonably large, it assumes evenly distributed peaks. Moreover, it is much less than can be obtained for that efficiency using conventional techniques that do not exhibit a limited elution range.

For conventional chromatography, R_S increases with increasing k'. However, with MECC the presence of the fourth factor in Eq. (3) alters this relationship, because the product of the last two parenthetical factors decreases at large k' values. These last two factors, and thus R_S, experience a maximum when k' is between one and five (Terabe et al., 1985a). The optimal value of k' is dependent on the breadth of the elution window. As the magnitudes of v_{eo} and $v_{e,m}$ approach the same value, but opposite in sign, the elution window is extended and the optimal k' is increased (Terabe et al., 1985a; Otsuka and Terabe, 1989). Unfortunately, under this condition, reproducible retention times may not be easily achieved.

Since R_S (as a function of k') experiences a maximum in MECC, it follows that reduction of the large k' values of the late eluting components would improve their resolution. As seen in Eq. (2), an increase in t_m relative to t_R would serve to decrease k' values. The time of elution of the micellar phase is determined by v_M, which, as mentioned earlier, is the sum of v_{eo} and $v_{e,m}$ (Terabe *et al.*, 1985a):

$$v_{eo} = \frac{-\varepsilon\, \zeta_c\, E}{\eta} \tag{5}$$

$$v_{e,m} = \frac{2\varepsilon\zeta_m}{3\eta} f(\kappa a) E \tag{6}$$

In these equations, ε is the permittivity of the mobile phase, ζ_c and ζ_m are (respectively) the zeta potentials for the column wall and the micelles, η is the viscosity of the mobile phase, E is the electric field strength, and $f(\kappa a)$ is dependent on the micelle shape and size. Because v_{eo} is generally larger and opposes $v_{e,m}$, reductions in ζ_c will decrease v_{eo} and v_M. This produces the desired increase in t_m relative to t_R. Moreover, as t_0/t_m is decreased, the magnitude of the optimal R_S is increased and shifted to larger k' values.

Terabe *et al.* (1985a) demonstrated that changes in SDS concentration above the CMC resulted in no significant deviation in v_{eo}. However, one method of reducing the zeta potential at the solution–column wall interface is to modify or change the wall surface. This can be accomplished in a number of ways. For example, reduction in the pH of the mobile phase (Otsuka and Terabe, 1989; Rasmussen and McNair, 1989) can serve to "titrate" the silanol sites on the capillary surface, reducing ζ_c. In particular, Otsuka and Terabe (1989) demonstrated that at solution pH values below approximately five, v_{eo} is reduced to a value less than $v_{e,m}$, causing v_M to change sign. A side effect of such a technique is that the acid/base functionalities of injected solutes can be altered, significantly changing their retention characteristics (Cole *et al.*, 1990). An alternative method is to silanate the fused silica surface of the capillary with a silanating agent such as trimethylchlorosilane, (TMCS) (Balchunas and Sepaniak, 1987), which reacts with the silanol sites on the capillary surface. Such a wall treatment produced a significant increase in t_m. Lux and co-workers (1990) have also applied polymer coatings of polymethylsiloxane, PMS, and polyethylene glycol, PEG, to the capillary wall, demonstrating the ability to either increase (with PMS) or decrease (with PEG) electroosmotic flow rates. Lukacs and Jorgenson (1985) reported smaller ζ_c and v_{eo} values in CZE using Teflon capillaries. Unfortunately, this was accomplished at the expense of band dispersion (presumably a result of the need for a detection flow cell) and thermal problems (due to the poor heat dissipation of Teflon).

The k' of a solute is the product of its thermodynamic partition coefficient, K, and the phase ratio, β (ratio of the stationary-phase volume to the mobile-phase volume), of the column.

$$k' = \beta K \tag{7}$$

In reverse-phase LC the incorporation of organic solvents in the mobile phase changes solute k' values, based on predictable changes in K. As stated above, MECC mimics reverse-phase LC in retention characteristics (notice the similarities in elution order for solutes separated using reverse-phase LC and SDS-MECC shown in Table II). In MECC, the retention situation is complicated, however, by the fact that organic solvents affect β as well as K. This is because β is the ratio of the micellar-phase volume to the mobile-phase volume, and organic solvents affect the CMC and thus the micellar volume. As a further complication, organic additives affect the viscosity and dielectric constants of the mobile phase and ζ_c, the net result being a change in the electroosmotic and electrophoretic flow rates (Fujiwara and Honda, 1987b).

Initial experiments with mixed organic-aqueous mobile phases have involved adding isopropanol, methanol, or acetonitrile to the mobile phase (Balchunas and Sepaniak, 1987, 1988; Balchunas *et al.,* 1988; Gorse *et al.,* 1988; Bushey and Jorgenson, 1989a,b). Particularly noteworthy is Bushey and Jorgenson's successful separation of dansylated methylamine and dansylated methyl-d_3-amine using a mobile phase containing 20% methanol (Bushey and Jorgenson, 1989a). These bulk additions of organic modifier to the mobile phase have produced significant reductions in k' values for late-eluting components, but at the expense of extremely long analysis times (i.e., very large elution windows were observed). This time factor is particularly problematic with alcohol organic modifiers, which strongly interact with the capillary wall and dramatically alter v_{eo}.

A logical extension of the use of organic modifiers to manipulate k' values is the application of solvent gradients. This allows early-eluting components (already possessing small k' values) to elute before substantial exposure to the organic modifier. Furthermore, gradients tend to dampen the effect of organic modifier on t_m, since v_M decreases only during later stages of the separation. Solvent stepwise gradients (wherein the voltage is turned off long enough for the inlet reservoir solution to be replaced with a new solution containing a higher concentration of organic solvent) were shown to have a marked effect on k' values of late-eluting components without the aforementioned excessive increase in elution range (Balchunas and Sepaniak, 1988; Balchunas *et al.,* 1988). Recently, the apparatus for performing continuous solvent

Table II   Comparison of Elution Order and Capacity Factor (k') for Reverse-Phase LC (RP-LC), SDS-MECC, and CTAC-MECC

	Elution order (k')				
Compound	RP-LC[a] (C18)	MECC[b] (SDS)	MECC (CTAC)	Gas phase dipole moment (D)	K_a^c
Aniline	1 (0.37)	1 (0.46)	1 (0.58)	1.5	10^{-5}
Phenol	2 (0.50)	2 (0.60)	3 (2.4)	1.5	10^{-10}
Nitrobenzene	3 (1.3)	3 (1.6)	2 (1.4)	4.2	—
Toluene	4 (4.8)	4 (4.4)	4 (3.2)	0.36	—

[a] RP-LC: 60% methanol, 40% aqueous (0.001 M Na$_2$HPO$_4$, 0.0006 M Na$_2$B$_4$O$_7$).

[b] MECC: 0.05 M surfactant, 0.01 M Na$_2$HPO$_4$, 0.006 M Na$_2$B$_4$O$_7$.

[c] K_a: of acid or conjugate acid.

gradients has been described (Sepaniak *et al.*, 1989), allowing for a virtually endless number of gradient shapes ranging from concave to convex. It has also been demonstrated that the shape and relative extremity of the gradient can influence the resolution and peak profiles (Sepaniak *et al.*, 1989) (see Fig. 2). An attempt has also been made to quantitate the effect of solvent gradients on MECC to allow a computational method for gradient-optimized separations (Powell and Sepaniak, 1990).

Organic modifiers can also influence selectivity. Gorse *et al.*, (1988) demonstrated that methanol and acetonitrile can favorably alter selectivity between toluene and nitrobenzene and between pyridine and phenol; in fact, the two solvents were shown to have differing influences on selectivity depending on solute polarity. Liu and co-workers have also shown beneficial changes in selectivity by using methanol and tetrahydrofuran (Liu *et al.*, 1989b). Moreover, Sepaniak and co-workers demonstrated a *reversal* in elution order for substituted phenolic compounds using an acetonitrile step gradient (Balchunas *et al.*, 1988). Other mobile-phase additives, such as tetraalkylammonium salts (Nishi *et al.*, 1989d) and divalent metal cations of copper, zinc, and magnesium (Cohen *et al.*, 1987) also influence selectivity.

Selectivity can also be improved by employing a micellar system that enhances differences in micelle–solute association. Surfactants such as sodium decyl sulfate

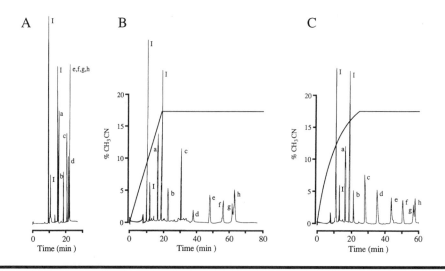

Figure 2   Effect of acetonitrile solvent gradients on MECC separations, comparing (A) no solvent modification of the mobile phase to (B) a linear gradient and (C) a convex gradient. Sample injected consisted of (a) NBD-*n*-propylamine, (b) NBD-*n*-butylamine, (c) NBD-*n*-pentylamine, (d) NBD-*n*-hexylamine, (e) NBD-*n*-heptylamine, (f) NBD-*n*-octylamine, (g) NBD-*n*-decylamine, (h) NBD-*n*-dodecylamine, and (I) impurities. Experimental conditions: 50-μm i.d. $\times$ 80-cm capillary; 24 kV applied voltage; mobile phase, 0.05 M SDS, 0.01 M Na$_2$HPO$_4$, and 0.006 M Na$_2$B$_4$O$_7$.

(STS), dodecyltrimethylammonium chloride (DTAC), and cetyltrimethylammonium chloride (CTAC) have been tested as pseudostationary phases for representative weakly acidic, weakly basic, polar, and nonpolar solutes (Burton *et al.*, 1987). In particular, CTAC, a cationic surfactant, displayed differing selectivity from SDS, an anionic surfactant. Table II illustrates this difference. Notice, in particular, the shift in elution order and k' for phenol between CTAC- and SDS-based micelles. This shift can be explained, given the acidic nature of phenol in terms of its increased affinity for the CTAC-micelle due to electrostatic interactions with the cationic head group of CTAC. A similar effect might have been observed for aniline using SDS, if the buffer system permitted the alkaline nature of aniline to be expressed. A phosphate/borate, pH ~ 8.5, system was employed in that work.

Terabe *et al.* (1985b) have also employed an *inclusion complex* mechanism, using an ionized form of β-cyclodextrin, to separate a number of organic isomers. Mixed micellar solutions, containing SDS and sodium octyl sulfate (SOS), have also demonstrated greater selectivity in the separation of catechols (Wallingford *et al.*, 1989). Furthermore, certain bile salts, a class of biological surfactants, have been used in MECC, demonstrating not only a general reduction in k' values due to their relatively polar nature but the possibility of chiral separations (Nishi *et al.*, 1989a; Terabe *et al.*, 1989b; Cole *et al.*, 1990; Nishi *et al.*, 1990b). As an example of the latter, consider the separation of isomeric binaphthyl compounds (useful in chiral syntheses) presented in Fig. 3. Dobashi and co-workers have likewise applied an enantioselective surfactant, *N*-dodecanoyl-L-valinate, to the separation of amino acid enantiomers, both as part of a mixed micellar system [with SDS (Dobashi *et al.*, 1989b)] and by itself (Dobashi *et al.*, 1989a). In combination, the successes of these various MECC systems indicate the versatility of the technique.

B. Efficiency

As mentioned earlier, MECC complements CZE by effectively addressing the separation of neutral compounds with high efficiency. On a fundamental level, the narrow bands obtained in MECC are the result of the relatively flat velocity profile characteristic of electroosmotic flow (Pretorius *et al.*, 1974). This differs from the hydrostatically pumped open capillaries that are employed in OCLC, where the resulting flow profile is parabolic. This radial mobile-phase velocity gradient often leads to excessive band broadening. To reduce this effect to acceptable levels, OCLC columns must be extremely narrow so that sufficient radial diffusion of injected solutes across the narrow flow profile can occur as they migrate down the capillary. Consequently, all solutes experience a net mobile-phase velocity that is the mean value in the capillary [*i.e.*, an "averaging-out effect" is observed (Sepaniak *et al.*, 1987)]. It has been stated that capillaries with i.d.s of 5 to 10 μm are required to yield high efficiency in OCLC (Knox and Gilbert, 1979). Use of such small diameters places stringent limitations on detection, injection, and column fabrication (Jorgenson and Guthrie, 1983; Sepaniak *et al.*, 1987). The "pluglike" flow profile resulting from the application of a large potential across 25- to 100-μm i.d. capillaries allows high efficiency in MECC, relaxing

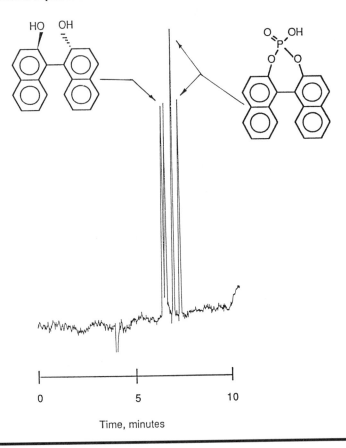

Figure 3   Enantiomeric separation of binaphthyl compounds using sodium deoxycholate. Experimental conditions: 50-μm i.d. × 65-cm capillary; 20 kV applied voltage; mobile phase, 0.01 M NaDC, 0.01 M Na$_2$HPO$_4$, and 0.006 M Na$_2$B$_4$O$_7$.

some of the instrumental constraints encountered when extremely narrow-bore capillaries are employed in OCLC. In fact, current trends in commercial instrumentation and product development indicate that capillary techniques such as CZE and MECC have the potential to become routine separation methodologies.

Under ideal conditions, the primary band-broadening mechanism in CZE is axial (or longitudinal) diffusion (Jorgenson and Lukacs, 1983). In MECC, the situation is more complex, owing to the distribution of a neutral solute between micellar and mobile phases. Contributions to band broadening stem not only from axial diffusion but also from various mass-transfer effects, arising from the type of surfactant employed and operating conditions unique to MECC (see the following). Factors important in determining column efficiency have been studied by several workers (Sepaniak and Cole, 1987; Davis, 1989; Terabe *et al.*, 1989a).

Table III shows the primary factors contributing to total plate height, H_T, in MECC, as well as the qualitative result of varying certain experimental parameters. The primary contributors to H_T discussed here are axial diffusion (H_a), micelle polydispersity (H_p), resistance to mass transfer (H_{mt}), and thermal gradients (H_t). Extra-column procedures such as injection and detection may also have substantial effects on efficiency (H_{ec}), but will be discussed later. H_T may be considered the sum of all these aforementioned contributions:

$$H_T = H_a + H_p + H_{mt} + H_t + H_{ec} \tag{8}$$

1. Axial Diffusion (H_a)

Axial diffusion can be a significant intracolumn band-broadening mechanism in MECC. As in CZE, the magnitude of the contribution to total plate height is directly proportional to the diffusion coefficient of the analyte in the mobile phase. The addition of a micellar phase causes the contribution from longitudinal diffusion to have dependencies on both the solute diffusion coefficient in the mobile phase and the effective diffusion coefficient while in the micellar phase (the latter is generally small). Solutes that have large k' values spend much of their time in the slow-diffusing micelle. However, they also require more time to elute, and hence, have greater opportunity to diffuse axially. The relative magnitudes of these offsetting effects determine the influence of k' on H_a (note ↑↓ in Table III). A similar situation exists for micelle concentration (i.e., an increase of micelle concentration causes solutes to spend less time in the mobile phase but results in longer retention times). In a more straightforward manner, small solute-diffusion coefficients (e.g., large solutes) and high operating voltages (i.e., faster flowrates) reduce H_a.

2. Micelle Polydispersity (H_p)

Micelle aggregation number is defined as the average number of surfactant monomers per micelle present in a micellar solution (Attwood and Florence, 1983). The constant exchange of monomer between micelle and bulk solution causes the size of a particular micelle to vary with time. The reported coefficient of variation (CV) in monomeric units per aggregate for SDS micelles is 20% (Attwood and Florence, 1983), indicating a significant degree of polydispersity. This micelle polydispersity has significant bearing on the suitability of a particular surfactant for use as a pseudostationary phase in MECC. Micelle electrophoretic mobility is determined by factors such as micelle size and shape [i.e., see $f(\kappa\alpha)$ term in Eq. (6)], which are influenced by aggregation number. Surfactant systems that inherently possess micelles with a wide range of sizes and shapes may not be suitable for use as pseudostationary phases in MECC. Solutes interacting with micelles exhibiting a range of migration velocities can produce broad bands. This effect is reduced as the rate of micelle–monomer exchange is increased, since this dynamic process tends to "average out" the effects of different micelle sizes encountered by a solute. The H_p is largely controlled by two experimental parameters: micelle concentration (see Table III) and operating temperature. Surfactant, hence micelle, concentration is particularly important in determining the separation perfor-

Table III Effect of Increased Experimental Parameters on Factors Contributing to Total Plate Height $(H_T)^a$

Experimental parameter	Longitudinal diffusion (H_a)	Thermal dispersion (H_t)	Micelle polydispersity (H_p)	Resistance to mass transfer (H_{mt})	
				Pseudostationary phase	Intracolumn (mobile phase)
Applied voltage	↓	↑	↑	↑	↑
Solute diffusion coefficient	↑	—	—	↑	↑
Capacity factor	↑↓	—	↑	↑	↑↓
Column diameter	—	↑↓	↓	—	↑
Micelle concentration	↑↓	↑	—	—	—
Power/length	—	↑	—	—	—

[a] ↑, an increase in plate height factor; ↓, a decrease in plate height factor; —, a negligible effect on total plate height factor; and ↑↓, an effect that is dependent on other factors.

mance in MECC. Table IV shows the variation of efficiency with SDS concentration for fluorescently labeled *n*-pentylamine. Near the CMC (0.005 *M*), the solute begins to show increased retention, indicating the formation of micelles (Sepaniak and Cole, 1987). However, efficiency and peak shape is poor, presumably owing in part to the adverse effects of micelle polydispersity. At this relatively low SDS concentration, the population of "small" micelles (i.e., those considerably smaller than the *N*-mer) within the micellar solution may be relatively high, leading to band dispersion. Increasing the surfactant concentration above the CMC alleviates this problem and, furthermore, increases the rate of micelle–monomer exchange.

The rate of micelle–monomer exchange is also increased by increasing the temperature. Improvements in efficiency have been observed (Balchunas and Sepaniak, 1988), and are illustrated in Fig. 4. Unfortunately, working at elevated temperatures may lead to increases in H_t (see the following). Moreover, studies of the direct influence of micelle polydispersity on efficiency are dependent on the inherently difficult process of accurately measuring micelle size distributions (Attwood and Florence, 1983). Nevertheless, Table IV and Fig. 4 clearly indicate the potential significance of the H_p term.

3. Mass Transfer Effects (H_{mt})

Band broadening due to the adverse effects of resistance to mass transfer in the pseudostationary and mobile phases, H_{mt}, can also occur in MECC. In general, low applied voltages (i.e., slow flow) and large solute diffusion coefficients minimize these adverse effects (see Table III). The pseudostationary-phase term, analogous to the conventional LC stationary-phase term, is often of little consequence in MECC because of the small size of micelles (approximately 40 Å for SDS), and the resulting rapid equilibration of nonpolar solutes within the hydrophobic interior of the micelle. In this case, the association between solute and micelle is based on weak dispersive interactions. With charged or amphiphilic solutes, strong electrostatic interactions with the

Table IV **Effect of Surfactant Concentration on Efficiency**[a]

SDS concentration (*M*)	Plate number
0.1	215,121
0.05	107,896
0.01	43,795
0.005	15,505

[a] Column, 25-μm i.d. × 50 cm, 40 cm to detection window; applied voltage, 30 kV; detection, Ar⁺ laser-based fluorescence; λ_{ex} 488 nm, λ_{em} 540 nm; mobile phase, 0.01 *M* Na₂HPO₄/0.006 *M* Na₂B₄O₇, pH 9; analyte, NBD-labeled *n*-pentyl amine.

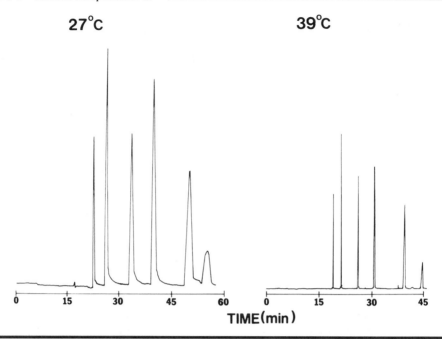

Figure 4 Effect of temperature on the MECC separation of selected NBD-amines. Components in order of elution are NBD-ethylamine, NBD-*n*-propylamine, NBD-*n*-butylamine, NBD-cyclohexylamine, NBD-*n*-hexylamine, and benzopyranoquinolizin-tetrahydro-(trifluoro-methyl) (Coumarin 153). SDS concentration relatively low (0.015 *M*) to illustrate relationship between temperature and efficiency. Experimental conditions: 50-μm i.d. × 65-cm capillary; 20 kV applied voltage; mobile phase, 0.015 *M* SDS and 0.005 *M* Na$_2$HPO$_4$.

charged head group of the surfactant can exhibit slow solute–micelle exchange kinetics and lead to band broadening, as is generally observed in ion exchange chromatography. For example, when obtaining the data for Table II (see Section II,A), it was observed that the efficiency for phenol was considerably less than that for the other solutes in the mixture when CTAC (a cationic surfactant) was used (Sepaniak *et al.*, 1987). The phenol is weakly acidic and partially ionized under the conditions employed, thereby permitting electrostatic interaction with the micelle.

Resistance to mass transfer within the mobile phase can also lead to band broadening in MECC. Two contributions, intermicelle and intracolumn, can be distinguished. Intermicelle effects arise from the slow diffusion between micelles. This is analogous to the resistance to mass transfer in the mobile-phase term for packed column LC, where rapid diffusion between particles is important. Even at moderate surfactant concentrations, the distance between micelles is extremely small (Sepaniak and Cole, 1987). Moreover, the pluglike flow within the capillary minimizes this source of dispersion. This indicates a fundamental reason for the characteristically high efficiencies in MECC, relative to packed-column LC. The micelles constitute a secondary phase that is more uniform in shape, more evenly dispersed (due to coulombic repulsion between micelles), and significantly smaller than conventional packed columns. Because of the near-perfect distribution of micelles and their fluidlike nature, eddy diffusion is also eliminated.

In OCLC, the contribution of intracolumn resistance to mass transfer to H_T is a consequence of the parabolic flow profile of the mobile phase (Pretorius *et al.*, 1974). Solutes at the center of the capillary move more rapidly than solutes near the walls. A similar effect occurs in packed-column LC, in which solutes between particles experience a local parabolic flow profile (see the preceding). The fluid nature of the primary and secondary phases in MECC indicates that slow intracolumn mass transfer can cause band broadening (as in OCLC) if radial velocity gradients exist in the capillary. Fortunately, the pluglike flow characteristic of electroosmotic flow minimizes this source of dispersion in MECC. Nevertheless, Van Deemter-like plots of plate height versus applied voltage indicate that this can still contribute to H_T (Sepaniak and Cole, 1987). It was found that at relatively high applied voltages (i.e., rapid flow, a condition that leads to significant band broadening in OCLC), the observed efficiency was dependent on capillary diameter (see Table III). Because of this and the thermal effects to be discussed, it is desirable to employ fairly narrow-bore capillaries in MECC (≤ 50 μm i.d.).

4. Thermal Effects (H_t)

Dissipation of Joule heat by the capillary is an important factor in determining efficiency. In the extreme case, boiling of the mobile phase within the capillary can occur. More commonly, radial thermal gradients are created in the capillary, since heat is dissipated at the capillary wall, leaving the center of the capillary at a higher temperature (Grushka *et al.*, 1989). Electrophoretic mobility increases with increasing temperature (approximately 2%/°C). Consequently, charged species in the center of the column move with greater electrophoretic velocity than those at the walls, distorting the pluglike flow profile, which, as mentioned earlier, can lead to band broadening. As discussed in Section I, CMC values often exhibit a significant temperature dependence. Thus, thermal gradients can produce a radial-phase ratio gradient, leading to further band broadening. Temperature may also influence micelle–monomer exchange kinetics, as discussed earlier. The effects of certain experimental parameters on H_t are summarized in Table III.

In most situations for which MECC is employed, high efficiency can be readily obtained. However, careful attention is required to strike a balance between conditions that minimize band broadening and those required for successful separation. The surfactant–micelle system should exhibit rapid exchange kinetics and/or minimal polydispersity. Surfactant concentration must be held at a level high enough to avoid the adverse effects of micelle polydisperity and low enough to allow adequate thermal dissipation. Likewise, columns must be selected with dimensions that minimize the effects of intracolumn resistance to mass transfer and thermal dispersion, while providing adequate detection sensitivity. Following these general guidelines should allow high-efficiency separations for many classes of compounds.

III. Experimental

Although the experimental apparatus for MECC is the same as that used for CZE, there are several subtle differences between the techniques that influence the selection of experimental operating conditions. In general, these differences result from (1) the increased conductivity of the mobile phase, (2) the different separation mechanism of MECC (i.e., chromatographic-based versus electrophoretic), and (3) the limited elution range. This section will discuss these differences and their effects on separation performance and operating procedures.

A. Capillary Conditioning and Modification

The implementation of CZE often entails washing the capillary with a base solution (i.e., 0.1 M KOH or NaOH), using a method first described by Lauer and McManigill (1986). This method can decrease the deleterious effects of capillary "aging," which can result in changes in sample–wall interactions and electroosmotic flow. Since MECC is often applied to the separation of small neutral molecules that do not have a tendency to interact with capillary surfaces, washing between each injection is not usually necessary. However, if the pseudostationary phase interacts with the walls, as seen with bile-salt surfactants, washing procedures may be required (Nishi *et al.*, 1989a). Furthermore, the use of a base wash as an initial capillary treatment is recommended to reduce changes in electroosmotic flow during initial use of a new capillary.

The majority of MECC separations have been performed in untreated fused-silica capillaries. These capillaries exhibit rapid electroosmotic flow and, consequently, result in a short elution range (when SDS micelles are employed) and a concomitant low peak capacity (see Section II,B). One way to increase the elution range is to modify the capillary wall with a variety of silanating reagents (Balchunas and Sepaniak, 1987; Lux *et al.*, 1990; see discussion in Section II,A). Silanation procedures block (or "cap") silanol groups on the capillary wall and reduce ζ_c [see Eq. (5)], thereby decreasing electroosmotic flow and increasing the elution range. Unfortunately, this method of extending the elution range effectively creates a "reversed phase" on the capillary wall that can produce a secondary retention mechanism and reduce efficiency. This effect was demonstrated by Balchunas and Sepaniak (1987). Although there are exceptions, most notably in the separation of proteins, we have generally observed higher efficiency using washed, unmodified capillaries.

B. Mobile-Phase Components and Considerations

In gas chromatography, the choice of mobile phase is not critical since it interacts only weakly with injected solutes. The choice becomes more important in supercritical fluid and liquid chromatography, where solute–mobile phase interactions are more significant and can be used to alter retention. With these methods, interactions with the stationary phase also strongly influence the separation. In MECC the choice of mobile-phase composition is crucial since both phases responsible for separation are fluid components. Thus, proper adjustment of mobile-phase parameters, such as (1) buffer composition, (2) surfactant type and concentration, (3) organic modifier concentration, and (4) other highly "specific" additives, is essential.

Higher buffer concentrations result in large electrophoretic currents and can cause band broadening due to Joule heating. Conversely, in our laboratory, we have observed poor retention-time reproducibility when low buffer concentrations (less than roughly 10 mM) are employed (possibly owing to an increase in surfactant–wall interactions under these conditions). Thus a compromise in buffer concentration is often required. The presence of free surfactant in MECC mobile phases further increases conductivity relative to CZE. Since low conductivity is desirable, surfactants with low CMCs are recommended. For example, sodium decyl sulfate (STS), which has a relatively high CMC (0.013 M), provides a longer elution range than the more commonly used SDS, but results in excessive Joule heating (Burton *et al.*, 1987). The effects of micelle polydispersity must also be considered. In the case of SDS, surfactant concentration must be held well above the CMC to reduce H_p (see Section II,B), exacerbating thermal problems. Because of the larger ionic strength of the mobile phase, Joule heating is generally more problematic in MECC than in CZE. As a rule of thumb, a 50-μm i.d. capillary can dissipate a "thermal load" of approximately one watt (i.e., the product of the current and the applied voltage) per meter (Burton *et al.*, 1987). Optimal field strengths for a 75-μm i.d. capillary are typically between 150 and 200 V/cm (Sepaniak and Cole, 1987). Smaller-diameter capillaries allow the use of considerably higher field strengths, without detrimental thermal effects.

The problem of Joule heating could be somewhat relieved by the use of low-conductivity organic zwitterion buffers. Although these have been extensively employed in CZE, they have been virtually ignored in MECC and should be investigated in the future. Traditionally, inorganic buffers have been employed in MECC; one of the most popular buffer systems is sodium phosphate-sodium borate. This buffer provides good reproducibility in retention times (see Section III,E) and capacity over a large pH range (pH 3 to pH 11). Other buffer components that have been employed include sodium acetate *tris*(hydroxymethyl)aminomethane (Tris), and sodium bicarbonate. Buffer composition can also influence separation selectivity. For example, borate-complex formation has been shown to influence selectivity in the separation of catecholamines (Wallingford and Ewing, 1988b).

The pH of the buffer can affect the elution range and alter the charge on a solute (see Section II,A). Typically a pH between 7 and 9 is employed to assure proper MECC flow characteristics. Buffer pH can also be used to affect changes in micelle–solute interactions by effectively titrating the acid or base functionalities on the compound. For example, chiral separations of amino acids using bile salts are not possible with an alkaline buffer, owing to ion repulsion with the micelle. However, in an acidic buffer (pH 4), ion repulsion is reduced, and separation is possible (Otsuka and Terabe, 1990). Because pH influences flow rates, and MECC is often used to separate weak acids and bases, the pH of the mobile phase is generally an important experimental factor.

The vast majority of MECC separations have employed SDS, because it provides selectivity similar to that of reverse-phase LC, is easy to use, and permits the separation of low to moderately hydrophobic compounds. Concentrations of SDS between 0.025 *M* and 0.1 *M* are usually employed. Concentrations in the upper part of this range result in larger k' values for a given solute (for hydrophobic solutes this is undesirable; see Section II,A) and smaller H_p (see Section II,B). For more hydrophobic compounds, the use of bile-salt surfactants is recommended because their use results in a general reduction in k'. Bile salts that have been investigated include sodium cholate, sodium deoxycholate, and sodium taurocholate. This class of surfactants also permits the use of higher organic modifier concentrations (up to 40% have been used in our laboratory) and can be used for chiral separations (see Section IV).

A few other surfactant systems have been employed to provide unique selectivities, usually based on some interaction with a functional group or other characteristic of the analyte (see Section II,A and Table V for specific details). Often these surfactants are not commercially available and must be synthesized. For example, amino acids have been reacted with dodecanoic acid *N*-hydroxysuccinimide ester to produce chiral alkyl surfactants (Dobashi *et al.*, 1989a). Cationic surfactants reverse the direction of electroosmosis (flow is toward the positive electrode) and usually require the reversal of electrode polarity (Burton *et al.*, 1987). The cationic surfactants most commonly employed are quaternary ammonium alkyl surfactants such as CTAC and DTAC (see Table I). These compounds can be used in the same concentration range as SDS and exhibit similar retention characteristics for nonpolar molecules. However, they can provide unique selectivities for weakly acidic charged species (see Table II and related discussion). Since relatively few of the common surfactants have been investigated as pseudostationary phases, other surfactants will probably be used in MECC.

In summary, desirable characteristics of potential MECC surfactants include (1) a low CMC, (2) an electrophoretic mobility slightly greater than that of SDS, and (3) good stability in the presence of organic solvent. A very low CMC (below 1 m*M*) would reduce the conductivity of the buffer and Joule heating. The greater electrophoretic mobility would result in a larger elution range and peak capacity. The values of v_{eo} and $v_{e,m}$ produced by commercial fused-silica capillaries and SDS result in elution ranges of typically 10 to 30 min for a 1-m capillary. A surfactant with a larger electrophoretic mobility could provide a more desirable, longer elution range (i.e., 10 to 60 min) without excessive analysis times. High concentrations of organic solvents inhibit the formation of SDS micelles. Greater micelle stability in mixed aqueous-organic mobile phases would allow higher organic modifier concentrations to be employed, and a greater degree of separation selectivity could possibly be achieved.

Other "specific" components can be added to MECC mobile phases to influence the separation. Separations are affected by directly altering the micelle–buffer equilibrium (via changes in K or β) or by creating a secondary equilibrium (see Section II,A).

Table V Separations of Interest Using MECC

Analyte	Surfactant(s)	Detection	Remarks	Reference(s)
Environmental interest				
Aflatoxins	SDS[a]	Laser-based fluorescence [Ar$^+$ (488 nm)]	10% acetonitrile in the mobile phase improves separation	Balchunas et al. (1988)
Phenolic compounds	SDS	UV-absorbance (270 nm)		Terabe et al. (1984)
	SDS	UV-absorbance (220 nm)		Otsuka et al. (1985a)
	SDS	UV-absorbance (254 nm)	Step gradient from 0% to 15% acetonitrile to help resolution	Balchunas et al. (1988)
Biological interest				
B vitamins	SDS	Laser-based fluorescence		Burton et al. (1986a)
	SDS	UV-absorbance (254 nm)		Fujiwara et al. (1988)
	SDS	Laser-based fluorescence [Ar$^+$ (488 nm) and HeCd (325 nm)]		Swaile et al. (1988)
	SDS	UV-absorbance (210 nm)	Tetraalkylammonium salts used to improve separation	Nishi et al. (1989d)
	SDS	UV-absorbance (210 nm)	pH effect on separation	Nishi et al. (1989b)
Antipyretic analgesic	SDS	UV-absorbance (214 nm)		Fujiwara and Honda (1987a)
Antibiotics	SDS	UV-absorbance (210 nm)	Cephalosporin antibiotics; tetraalkylammonium salts used to improve separation	Nishi et al. (1989d)
	SDS, LMT[b]	UV-absorbance (210 nm with SDS, 220 nm with LMT)	Penicillin and cephalosporin antibiotics	Nishi et al. (1989c)

(Continued)

Table V Separations of Interest Using MECC (*Continued*)

Analyte	Surfactant(s)	Detection	Remarks	Reference(s)
Chiral drugs	Bile salts	UV-absorbance (210 nm or 220 nm)	Sodium cholate and derivatives used for chiral separation of diltiazem, trimetoquinol, and carboline isomeric drugs	Nishi *et al.* (1989a)
Cold medicine	Bile salts	UV-absorbance (210 nm or 220 nm)	Sodium cholate and derivatives used for chiral separation	Nishi *et al.* (1990b)
Amino acids	SDS, DTAB[c]	UV-absorbance (260 nm)	Phenylthiohydantoin-amino acid derivatives separated	Otsuka *et al.* (1985b)
	SDS	UV-absorbance (254 nm)	Dinitrophenylhydrazine-amino acids separated	Swaile *et al.* (1988)
	Bile salts	UV-absorbance (210 nm or 220 nm)	Sodium cholate and derivatives used for chiral separation of racemic dansylated-amino acids	Terabe *et al.* (1989a)
	SDS and SDVal[d]	UV-absorbance (254 nm)	A mixed micellar phase of SDS and SDVal was used to separate the *N*-(3,5-dinitrobenzoyl) *O*-isopropyl ester derivatives of enantiomeric amino acid pairs	Dobashi *et al.* (1989b)
	SDVal	UV-absorbance (230 nm or 260 nm)	*N*-(3,5-dinitrobenzoyl) *O*-isopropyl ester derivatives of enantiomeric amino acid pairs were separated	Dobashi *et al.* (1989a)
	SDS	Fluorescence [excitation, 365 nm; detection, 418 cut-off filter]	Mixture of *o*-phthalaldehyde-derivatized amino acids separated in the presence of 15% methanol, 1% tetrahydrofuran	Liu *et al.* (1989a)

Analytes	Surfactant	Detection	Comments	References
Catechols	SDS	Off-column amperometric		Wallingford and Ewing (1988a); Wallingford and Ewing (1988b)
	SDS and SOS[e]	Off-column amperometric	A mixed micellar solution of SDS and SOS demonstrated improved selectivity and efficiency	Wallingford et al. (1989)
Bases, nucleosides, oligonucleotides	SDS	UV-absorbance (260 nm)	Addition of divalent metals (Cu(II), Zn(II), Mg(II)) improved separation of oligonucleotides	Cohen et al. (1987)
Nucleosides, nucleotides	SDS	UV-absorbance (256 nm)		Row et al. (1987)
		UV-absorbance (? nm)		Griest et al. (1988)
Nucleosides	SDS, DTAB, CTAB[f]	UV-absorbance (254 nm)	An anionic surfactant (SDS) and two cationic surfactants (DTAB and CTAB) were each used to separate phosphorylated nucleoside mixture	Liu et al. (1989b)

[a] Sodium dodecyl sulfate.
[b] Sodium N-lauroyl-N-methyltaurate.
[c] Dodecyltrimethylammonium bromide.
[d] Sodium N-decanoyl-L-valinate.
[e] Sodium octyl sulfate.
[f] Hexadecyltrimethyl-ammonium bromide.

Ions such as Mg(II), Zn(II), Cu(II), and tetraalkylammonium will interact with the charged head group of anionic surfactants and alter the solute–micelle interactions of charged species. Organic solvents can also be added to the buffer to affect selectivity, as in reverse-phase LC (see Section II,A). In reverse-phase LC, it is common practice to optimize mobile-phase composition through considerations of specific solute–mobile-phase interactions (dipole–dipole, hydrogen-bonding, etc.). Similar considerations in MECC are possible. However, large changes in selectivity are usually not observed, owing to restrictions on the amount of solvent that can be added to the buffer without preventing micelle formation (Gorse *et al.*, 1988).

The major advantages of using mixed aqueous-organic mobile phases in MECC (i.e., extending elution range and reducing k') have been discussed previously but are elaborated herein from an experimental point of view. The degree to which the addition of organic solvents extends the elution range is generally larger for solvents that dramatically alter ζ_c. Alcohols generally extend elution range greatly. A rough comparison of the general effects of organic solvents on k' can be made by considering the polarity of the organic solvents (Snyder and Kirkland, 1979) and ignoring their effects on the CMC of the surfactant. For example, acetonitrile is a less-polar solvent (i.e., a stronger reverse-phase eluant) than methanol and generally reduces solute k' values more significantly than does MECC. The choice of organic modifier is based largely on the nature of the sample to be separated. If it contains a large number of components, then an organic modifier that extends the elution range, such as methanol, should be employed. However, if the sample is relatively simple, then an organic modifier such as acetonitrile or dioxane, which reduces k' values without excessive changes in elution range, should be used. If a combination of these two effects is desired, isopropanol could be used to both extend elution range and reduce k'. Furthermore, the use of ternary systems (Balchunas *et al.*, 1988) and step- or continuous-gradient systems might be required to optimize separation and analysis time (Powell and Sepaniak, 1990).

C. Injection

Samples are injected into CZE/MECC capillaries using either electromigration (Burton *et al.*, 1986b) or hydrostatic injection. Hydrostatic injection is performed either by aspirating at the outlet of the capillary or by applying positive pressure to the sample solution at the capillary inlet (sometimes accomplished by raising the inlet of the capillary, relative to the outlet, while it resides in the sample). Hydrostatic injection uniformly injects all sample components. Electromigration is accomplished by placing the inlet of the capillary in the sample and applying an electric field. Sample components migrate into the capillary at rates determined by their individual electrokinetic and solute–micelle characteristics; consequently, sample component discrimination can occur.

The rate at which a sample component migrates into the capillary during the injection determines the effective injection length, which, in turn, influences both efficiency and peak shape. These effects have been studied by Burton *et al.* (1986b).

Their work showed that long injection plugs resulted in distorted peak shapes. The contribution of the injection length can be determined by the equation

$$\sigma_{inj}^2 = (1/12)L_{inj}^2 \tag{9}$$

where σ_{inj}^2 is the contribution of the injection plug length to the total peak variance, and L_{inj} is the length of the injected plug (Terabe *et al.*, 1989a). Terabe *et al.* (1989a) calculated that an injection plug length of 0.79 mm (a 1.5 nl injection volume) would contribute less than 5% to the overall plate height when using a 50-μm i.d. $\times$ 50-cm capillary. Although short injection plugs are required to minimize injection-related band dispersion, the use of longer injection plugs can be advantageous (i.e., improve detection) when injection-related contributions to H_T are negligible or when resolution is not a critical concern.

D. Detection

The diminutive size of the capillaries employed in CZE/MECC complicates detection. Because solute band volumes are generally below 10 nl and detection-zone residence times are only a few seconds, on-column flow cells with short detection zones and rapid detector response times are needed. If the length of the detection zone or the response time constant is too large, a significant amount of extra-column band broadening can result. To remedy this situation, a variety of detection schemes have been adapted for CZE; however, the detection schemes for MECC to date have been limited to ultraviolet (UV) absorbance, fluorescence, and amperometry. This is probably a result of the more complex mobile phases employed in MECC. For example, indirect detection (Kuhr and Yeung, 1988) based on analyte charge displacement cannot be efficiently used in MECC. The presence of charged surfactant molecules in the mobile phase will reduce the transfer ratio for an analyte molecule, either by ion pairing with the analyte or by replacing the chromophore that is to be displaced. However, it may be possible to perform other indirect detection methods. Also, the mass spectrometric detectors adapted for CZE have not been applied to MECC; however, this may be possible if the pseudostationary phase does not produce excessive chemical noise owing to micelle–surfactant fragmentation.

The most commonly used detection scheme for MECC and CZE is UV absorbance (Green and Jorgenson, 1989; Sepaniak *et al.*, 1989; Vindevogel *et al.*, 1990). Since on-column detection is generally required, sensitivity with this method is limited by the path length of the detection region (usually considered to be approximately the i.d. of the capillary). The diminutive size and shape of the capillary flow cell can also result in large amounts of stray light that further reduce the sensitivity of this method. Often CZE is applied to the separation of large macromolecules with very large molar absorptivities (e.g., proteins and oligonucleotides), resulting in reasonably good detectability. However, MECC is typically applied to the separation of relatively small hydrophilic molecules with molar absorptivities of less than 10^4. A compound with a molar absorptivity of 5000 (in a 50-μm i.d. $\times$ 500-mm capillary) would result in a minimum injectable concentration of 5.8 μM (S/N = 2), if a Gaussian peak exhibiting

100,000 theoretical plates and a base-line noise of 5×10^{-5} AU were observed. Although this provides good mass sensitivity (9 femtomole for a 1.5 nl injection volume), the relatively poor concentration-based sensitivity limits the practical utility of this detection method in MECC.

Whereas it is not so universal as UV-absorbance detection, fluorometric detection can provide much greater sensitivity, since large excitation powers can compensate for short optical path lengths and produce large fluorescence signals (Burton *et al.*, 1986a; Balchunas and Sepaniak, 1987; Burton *et al.*, 1987; Sepaniak and Cole, 1987; Balchunas *et al.*, 1988; Swaile *et al.*, 1988; Sepaniak and Swaile, 1989; Sepaniak *et al.*, 1989). Both conventional and laser sources have been employed; however, laser excitation is preferred, because of the ability to tightly focus the laser output into the capillary. This not only provides large excitation powers, but also results in very short detection zones, often on the order of 20 μm in length. Detection zones of this size are much shorter than typical injection lengths and obviously result in negligible detector-related contributions to H_{ec}. The utility of this mode of detection is illustrated in Fig. 5 by the MECC separation and laser (Ar$^+$ ion) fluorometric detection of a number of amino derivatized with 7-chloro-4 nitrobenzo-2,1,3-oxadiazole (NBD-

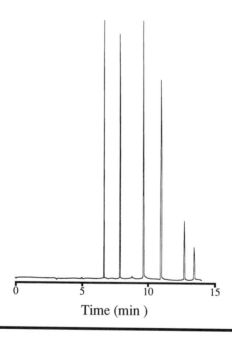

Time (min)

Figure 5 Separation of NBD-derivatized alkyl amines (same mixture as Fig. 4) as an illustration of the high sensitivity and efficiency afforded by laser-based detection (Ar$^+$ laser, 488 nm). Limit of detection for NBD-ethylamine was 10 fg injected. Experimental conditions: 50-μm i.d. $\times$ 65-cm capillary; 20 kV applied voltage; mobile phase, 0.025 M SDS, 0.005 M Na$_2$HPO$_4$, and 10% (v/v) 2-propanol.

Cl). This mode of detection exhibits excellent detectability [limits of detection (LODs) in the low femtogram injected range (Balchunas *et al.*, 1988)] and separation efficiency.

The applicability of this method is limited to compounds that naturally fluoresce and those that can be conveniently labeled. The limited number of easily accessed laser lines also reduces the usefulness of the method. Although it is possible to access virtually all wavelengths between the vacuum UV and the near infrared (IR) with tunable dye laser systems, their complexity and cost are distinct disadvantages. The use of fluorescent labels to derivatize analytes can increase the utility of laser fluorometric detection, since the spectroscopic characteristics of the label can be matched to the outputs of simple, inexpensive laser systems. Labeling, however, can require time-consuming sample treatment and result in more complex chromatograms.

The solubilization of the fluorophore by the micellar phase can enhance the fluorescent signal (Balchunas *et al.*, 1988). This detectability improvement occurs only for solutes that associate with the micelle. Proper choice of surfactant counterion is necessary owing to possible heavy atom quenching effects. This has been observed in our laboratory when bromide is the counterion (e.g., with CTAB).

An amperometric detection scheme for MECC has been described by Wallingford and Ewing (1988a,b; Wallingford *et al.*, 1989). Although a post-capillary detection scheme, it resulted in excellent sensitivity and minimal detector-related band broadening. This is probably a result of a novel capillary–detector connection method in which a detection capillary is attached to the separation capillary via a porous glass tube that allows electrical connection to the ground electrode. Amperometric detection is then performed using a carbon fiber placed in the detection capillary. A detection limit of less than 1 femtomole was achieved in a 26-μm i.d. capillary (Wallingford and Ewing, 1988a). Since the sensitivity of electrochemical detection is not path-length dependent, it is compatible with very small i.d. capillaries and should play a large part in the future development of MECC and CZE.

E. Analytical Figures of Merit

For any separation method to be used for the analysis of complex unknown samples, a high degree of reproducibility must be achieved for both band velocity and signal. In MECC, the two parameters that most greatly degrade reproducibility are variances in injection volume (see previous discussion) and electroosmotic velocity. Small changes in electroosmotic flow can reduce retention reproducibility; however, for isocratic conditions, the relative standard deviations (RSDs) are typically below 2% (Fujiwara and Honda, 1987a; Otsuka *et al.*, 1987; Fujiwara *et al.*, 1988). The use of gradient elution decreases the reproducibility of this method somewhat, with RSDs near 2% for step gradients (Balchunas and Sepaniak, 1988) and around 5% for continuous gradients (Powell and Sepaniak, 1990).

Electroosmosis also affects quantitative measurement since the peak areas for concentration-sensitive detectors such as spectrophotometry and fluorimetry are dependent on band velocity. On-column detection in MECC further exacerbates this problem. Since the velocity of the analyte in the flow cell is dependent on K, β, and

v_M, these parameters will affect peak area. This could account for the slightly poorer reproducibility for peak area, relative to peak height, reported in an early study of quantitation (Otsuka *et al.*, 1987). An internal standard is often employed to improve reproducibility (Fujiwara and Honda, 1987a; Fujiwara *et al.*, 1988). By using the analyte peak area : internal standard peak area ratio for quantitation, it is possible to normalize for flow rate and injection volume variances. Typical RSDs for quantitation using the internal-standard method are below 2%. This level of reproducibility is adequate for the routine analysis of many compounds. The increased development and use of automated CZE/MECC instrumentation should further improve reproducibility, by reducing variation in injection volume, mobile-phase composition, and random operator errors.

IV. Applications

Although it is a relatively new separation methodology, MECC has been successfully applied in numerous analyses. As listed in Table V, MECC has been used in the analysis of compounds of environmental interest, pharmaceuticals, and nucleic acids, as well as neurotransmitters and amino acids. This diverse array of analytes, along with anticipated advances in selectivity and detectability, suggests further potential for MECC in analytical separations.

Certainly, a major hallmark of capillary/electrokinetic techniques is high efficiency. This high efficiency, coupled with the chromatographic partitioning afforded by the micellar pseudostationary phase, results in a separation technique that can separate both charged and neutral compounds (Burton *et al.*, 1986a). The low injection volume (on the order of nanoliters) also contributes to its applicability in bioanalyses. In fact, CZE has been applied to the cytoplasmic analysis for a single nerve cell (Olefirowicz and Ewing, 1990).

Currently, MECC is suitable only for compounds of at least a moderately hydrophilic nature. It is also limited to relatively high sample concentrations (i.e., not appropriate for trace analysis). Also, the limited elution range of MECC renders the technique inappropriate for extremely complex samples, unless some type of selective detection is used. Within these limits, MECC should be readily applicable to analyses such as quality control for pharmaceuticals (i.e., drugs, bioactive peptides).

An area of particular interest in MECC is that of chiral separations. The separation of optical isomers is one of the more important and challenging problems in modern separation science. The assessment of optical purity can be of critical importance in the pharmaceutical and food industries because of the different physiological responses that different enantiomers can elicit. These types of separations are usually carried out by the use of a stationary phase possessing molecular asymmetry. Chiral stationary phases have been used in both gas and liquid chromatography and have proven successful in a variety of separations (Armstrong, 1987). Chiral separations using MECC offer the advantages of economy and versatility over the conventional HPLC system.

In MECC, two general approaches are used to separate enantiomers. The simplest method involves the use of a chiral surfactant. Bile-salt surfactants have been employed in this manner. Amino acid, pharmaceutical, and binaphthyl enantiomers have all been separated by MECC, using bile-salt surfactant systems (Nishi *et al.,* 1989a; Terabe *et al.,* 1989b; Cole *et al.,* 1990). Possessing a structure that is steroid-like, this naturally occurring class of surfactant forms unusual aggregates that are helical in composition (Campanelli *et al.,* 1989; Epiosito *et al.,* 1989). Studies in our laboratory with binaphthyl compounds indicate that enantiomeric compounds possessing some degree of rigidity or planar character are most effectively separated with bile-salt systems (Cole *et al.,* 1990).

An alternate method of achieving chiral recognition in MECC is the use of chiral additives that form mixed micelles. Dobashi and co-workers employed a surfactant that consisted of an *n*-alkyl tail and a valine head group (SDVal) to separate amino acid enantiomers (Dobashi *et al.,* 1989a,b). This approach proved to be more efficient when mixed with SDS surfactant; mixed micelle formation evidently allowed for more facile kinetics between solute and micelle than did the SDVal surfactant system by itself (Otsuka and Terabe, 1990). Neutral chiral additives have also been utilized with some success. Otsuka and Terabe (1990) have employed digitonin-SDS micelles to separate enantiomers of amino acid.

V. Conclusion

Micellar electrokinetic capillary chromatography provides the benefits of CZE separations (high efficiencies and small injected volumes) for neutral solutes. As demonstrated by the listing in Table V, MECC has the potential for a wide variety of applications. Future improvement of the technique must include expansion of its ability to separate hydrophobic solutes and lower limits of detection. Furthermore, the use of novel surfactant systems should also be investigated for their potential enhancements of selectivity.

Acknowledgments

The authors wish to thank Shannon Gerhardt for her valued assistance in the editing of this chapter. The research presented herein was sponsored by the Division of Chemical Sciences, Office of Basic Sciences, U.S. Department of Energy, under Grant DE-FG05-86ER13612 with the University of Tennessee.

References

Armstrong, D. W. (1987). *Anal. Chem.* **59**, 84A–91A.
Attwood, D., and Florence, A. T. (1983). *In* "Surfactant Systems: Their Chemistry, Pharmacy, and Biology." Chapman and Hall, London.
Balchunas, A. T., and Sepaniak, M. J. (1987). *Anal. Chem.* **59**, 1466–1470.
Balchunas, A. T., and Sepaniak, M. J. (1988). *Anal. Chem.* **60**, 617–621.

Balchunas, A. T., and Swaile, D. F., Powell, A. C., and Sepaniak, M. J. (1988). *Sep. Sci. Technol.* **23**, 1891–1904.

Burton, D. E., Sepaniak, M. J., and Maskarinec, M. P. (1986a). *J. Chromatogr. Sci.* **24**, 347–351.

Burton, D. E., Sepaniak, M. J., and Maskarinec, M. P. (1986b). *Chromatographia* **21**, 583–586.

Burton, D. E., Sepaniak, M. J., and Maskarinec, M. P. (1987). *J. Chromatogr. Sci.* **25**, 514–518.

Bushey, M. M., and Jorgenson, J. W. (1989a). *J. Microcol. Sep.* **1**, 125–130.

Bushey, M. M., and Jorgenson, J. W. (1989b). *Anal. Chem.* **61**, 491–493.

Campanelli, A. R., De Sanctis, C., Chiessi, E., D'Alagni, M., Giglio, E., and Scaramuzza, L. (1989). *J. Phys. Chem.* **93**, 1936.

Cline Love, L. J., Habarta, J. G., and Dorsey, J. G. (1984). *Anal. Chem.* **56**, 1132A–1148A.

Cohen, A. S., Terabe, S., Smith, J. A., and Karger, B. L. (1987). *Anal. Chem.* **59**, 1021–1027.

Cole, R. O., Sepaniak, M. J., and Hinze, W. L. (1990). *J. High Res. Chromatogr.* **13**, 579–582.

Cole, R. O., Gorse, J., Oligis, K., and Sepaniak, M. J. (1991). *J. Chromatogr.* **557**, 113–123.

Davis, J. M. (1989). *Anal. Chem.* **61**, 2455–2461.

Dobashi, A., Ono, T., Hara, S., and Yamaguchi, J. (1989a). *J. Chromatogr.* **480**, 413–420.

Dobashi, A., Ono, T., Hara, S., and Yamaguchi, J. (1989b). *Anal. Chem.* **61**, 1984–1986.

Epiosito, G., Giglio, E., Pavel, N. V., and Zanobi, A. (1989). *J. Phys. Chem.* **480**, 356.

Ewing, A. G., Wallingford, R. A., and Olefirowicz, T. M. (1989). *Anal. Chem.* **61**, 292A–303A.

Fujiwara, S., and Honda, S. (1987a). *Anal. Chem.* **59**, 2773–2776.

Fujiwara, S., and Honda, S. (1987b). *Anal. Chem.* **59**, 487–490.

Fujiwara, S., Iwase, S., and Honda, S. (1988). *J. Chromatogr.* **447**, 133–140.

Gorse, J., Balchunas, A. T., Swaile, D. F., and Sepaniak, M. J. (1988). *J. High Res. Chromatogr.* **11**, 554.

Green, J. S., and Jorgenson, J. W. (1989). *J. Liq. Chromatogr.* **12**, 2527–2561.

Griest, W. H., Maskarinec, M. P., and Row, K. H. (1988). *Sep. Sci. Technol.* **23**, 1905–1914.

Grushka, E., McCormick, R. M., and Kirkland, J. J. (1989). *Anal. Chem.* **61**, 241–246.

Guiochon, G. (1981). *Anal. Chem.* **53**, 1318–1325.

Guiochon, G., and Martin, M. (1984). *Anal. Chem.* **56**, 614–620.

Hinze, W. L. (1987). *In* "Organized Surfactant Assemblies in Separation Science" (M. J. Comstock, ed.), Vol. 342, pp. 2–82. American Chemical Society, Washington, D.C.

Johnson, E. L., and Stevenson, R. (1978). *In* "Basic Liquid Chromatography," pp. 37. Varian Associates, Palo Alto, California.

Jorgenson, J. W., and Guthrie, E. J. (1983). *J. Chromatogr.* **255**, 335–348.

Jorgenson, J. W., and Lukacs, K. D. (1981). *Anal. Chem.* **53**, 1298–1302.

Jorgenson, J. W., and Lukacs, K. D. (1983). *Science* **222**, 266–272.

Knox, J. H. (1980). *J. Chromatogr. Sci.* **18**, 453–472.

Knox, J. H., and Gilbert, M. T. (1979). *J. Chromatogr.* **186**, 405–418.

Kuhr, W. G., and Yeung, E. S. (1988). *Anal. Chem.* **60**, 2642–2646.

Lauer, H., and McManigill, D. (1986). *Anal. Chem.* **58**, 166–170.

Liu, J., Banks, J. F., and Novotny, M. (1989a). *J. Microcol. Sep.* **1**, 136–141.

Liu, J., Cobb, K. A., and Novotny, M. (1989b). *J. Chromatogr.* **468**, 55–65.

Lukacs, K. D., and Jorgenson, J. W. (1985). *J. High Res. Chromatogr.* **8**, 407–411.

Lux, J. A., Yin, H., and Schomburg, G. (1990). *J. High Res. Chromatogr.* **13**, 145–147.

Nishi, H., Fukuyama, T., Matsuo, M., and Terabe, S. (1989a). *J. Microcol. Sep.* **1**, 234–241.

Nishi, H., Tsumagari, N., Kakimoto, T., and Terabe, S. (1989b). *J. Chromatogr.* **465**, 331–343.

Nishi, H., Tsumagari, N., Kakimoto, T., and Terabe, S. (1989c). *J. Chromatogr.* **477**, 259–270.

Nishi, H., Tsumagari, N., and Terabe, S. (1989d). *Anal. Chem.* **61**, 2434–2439.

Nishi, H., Fukuyama, T., Matsuo, M., and Terabe, S. (1990a). *J. Chromatogr.* **515**, 233–243.

Nishi, H., Fukuyama, T., Matsuo, M., and Terabe, S. (1990b). *J. Chromatogr.* **498**, 313–323.

Novotny, M. (1988). *Anal. Chem.* **60**, 500A–510A.

Olefirowicz, T. M., and Ewing, A. G. (1990). *Anal. Chem.* **62**, 1872–1876.

Otsuka, K., and Terabe, S. (1989). *J. Microcol. Sep.* **1**, 150–154.

Otsuka, K., and Terabe, S. (1990). *J. Chromatogr.* **515**, 221–226.

Otsuka, K., Terabe, S., and Ando, T. (1985a). *J. Chromatogr.* **332**, 219–226.

Otsuka, K., Terabe, S., and Ando, T. (1985b). *J. Chromatogr.* **348**, 39–47.

Otsuka, K., Terabe, S., and Ando, T. (1987). *J. Chromatogr.* **396**, 350–354.

Powell, A. C., and Sepaniak, M. J. (1990). *J. Microcol. Sep.* **2**, 278–284.

Pretorius, V., Hopkins, B., and Schieke, J. (1974). *J. Chromatogr.* **99**, 23–30.

Rasmussen, H. T., and McNair, H. M. (1989). *J. High Res. Chromatogr.* **12**, 635–636.

Row, K. H., Griest, W. H., and Maskarinec, M. P. (1987). *J. Chromatogr.* **409**, 193–203.

Sepaniak, M. J., and Cole, R. O. (1987). *Anal. Chem.* **59**, 472–476.

Sepaniak, M. J., and Swaile, D. F. (1989). *J. Microcol. Sep.* **1**, 155–158.

Sepaniak, M. J., Burton, D. E., and Maskarinec, M. P. (1987). *In* "Micellar Electrokinetic Capillary Chromatography" (M. J. Comstock, ed.), Vol. 342, pp. 142–151. American Chemical Society, Washington, D.C.

Sepaniak, M. J., Swaile, D. F., and Powell, A. C. (1989). *J. Chromatogr.* **480**, 185–196.

Snyder, L. R., and Kirkland, J. J. (1979). *In* "Introduction to Modern Liquid Chromatography" Wiley, New York.

Swaile, D. S., Burton, D. E., Balchunas, A. T., and Sepaniak, M. J. (1988). *J. Chromatogr. Sci.* **26**, 406–409.

Terabe, S., Otsuka, K., Ichikawa, K., Tsuchiya, A., and Ando, T. (1984). *Anal. Chem.* **56**, 111–113.

Terabe, S., Otsuka, K., and Ando, T. (1985a). *Anal. Chem.* **57**, 834–841.

Terabe, S., Ozaki, H., Otsuka, K., and Ando, T. (1985b). *J. Chromatogr.* **332**, 211–217.

Terabe, S., Otsuka, K., and Ando, T. (1989a). *Anal. Chem.* **61**, 251–260.

Terabe, S., Shibata, M., and Miyashita, Y. (1989a). *J. Chromatogr.* **480**, 403–411.

Terabe, S., Miyashita, Y., Shibata, O., Barnhart, E. R., Alexander, L. R., Patterson, D. G., Karger, B., Hosoya, K., and Tanaka, N. (1990). *J. Chromatogr.* **516**, 23–31.

Vindevogel, J., Sandra, P., and Verhagen, L. C. (1990). *J. High Res. Chromatogr.* **13**, 295–298.

Wallingford, R. A., and Ewing, A. G. (1988a). *Anal. Chem.* **60**, 258–263.

Wallingford, R. A., and Ewing, A. G. (1988b). *J. Chromatogr.* **441**, 299–309.

Wallingford, R. A., Curry, P. D., and Ewing, A. G. (1989). *J. Microcol. Sep.* **1**, 23–27.

Willard, H. H., Merrit, L. L., Dean, J. A., and Settle, F. A. (1981). *In* "Instrumental Methods of Analysis," p. 445. Wadsworth Publishing, Belmont, California.

CHAPTER

7

Isoelectric Focusing in Capillaries

Stellan Hjertén

I. Introduction: A Comparison between Isoelectric Focusing and Other Electrophoresis Methods

The large number of electrophoresis methods described in the literature makes it difficult, particularly to the newcomer in the field, to distinguish the characteristic feature of each individual method. However, understanding of an electrophoresis method is facilitated if one is aware of the fact that it can be classified into one of the following categories: moving-boundary electrophoresis, zone electrophoresis, displacement electrophoresis, and isoelectric focusing. All these categories have counterparts in chromatography and centrifugation (see Section V). Since this chapter deals exclusively with isoelectric focusing, i.e., electrophoretic separation of ampholytes in a pH gradient, I shall mention only the counterparts of this technique: chromatofocusing (chromatographic separation of ampholytes in a pH gradient) and isopycnic centrifugation (separation by sedimentation in a density gradient).

In moving-boundary electrophoresis (Fig. 1) and in zone electrophoresis, the boundaries will become more blurred the longer the run time, whereas in isoelectric focusing and displacement electrophoresis, they will remain sharp as soon as the steady state has been attained. In comparison with isoelectric focusing, displacement electrophoresis (isotachophoresis) has the disadvantage that there is no space between the solute zones. Considering these principal differences between different electrophoretic separation methods (see also Section II), it is not surprising that those based on isoelectric focusing often give higher resolution of proteins than do other electrophoresis methods.

Capillary Electrophoresis
Copyright © 1992 by Academic Press, Inc. All rights of reproduction in any form reserved.

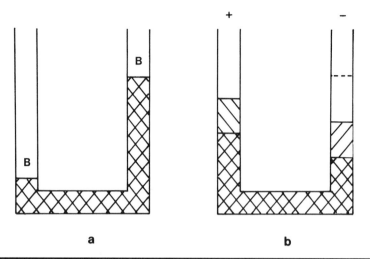

Figure 1 The principle of moving-boundary electrophoresis. The sample, containing the proteins p1 (\\\) and p2 (///), has been dialyzed against the buffer B. The figure illustrates the start (a) and the end (b) of a run. The convection is suppressed since the density below each boundary is higher than that above. Observe that the method permits only a fraction of proteins p1 and p2 to be isolated in pure form (p1 in the left limb of the U-cell and p2 in the right limb). The isoelectric point (pI) of a protein can be established by determinations of the mobility (μ) of the protein at some pH values around pI and extrapolation to $\mu = 0$.

It should also be mentioned that in moving-boundary and zone electrophoresis, the solutes can be characterized (identified) by their mobilities — in displacement electrophoresis, by their concentrations in the zones, and in isoelectric focusing, by their isoelectric point (pI) values. A disadvantage with isoelectric focusing [and to some extent with displacement electrophoresis (Johansson *et al.*, 1987)] is the risk of precipitation of some proteins (see Section VIII).

II. Different Ways to Suppress Convection in Carrier-Free Electrophoresis, Including Isoelectric Focusing

The main problem in designing a free electrophoresis method is how to eliminate convection, i.e., the movement that takes place in a liquid layer when the density at the top is higher than that at the bottom. It may, therefore, be appropriate to discuss briefly different ways to suppress convection in electrophoresis, including isoelectric focusing.

In the moving-boundary method, for which Tiselius (1930, 1937) was awarded the Nobel prize in 1948, the convection is strongly suppressed, since each sample (protein) phase has a concentration, and thereby density, that is higher than that in the phase above, as shown in Fig. 1. The isoelectric point of a protein can be determined from a plot of its mobility (μ) against pH and extrapolation to $\mu = 0$. Figure 1 also illustrates that only a *partial* separation can be accomplished. In addition, the amount of protein required for an analysis is as much as 50 to 100 mg (in a microversion of the moving-boundary method, the amount can be reduced to 0.1 mg; Hjertén, 1967). Therefore, Hjertén (1958, 1967, 1976) introduced a carrier-free capillary electrophoresis method, which permits analysis of much smaller amounts of protein and other substances (on the μg scale) and affords *complete* separation of the solutes into discrete zones (Fig. 2, case a). The diameter of the electrophoresis tube is $1-3$ mm. The capillary is slowly rotated around its long axis for elimination of disturbances caused by convection. As an alternative of this method for stabilization against convection, one can use a nonrotating capillary if its diameter is small enough (Fig. 2, case b), as was pointed out as early as 1967 (Hjertén). It must be stressed, however, that stationary

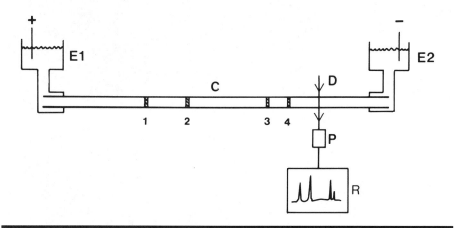

Figure 2   The principle of capillary electrophoresis. C, capillary; E1, E2, electrode vessels; D, detecting light beam; P, photo detector; R, recorder; 1, 2, 3, 4, zones (the sample components).

Case (a). When the diameter is $0.5-3$ mm, the capillary is rotated around its long axis (40 rpm) for stabilization against convection. The technique permits automatic scanning of the capillary at predetermined intervals. The course of the separation can therefore be followed. The resulting relatively large number of measuring points during a run gives very accurate determinations of mobilities and peak areas. Notice that stationary chambers with a rectangular cross-section also permit scanning (see Section XIII and Fig. 11).

Case (b). In capillaries with diameters less than 0.1 mm, the convection is negligible also when the capillary is not rotated. The zones 1, 2, 3, and 4 are detected when they pass the stationary detector ($D-P-R$). Only one value of the migration time and one of the peak area are obtained for each zone [if the scanning technique employed in (a) is utilized in (b), a very high noise : signal ratio will be obtained].

capillaries with diameters greater than 0.1 to 0.2 mm seldom offer a satisfactory zone stabilization, particularly not when the solutes consist of proteins, since solutions of these macromolecules have a higher density than those of low-molecular-weight substances (for instance, peptides). In capillaries where the diameter of the electrophoresis tube is 0.25–0.40 mm (Virtanen, 1974; Mikkers *et al.,* 1979), one can therefore expect some degree of convective-zone broadening, which is strongly reduced when the diameter is reduced to 0.075 mm (Jorgenson and deArman Lukacs, 1981), although the concentration limit of the detection of the solutes increases. These considerations of the degree of convection in capillaries of different diameters refer to free zone electrophoresis. In free isoelectric focusing (the principle of which is outlined in Fig. 3) and displacement electrophoresis, the convection is counteracted to some extent by the automatic zone sharpening that is an attractive feature of these methods. These two methods have also in common that the separation pattern does not change (or relatively little) with time once the "steady state" has been attained. In gel-filled capillaries, the convection is completely eliminated.

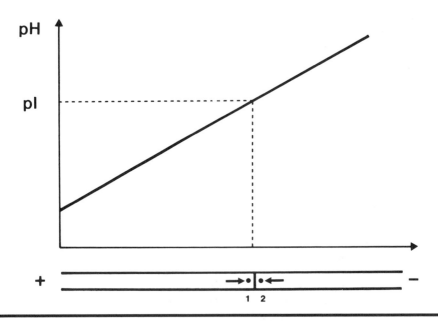

Figure 3   The principle of isoelectric focusing. A protein with the isoelectric point pI has a negative (positive) net charge at pH above (below) the pI. Protein molecules distributed over the whole pH gradient will accordingly be focused at a pH = pI (where the net charge is zero). Protein molecules that by diffusion leave the main zone (for instance, the molecules 1 and 2) will for the same reason migrate back into the zone. An analogous zone-sharpening effect occurs in displacement electrophoresis, where it is accomplished by a gradient in conductivity.

III. The Creation of a pH Gradient from a Mixture of Carrier Ampholytes with Different pI Values

Let us assume that we have chosen an ampholyte mixture that, according to the manufacturer, will give a gradient in pH from 3 to 10. The pH of such a mixture is about 7.8. The whole electrophoresis tube is filled with this mixture diluted with water to an ampholyte concentration of about 1%. Ampholytes with pI > 7.8 will be positively charged and migrate toward the negative pole when a voltage is applied, whereas those with pI values < 7.8 will acquire a negative charge and migrate toward the positive pole. The pH will accordingly decrease at the anodic section of the electrophoresis tube and increase at the cathodic section (the lowest pH in the tube is thus always found at the anodic end). This change in pH will continue until each ampholyte species has come to a final position, where it is noncharged. The pH in this position is equal to the pI of the particular ampholyte species. The more species of the carrier ampholytes and the more uniformly they are spaced at the steady state, the smoother will become the pH gradient. Owing to diffusion, the zones of the different species of ampholytes will not become infinitesimally thin, but will have a finite width and will thus not be surrounded by zones of pure water, although the concentrations of the ampholytes between their concentration maxima will become relatively low (cf. Fig. 8). The ohmic resistance (R) is accordingly high between these maxima, which means that the heat developed ($= R \cdot I^2$, where I is the current) is high, causing so-called "hot spots" to be generated.

The pH in the anode vessel must be lower than the pI of the most acidic ampholytes to prevent them from migrating into the anolyte (in the chosen example, lower than 3); if an ampholyte happened to enter the anolyte, it would be positively charged and return into the electrophoresis tube. For similar reasons, the catholyte have a pH > 10.

IV. The Principle of Isoelectric Focusing

In a focusing experiment the capillary tube contains not only carrier ampholytes but also proteins. Since proteins are amphoteric, they will focus at their pI values in narrow zones in the same way as do the individual carrier ampholytes (see Section III).

With reference to Fig. 3, the zone-sharpening effect, characteristic of isoelectric focusing, will now be discussed. If a focused protein molecule (2 in Fig. 3) diffuses into the region where the pH is higher than the pI of the protein, the protein molecule will acquire a negative net charge and therefore migrate back into the focused zone. In an analogous way, a protein molecule (1 in Fig. 3) diffusing into the pH region where the pH is lower than its pI will become positively charged and return to the focused zone. This self-sharpening focusing effect gives rise to very narrow zones, making isoelectric focusing a high-resolving separation method. Observe that the whole electrophoresis chamber can be filled with the protein sample at the start of a run. Isoelectric focusing therefore permits analysis of very dilute samples, which renders the method especially attractive. The obvious problem in high-performance capillary electrophoresis with detection of very dilute samples is thus eliminated when this technique is utilized in the isoelectric focusing mode.

In this discussion, it has been assumed that the sample consists of proteins. Isoelectric focusing is, however, applicable to all kinds of ampholytes, including peptides (Righetti and Hjertén, 1981), viruses (Talbot, 1975), and bacteria (Boltz and Miller, 1978; McGuire *et al.*, 1980), provided that one can create a pH gradient that covers the pI values of the solutes.

The first isoelectric focusing experiments were described by Kolin (1954). He used a pH gradient formed by ordinary buffer constituents. Since such gradients are not stable in an electrical field, Svensson (1961) introduced "natural pH gradients," created by means of a series of ampholytes, as discussed in Section III. Svensson showed theoretically that for these carrier ampholytes to function satisfactorily, they should have good buffering capacity and conductivity at their isoelectric points (Svensson, 1962). One should observe, however, that the pH gradients in practice move somewhat with time. It was not until 1987 that a mathematical explanation was reported (Hjertén *et al.,* 1987). It should be emphasized that Vesterberg (1969) made an extremely important contribution to the development of isoelectric focusing when he succeeded in preparing carrier ampholytes with the properties desired. These ampholytes consist of a mixture of compounds (containing amino and carboxylic groups) with slightly different pI values. For a review of isoelectric focusing techniques, see Righetti *et al.,* (1980) and Righetti (1984).

V. Terminology

An equation has been derived that applies to all methods based on differential migration (Hjertén, 1990) and is therefore valid for electrophoresis, chromatography, and centrifugation. Accordingly, any chromatographic method has, for instance, an electrophoretic counterpart. Such analogous methods should therefore be given analogous notations in order to emphasize the similarities between the methods.

The electrophoretic counterpart of high-performance liquid chromatography (HPLC) is electrophoresis in narrow-bore tubes. It is, accordingly, natural to call this method high-performance (capillary) electrophoresis, HP(C)E, a notation used in the title of the annual symposia devoted to this technique. Being analogous methods, HPLC and HPCE are both characterized by high resolution and short analysis times. The analogy between isoelectric focusing, the particular electrophoresis technique described in this chapter, and chromatofocusing and isopycnic centrifugation is obvious.

Discussing terminology, it may be of interest to refer to papers by Lederer (1979), Mikkers *et al.* (1979), and Hjertén (1983) with the titles "High-Performance Paper Electrophoresis"; "High-Performance Zone Electrophoresis"; and "High-Performance Electrophoresis: The Electrophoretic Counterpart of High-Performance Liquid Chromatography," respectively.

VI. Elimination of Electroendosmosis and Adsorption

A. Theoretical Considerations

In HPCE the capillaries are made from quartz, fused silica, or glass. Glass capillaries can be used in focusing experiments if they are sufficiently thin-walled to transmit light at a wavelength of 280 nm. Shorter wavelengths cannot be employed in isoelectric focusing since the light absorption of the carrier ampholytes gives a very disturbing background on UV-detection of the mobilized, focused protein zones (Fig. 8). The glass tubes have the advantage (in comparison with fused-silica tubing) that they are less brittle and therefore do not require an outer coating of polyimide.

Both quartz, fused silica, and glass have negative surface charges. The positive ions in the buffer are attracted to these charges, and the negative ones are repelled. Close to the capillary wall (in the so-called diffuse part of the double layer), there will accordingly be more positive than negative ions (electroneutrality will accordingly not prevail in the double layer). In an electrical field, the hydrated, positively charged surface layer will move toward the negative pole. This flow of liquid is called electroendosmosis. Because of the friction between adjacent liquid layers, all layers will move at the same speed, except the layers close to the capillary wall (at the wall the velocity is zero); see Fig. 4b. There is some evidence, however, that electroendosmosis is not such a perfect plug flow (Tsuda *et al.*, 1982). The zones become deformed parabolically if a hydrodynamic flow appears in the electrophoresis tube (see Fig. 4d), for instance, when the levels of the solutions in the electrode vessels differ.

Theoretically, the electroendosmotic flow should be zero if the inside surface of the electrophoresis capillary were neutral, but such surfaces probably do not exist (all plastic materials we have tested have exhibited electroendosmosis). However, the following formula (Hjertén, 1967) shows how the electroendosmosis can be suppressed without the requirement that the surface charge be zero.

$$\mu_{eo} = \frac{\varepsilon}{4\pi} \int_0^\zeta \frac{1}{\eta} d\,\psi(x) \tag{1}$$

where ε is the dielectric constant, $\psi(x)$ is the electric potential at the distance x from the capillary wall, ζ is the zeta potential, i.e., the potential in the double layer at the so-called *slipping plane*. This formula differs from the Helmholtz equation for electroendosmosis in the sense that η is the viscosity in the double layer (not that in the bulk of the buffer). In other words, by coating the surface of the capillary with an hydrophilic, nonionic polymer, the viscosity in the double layer will be high, and the electroendosmosis is virtually eliminated. In analogy with electrophoretic mobility, the electroendosmotic mobility, μ_{eo}, by definition is equal to v_{eo}/F, where v_{eo} is the velocity of the electroendosmotic flow (cm/sec) and F is the electrical field strength (volt/cm).

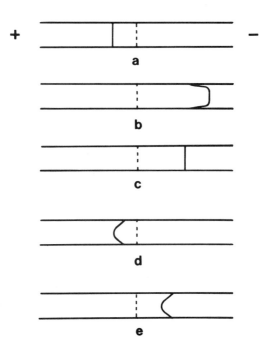

Figure 4 Deformation of an electrophoresis zone caused by electroendosmosis and a hydrodynamic flow. The figure illustrates the position and the shape of a zone affected by (a) electrophoresis only (the dotted line refers to the starting zone); (b) electroendosmosis only (the deformation occurs only in the very thin Helmholtz double layer, close to the tube wall; (c) a combination of a and b (the deformation in the double layer has been neglected); (d) hydrodynamic flow; (e) a combination of c and d. We have assumed here that electroendosmosis only displaces an electrophoresis zone without affecting its shape (with exception of the deformations very close to the tube wall). This assumption can, however, be questioned (Tsuda *et al.*, 1982).

A further advantage of coating the capillary with a neutral, hydrophilic polymer is that the interaction of solutes with the silanol groups in the capillary wall will be strongly suppressed [this interaction is not only of electrostatic but also of hydrophobic nature, since proteins adsorbed to naked silica beads in a chromatographic column cannot be completely desorbed by increasing the ionic strength of the buffer; electrostatic interactions can be eliminated on an increase in the buffer concentration, but the hydrophobic interactions will then increase (Hjertén, 1981)]. However, after repeated runs (days or weeks), proteins can become adsorbed to the coated capillary, which is expected from the general observation that no bed material used in chromatography is completely free from adsorption. It is easy to decide when protein adsorption occurs, since the peaks become broad, and the migration velocities, irreproducible [the capillary tube is flushed between the runs with 0.5 mM NaOH and with 0.02 M sodium phosphate, pH 2.5, for desorption of the proteins; the coating described by Hjertén (1985) is stable at pH 11 for at least 4 weeks].

B. Coating Procedure

The coating procedure described subsequently is the same as that reported earlier (Hjertén, 1985), with some modifications. The reaction scheme is outlined in Fig. 5. The glass or fused-silica tube is filled with 0.1 M sodium hydroxide. After 3 hr this solution is removed by rinsing with water. Following a 1-min contact with 0.1 M HCl and subsequent rinsing with water, the tube is ready for covalent attachment of linear polyacrylamide. This pretreatment of the inside of the tube cleans the inside of the tube and uncovers silane groups.

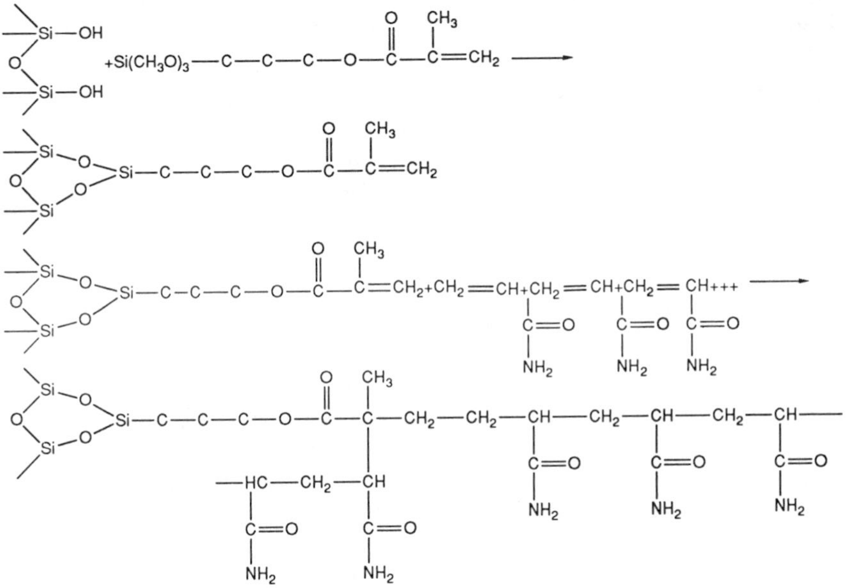

Figure 5   The coating procedure. To virtually eliminate electroendosmosis (and adsorption), the inside of the capillary should be coated with a thin layer of a hydrophilic noncharged polymer, as is evident from Eq. (1). The reaction scheme illustrates only one of many possible methods. γ-Methacryloxypropyltrimethoxysilane is first coupled to the silanol groups in the glass or fused-silica tubing via the trimethoxy groups. The subsequent reaction with acrylamide gives a covalently attached monolayer of linear polyacrylamide. It is important that the polymer layer be very thin to avoid tailing of a solute zone because of slow diffusion of the solute into and out of the layer.

The tube is filled with a mixture of 30 μl of γ-methacryloxypropyltrimethoxysilane (Bind-Silane, LKB, Bromma, Sweden) and 1 ml of a 60% (v/v) acetone solution. An efficient reaction between the methoxy groups and the silanol groups at the tube wall is accomplished if the coupling takes place at room temperature overnight. A 3% deaerated solution of acrylamide in water, supplemented with ammonium persulfate (2 mg/ml) and N,N,N',N'-tetramethylethylenediamine (TEMED) (0.8 μl/ml), is introduced into the tube to wash off the excess silane solution and is then allowed to polymerize overnight and, at the same time, to react with the double bond in the covalently attached silane. The ends of the tube are closed with modeling clay (plasticine) in this reaction step to prevent evaporation of solvent. The same precautionary measure is taken in the silanization step.

The smaller the diameter of the capillary, the larger the risk of adsorption of the solutes to the inner surface of the capillary, since the ratio between the area of this surface and the volume of the capillary is inversely proportional to its diameter. Many attempts have therefore been made to decrease the solute adsorption in HPCE. However, in isoelectric focusing, it is necessary not only to suppress adsorption but also to eliminate electroendosmotic flow. Otherwise the carrier ampholytes and the proteins will be displaced out of the capillary. Accordingly, methods that give only a partial reduction in electroendosmosis (and adsorption) cannot be employed with success in isoelectric focusing. Such unusable methods are those based on the addition of certain positive ions to the carrier ampholytes to "block" the negative silanol groups and the coupling of low-molecular-weight compounds to the inner surface of the capillary. The former alternative is precluded also because it causes a drift of the pH gradient (see Section IX).

VII. Determination of the pI of a Protein

The on-tube detection technique, described herein, does not permit a direct measurement of the pH of the focused protein zones. To determine the pI of a protein, one is therefore forced to use standard proteins with known pI values. These proteins and the sample proteins to be analyzed are included in the carrier ampholyte solution. The migration times for the standard proteins in the mobilization step are plotted against their pI values. From the migration times of the sample proteins, one can obtain their pI values from this calibration curve. A typical calibration is shown in Fig. 6 [for further experimental details, see Hjertén et al. (1986)].

VIII. Suppression of Precipitation of Proteins

A protein has a negative (positive) net charge at a pH above (below) its isoelectric point. At such pH values, the protein molecules repel each other, which decreases the risk of aggregation (which may lead to precipitation). However, it is not surprising that certain proteins have a tendency to precipitate when pH approaches the pI, where the

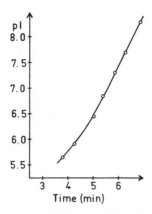

Figure 6 Calibration curve for the determination of the pI of an unknown protein. The protein, the pI of which is to be determined, is mixed with the carrier ampholytes and standard proteins with known pI values. Following focusing and drawing the calibration curve, the migration time of the protein of interest permits estimation of its pI. Reproduced with permission (Hjertén *et al.*, 1986).

net charge is zero. When subjecting such proteins to isoelectric focusing, the amount applied should be relatively low to suppress the risk of the formation of precipitates. Fortunately, a precipitate gives rise to an extremely narrow peak that therefore is easy to distinguish from a peak corresponding to a protein zone. When the protein molecules precipitate, they probably aggregate by hydrophobic interactions. It is therefore logical to suppress the precipitation by supplementing the carrier ampholytes with agents known to decrease hydrophobic interactions, for instance, ethylene glycol [10–40% (v/v)] or detergents [1–4% (w/v)] (Hjertén *et al.*, 1982, 1987). The detergents must be of the nonionic type and devoid of disturbing UV-absorption. Examples of such detergents are G 3707 (available from Atlas Chemie, Everberg, Belgium) and reduced Triton X-100 (Sigma Chemical Company, St. Louis, Missouri). Both ethylene glycol and neutral detergents are mild agents and seldom change the pI of a water-soluble protein significantly. In fact, ethylene glycol often has a stabilizing effect on the structure of a protein (Hjertén *et al.*, 1982). Urea at high concentrations (6–8 M) can prevent precipitation, but denatures the proteins (Righetti *et al.*, 1980). Intrinsic membrane proteins, with their pronounced hydrophobic surface structure and attendant risk of precipitation, can seldom be analyzed successfully by isoelectric focusing.

When proteins are not suitable for study by isoelectric focusing in free solution owing to precipitation, one should try to perform the runs in a polyacrylamide gel of low concentration, since experience shows that the tendency of precipitation then is less, perhaps because the pore structure of the gel prevents the proteins from coming too close to each other.

In particularly difficult cases isoelectric focusing should be replaced by zone electrophoresis in a pH gradient (see Section XV). This method also gives an automatic zone sharpening, but the risk of precipitation is much lower because the protein will never reach a pH equal to its pI.

IX. Cathodic and Anodic Drift

The compounds used in isoelectric focusing to create a pH gradient are ampholytes and thus contain acidic (A^-) and basic (BH^+) groups ($A^- = COO^-$, SO_3^-; $BH^+ = NH_3^+$).

Let us consider an experiment (without applied sample) in which carrier ampholytes are used. The electroneutrality condition can then be formulated

$$C_{H^+} + \Sigma C_{BH^+} = C_{OH^-} + \Sigma C_{A^-} \tag{2}$$

where C is the concentration in equivalents/liter.

The net charge of the ampholytes, expressed as $\Sigma C_{BH^+} - \Sigma C_{A^-}$, is obtained from Eq. (2):

$$\Sigma C_{BH^+} - \Sigma C_{A^-} = C_{OH^-} - C_{H^+} \tag{3}$$

$$\Sigma C_{BH^+} - \Sigma C_{A^-} = \frac{K_w}{C_{H^+}} - C_{H^+} \tag{4}$$

where K_w = the dissociation constant of water ($\approx 10^{-14}$).

The net charge of the ampholytes differs accordingly from zero, except when $C_{OH^-} = C_{H^+}$ (pH 7). We thus arrive at the important conclusion that most carrier ampholytes at the steady state are either basic, and therefore migrate toward the cathode, or acidic, and migrate toward the anode [see Fig. 7, which is based on Eq. (4)]. One can thus expect a drift of the pH gradient toward both the anode and the cathode, which, in fact, has been observed (Chrambach *et al.*, 1973). This anodic and cathodic drift is also in agreement with the observation that measurements of pH, following a focusing based on ampholytes that cover nominally the pH range from 3 to 10, often show that the actual pH gradient is shorter (for instance, 4–9). It should be emphasized, however, that the electrophoretic velocities of the carrier ampholytes (the drift) is determined not only by their charge, but also by their sizes and the field strength, which is inversely proportional to the conductivity. The magnitude of (some of) these parameters may very well be such that the cathodic drift is more rapid than the anodic drift — which is observed in most isoelectric focusing experiments. In cases in which electroendosmosis has not been eliminated, the attendant displacement of the carrier ampholytes toward the cathode (Fig. 4b) may be stronger than the electrophoretic anodic drift, and therefore, only a net migration of ampholytes toward the cathode is observed.

Since the charges of the silanol groups in fused silica tubing are pH dependent, the electroendosmotic flow in a noncoated tubing is different in different parts of the pH gradient, which may cause local circulation of ampholytes and thus impair the res-

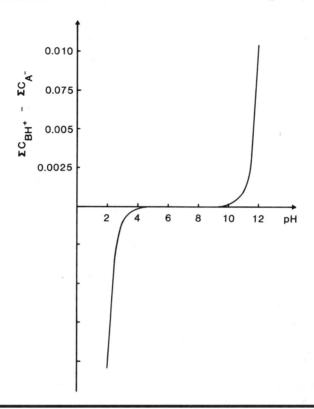

Figure 7 Cathodic and anodic drift. The curve is a graphic presentation of Eq. (4). $\Sigma C_{BH^+} - \Sigma C_{A^-}$ is the net charge of the carrier ampholytes. Since the net charge evidently is positive at basic pH and negative at acidic pH, a pH gradient formed by ampholytes is not entirely stable but drifts toward both the cathode and the anode. The electrophoretic cathodic drift is often strengthened by a superimposed electroendosmotic flow in the same direction.

olution (Öfverstedt *et al.*, 1981). Methods based on electroendosmosis for the mobilization of focused protein zones (see the next section) cannot therefore give high resolution over the entire pH gradient. The electroendosmotic migration of the ampholytes (most often toward the cathode) can be eliminated by coating the inside of the tube with a polymer (see Section VI).

X. Mobilization of the Focused Protein Zones

Since solutes can be monitored only if they pass the stationary UV-detector in an HPCE apparatus, one has to find a method to mobilize the focused, almost stationary protein zones without impairing the resolution obtained in the focusing step. One way to do this is to change the net charge ($\Sigma C_{BH^+} - \Sigma C_{A^-}$) of all species of the focused carrier ampholyte to such an extent that they rapidly migrate electrophoretically out of the tube. Equation (3) shows that this can be accomplished if we could modify the equation in an appropriate way so that its right member becomes sufficiently positive (corresponding to cathodic migration) or sufficiently negative (anodic migration). This requirement can be fulfilled theoretically if we reformulate Eq. (3) as follows:

$$\Sigma C_{BH^+} - \Sigma C_{A^-} = C_{OH^-} - C_{H^+} - C_{X^{+n}} \tag{5}$$

$$\Sigma C_{BH^+} - \Sigma C_{A^-} = C_{OH^-} - C_{H^+} + C_{Y^{-m}} \tag{6}$$

where X^{+n} is a cation with the valency $+n$, and Y^{-m}, an anion with the valency $-m$. If we add X^{+n} in the form of a salt to the anolyte (anodic migration) or Y^{-m} as a salt to the catholyte (cathodic migration), Eqs. (5) and (6), respectively, will be satisfied as soon as the ions X^{+n} and Y^{-m} have migrated electrophoretically into the tube. This theoretical approach works very well in practice, not only when a focusing experiment is performed in free solutions (see Figs. 8, 9a, and 10), but also in gels (Hjertén and Zhu, 1985). For instance, when sodium chloride is added to the anolyte, the pH gradient (and the protein zones) move toward the anode and, when added to the catholyte, they move to the cathode.

It is important that the anolyte in an isoelectric focusing experiment (often a sodium hydroxide solution) be fresh. Otherwise the uptake of carbon dioxide from the air can be so high that the carbonate formed may cause a cathodic drift. For this reason a solution of barium hydroxide is preferable to one of sodium hydroxide, since the concentration of carbonate can never be high in the former solution because of the low solubility of barium carbonate.

XI. The Practical Performance of a Run

The protein zones can be detected at 280 nm but not at shorter wavelengths where the carrier ampholytes have a strong absorption (see Fig. 8). At this wavelength, one can with advantage use cheap, thin-walled tubes of glass (Hjertén, 1983). Suitable dimensions of the tubes are 0.05–0.10 mm (i.d.) × 100–200 mm (length). The carrier ampholytes and the sample are mixed so that the final concentration of the ampholytes is about 1% [the sample should not contain salts, since they will cause a drift of the pH gradient, according to Eqs. (5) and (6)]. The glass tube, which must be coated with a polymer (see Section VI), is cut off at the ends with scissors to get a uniformly polymerized coating along the whole length of the tube. The tube is dipped into the sample solution containing the sample and the carrier ampholytes; the solution is sucked up automatically into the tube by the capillary forces (alternatively, a water pump or a syringe can be used). One end of the tube is then pressed into a 1% agarose gel, prepared in 0.02 M sodium hydroxide or 0.02 M phosphoric acid (after the dissolution of agarose in these hot basic or acidic solutions, the temperature is immediately lowered by water cooling to avoid strong degradation of the agarose). The gel plug thus inserted into the tube end prevents zone-deforming hydrodynamic flow in the tube during the subsequent focusing (see Fig. 4d). A voltage of 4000 to 6000 V is applied.

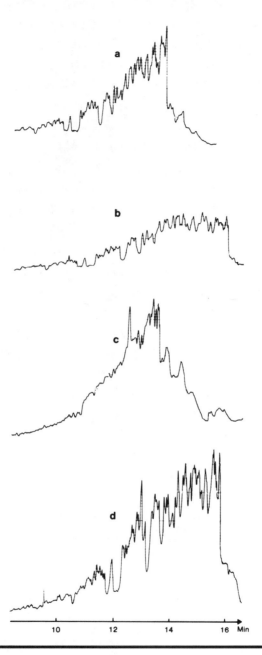

10   12   14   16   Min

Figure 8 Isoelectric focusing of different commercially available carrier ampholytes. Dimensions of the fused-silica tubing: 0.05 (i.d.) × 130 mm. Voltage at focusing and mobilization, 4000 V; focusing time, 6 min; anolyte during focusing and mobilization, 0.02 M phosphoric acid and 0.02 M sodium hydroxide, respectively; catholyte, 0.02 M sodium hydroxide; detection wavelength, 220 nm. (a) Ampholine, pH 3.5–10. (b) Bio-Lyte, pH 3–10. (c) Pharmalyte, pH 3–10. (d) A mixture of a, b, and c.

a

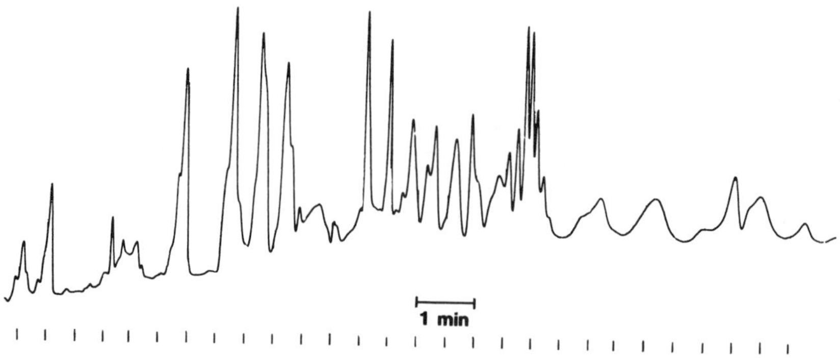
1 min

b

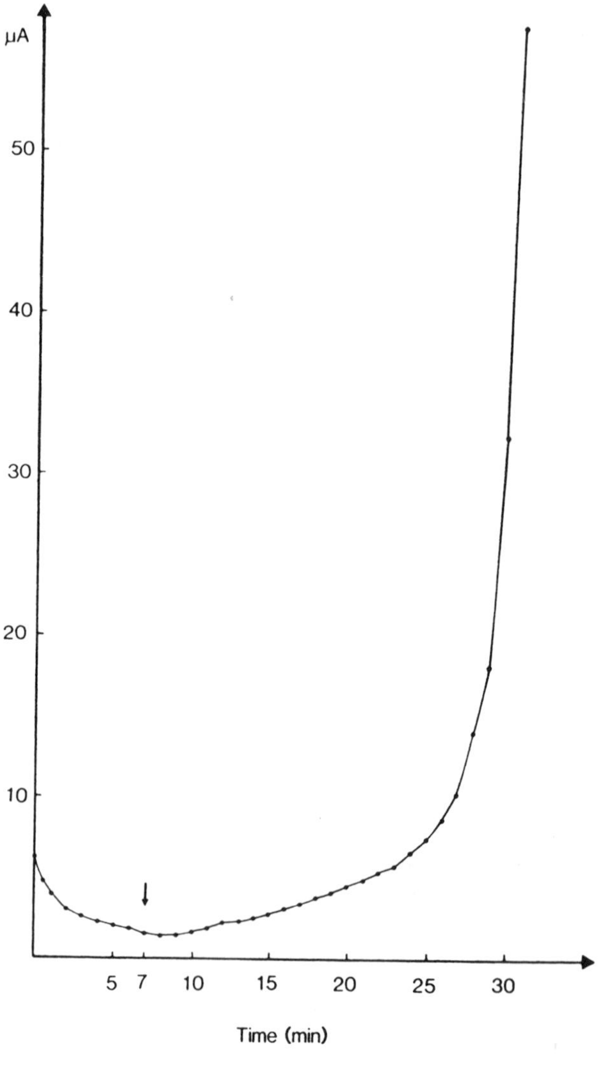
μA
50
40
30
20
10
5
7
10
15
20
25
30
Time (min)

When the steady state has been attained, which occurs when the current has dropped to about 10 to 25% of the value at the start (see Fig. 9b), the voltage is switched off. For anodic mobilization, the anolyte is replaced, for instance with 0.02 M sodium hydroxide or a solution of 0.02 M phosphoric acid containing 0.02–0.04 M sodium chloride [Na^+ is the mobilizing ion X^{+n} in Eq. (5)]. When the voltage is turned on, the focused protein zones migrate toward the anode and are thus detected (Figs. 9a and 10). For cathodic mobilization generated by chloride ions [Y^{-m} in Eq. (6)], the catholyte is replaced by a 0.02 M sodium hydroxide solution containing 0.02–0.04 M sodium chloride (one can also use 0.02 M phosphoric acid, but its pH-buffering capacity can affect the appearance of the pH gradient). For other compositions of the mobilizing solutions, see Hjertén *et al.* (1987).

The gel plug may be omitted when the diameter of the capillary is smaller than 0.03 mm, since the risk is small of hydrodynamic flow in the capillary caused by the siphoning effect. "Electroendosmosis-free" agarose should be used for the preparation of the gel plug. However, some electroendosmotic flow in the gel will not deform the zones significantly, because of the automatic zone sharpening obtained in isoelectric focusing. Many practical issues on isoelectric focusing have been discussed by Zhu *et al.* (1989).

XII. Analysis and Comparison of Different Carrier Ampholytes

The coated glass capillary was filled with a 1% (v/v) solution of Ampholine (pH 3.5–10) from LKB, Bromma, Sweden. The capillary had the dimensions 0.05 (i.d.) × 130 mm (the length up to the detector: 118 mm). The focusing (without sample) was conducted at 4000 V (6 min). The mobilization was performed at the same voltage with the 0.02 M phosphoric acid in the anolyte replaced by 0.02 M sodium hydroxide. The UV pattern at 220 nm is shown in Fig. 8a.

The experiment was then repeated with Bio-Lyte (pH 3–10), and Pharmalyte (pH 3–10), carrier ampholytes from Bio-Rad (Richmond, California) and Pharmacia (Uppsala, Sweden), respectively (Figs. 8b and c). Surprisingly, Ampholine and Bio-Lyte gave rather similar UV patterns (although that corresponding to Ampholine is more compressed), whereas Pharmalyte had a quite different pattern. It may be an advantage to create the pH gradient with a mixture of carrier ampholytes manufactured

Figure 9   High-performance isoelectric focusing of carbamylated carbonic anhydrase in a fused-silica tubing. Dimensions of the fused-silica tubing, 0.1 (i.d.) × 120 mm; voltage, 3000 V in both the focusing and the mobilization steps; carrier ampholyte, 1% Bio-Lyte 3/10, containing 8 M urea. (a) Electropherogram. (b) A plot of current against focusing and mobilization time. The arrow indicates the start of the mobilization.

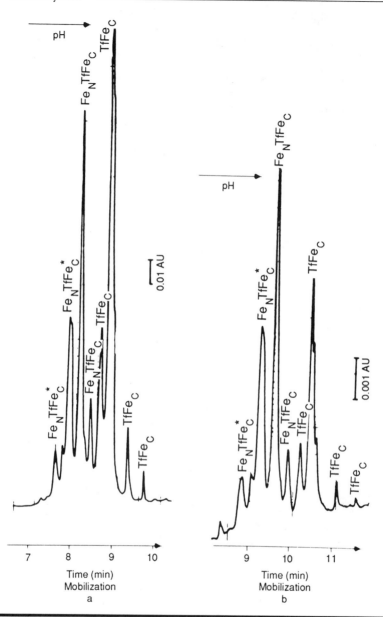

Figure 10   High-performance isoelectric focusing of transferrin in a glass tube. Dimensions of the glass tube, 0.1 (i.d.) × 140 mm; voltage, 4000 V for both focusing and mobilization; carrier ampholyte, 2% Bio-Lyte 5/7 (containing about 1 μg of transferrin). The protein and iron contents in the focused zones were monitored at 280 (a) and 460 (b) nm, respectively. Reproduced with permission, Kilár and Hjertén (1989).

by different companies (Fig. 8d), since the pH gradient will be smoother, the more ampholyte species there are in the capillary.

XIII. Apparatus for Focusing in a Chamber of Rectangular Cross-Section, Permitting Scanning for the Focused Proteins at Different Times and at Different Wavelengths

The chamber was built up of a bottom quartz microscope slide ($0.5 \times 24 \times 75$ mm) and an upper glass microscope slide ($0.12 \times 24 \times 60$ mm) and two greased polyvinyl chloride (PVC) strips ($0.14 \times 3 \times 60$ mm) at the edges of the slides (Hjertén et al., 1988). Since the two slides had different lengths, only part of the quartz slide was covered by the glass slide. The open sections thus formed at the ends of the quartz slide were utilized as electrode vessels. The electrophoresis chamber was "sealed" at the openings with some drops of a high-viscous polymer solution [10% (w/v) linear polyacrylamide dialyzed overnight against water]. Electric contact between the electrophoresis chamber and the electrodes was obtained via bridges of Wettex cloth moistured with 0.1 M sodium hydroxide (cathode) and 0.1 M phosphoric acid (anode). The focusing apparatus, with the slides clamped together in a PVC frame, was placed on the horizontal quartz plate of a scanner, originally designed for UV detection of the protein zones in polyacrylamide gel electrophoresis (without staining) (Fries and Hjertén, 1975). The scanner was constructed at the Institute of Biochemistry by Per-Axel Lidström and Hans Pettersson. The rectangular electrophoresis (focusing) chamber has the advantage over glass tubes or fused silica tubing, in that the optical surfaces are well defined, and therefore permit scanning with a satisfactory noise : signal ratio. If necessary this ratio can be rendered lower by scanning at two wavelengths, one where the proteins absorb light and one where they do not; the ratio between the absorption values at these wavelengths is recorded (Hjertén, 1967; Fries and Hjertén, 1975). In this connection, it should be mentioned that the free electrophoresis apparatus, based on rotation of the horizontal quartz capillary, also allows scanning of the capillary for detection of focused protein zones (Hjertén, 1967, 1976).

The advantage of scanning equipment in comparison with a stationary detector is that one can follow the separation during the whole focusing process. With some modifications, the rectangular chamber permits the application and analysis of more than one sample in a run, which is an obvious advantage over focusing chambers in the form of capillary tubing.

Thormann et al. (1987) used capillaries of rectangular cross-section and followed the variations in field strength during the focusing with the aid of a linear array of potential gradient sensors. Capillaries of the same shape have been utilized by Tsuda et al. (1990) to increase the light path, i.e., the sensitivity in HPCE, by letting the detecting light beam pass through the length (1 mm) of the cross-section instead of its depth (0.05 mm). Since the zones are not perfectly straight across such a relatively long distance (1 mm), the resolution will be somewhat less than that in capillaries with circular cross-section and diameters less than 0.1 mm.

XIV. Resolution and Optimal Field Strength

In all electrophoresis methods, including isoelectric focusing, the zone broadening caused by diffusion decreases when the voltage is increased (since the run time decreases), whereas the broadening due to the Joule heat increases. There must accordingly exist a field strength at which the total zone broadening generated by both diffusion and Joule heat is at a minimum. This optimal field strength can easily be calculated for zone electrophoresis experiments (Foret *et al.*, 1988; Hjertén, 1990). However, for isoelectric focusing, the situation is more complicated, owing to the zone-sharpening effect, and will not be discussed here. The optimal voltage (field strength) may be determined experimentally from a series of runs at different voltages (in isoelectric focusing, the field strength is not constant in the capillary, and therefore it may be appropriate to give the voltage and the length of the capillary instead of an average field strength, as shown in the figure legends).

XV. Applications

A. Isoelectric Focusing of Carbamylated Carbonic Anhydrase in a Fused-Silica Tubing

The sample, 150 μg of carbamylated carbonic anhydrase, was a gift from Jorge Lizana, Pharmacia. It was dissolved in 15 μl of water and mixed with 200 μl of a 1% (v/v) solution of Bio-Lyte (3/10), supplemented with 8 M urea.

The polymer-coated, fused-silica tubing [0.1 (i.d.) $\times$ 120 mm] was filled with this solution and then closed with a gel plug (see Section XI). A voltage of 3000 V was used for the focusing step (7 min). The same voltage was used for the mobilization, which was performed by replacing the anode solution by 0.02 M sodium hydroxide, containing 8 M urea. The photo of the electropherogram in Fig. 9a indicates the high resolution often obtained in isoelectric focusing experiments. A plot of current against time is shown in Fig. 9b for both the focusing and the mobilization steps.

B. Isoelectric Focusing of Transferrin in a Glass Capillary

The glass tube [0.1 (i.d.) $\times$ 0.3 (o.d.) $\times$ 140 mm] was treated with γ-methacryloxy-propyl-trimethoxysilane and acrylamide to eliminate adsorption and electroendosmosis, as described in Section VI.

Human transferrin was dissolved in the carrier ampholyte (2% Bio-Lyte 5/7) to a concentration of 1 μg/μl. The focusing took 6 min at a voltage of 4000 V with

0.02 *M* sodium hydroxide as catholyte and 0.02 *M* phosphoric acid as anolyte. The tube was closed at the anodic end with a 1% agarose gel, prepared in the 0.02 *M* phosphoric acid. The mobilization (4000 V) was achieved by replacing the phosphoric acid solution with 0.02 *M* sodium hydroxide. In this step the current increased, and when it was 50 μA and all transferrin species had left the capillary, the experiment was stopped (compare Fig. 9b). The detection was made at 280 nm for protein (Fig. 10a) and at 460 nm for iron (Fig. 10b) at a distance of 15 mm from the cathodic end of the glass tube. For more details of this experiment, see Kilár and Hjertén (1989).

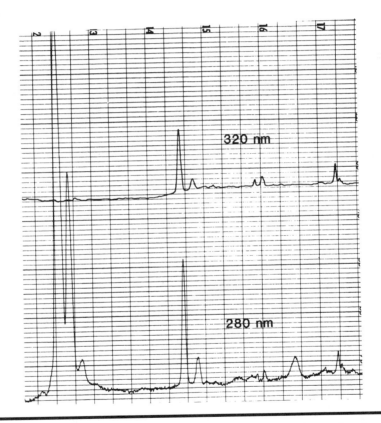

Figure 11   High-performance isoelectric focusing of chymotrypsinogen A and myoglobin in a rectangular focusing chamber. The inner dimensions of the chamber, 0.14 × 18 × 60 mm; voltage during focusing, 200 V; carrier ampholyte, 4% Pharmalyte, pH 3–10, containing 5 μl TEMED per ml to get a more stable pH gradient at the extremely basic pH values (Guo and Bishop, 1982). The chamber was scanned at 280 nm for protein and at 320 nm specifically for myoglobin. The scanning technique has several advantages: for instance, that one can follow the course of the focusing and that the mobilization step, required when the detector and the electrophoresis chamber are stationary, can be excluded.

C. Isoelectric Focusing of a Mixture of Chymotrypsinogen A and Myoglobin in a Rectangular Focusing Chamber

Before mounting in the PVC frame (see Section XIII), the quartz and glass plates were coated with methyl cellulose to eliminate electroendosmosis and adsorption (Hjertén, 1967).

Chymotrypsinogen A (500 µg; pI, 9.6) and myoglobin (35 µg; pI, 7.2) were dissolved in 150 µl of a 4% (v/v) solution of Pharmalyte, pH 3–10, containing 5 µl of TEMED per ml. TEMED facilitated the focusing of the extremely basic protein chymotrypsinogen A (Guo and Bishop, 1982). After the filling of the electrophoresis chamber with this sample–ampholyte solution, 200 V was applied. The course of the focusing was followed by scanning at 280 nm without switching off the voltage. After 45 min an almost stationary UV pattern was obtained, and another scanning at 320 nm was done (Fig. 11). Since both proteins absorb light at 280 nm, and only myoglobin, at 320 nm, a comparison of the two scanning patterns shows which peaks correspond to chymotrypsinogen A and which, to myoglobin.

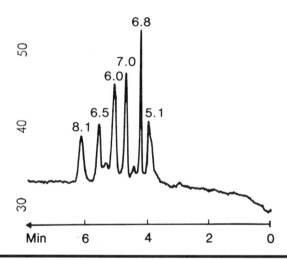

Figure 12 Zone electrophoresis in a pH gradient. Dimensions of the glass tube, 0.1 (i.d.) × 90 mm. Detection was made 1.5 cm from the anodic end of the tube. Voltage, 2000 V. Carrier ampholyte, 5% Bio-Lyte 8/10 (the pH was around 8.5 at the position where the light beam passed through the electrophoresis tube). Anolyte, 0.01 M N-2-hydroxyethylpiperazine propane sulfonic acid (pH 7.0). Catholyte, 0.01 M ethanolamine (pH 11.0). Sample, β-lactoglobulin (pI 5.1), human hemoglobin (pI 6.8), equine myoglobin (pI 7.0), bovine carbonic anhydrase (pI 6.0), human carbonic anhydrase (pI 6.5), and whale myoglobin (pI 8.1). Detection wavelength, 280 nm. Upon zone electrophoresis in a pH gradient the proteins migrate under zone sharpening as in isoelectric focusing, but do not reach the pH values corresponding to their pI values. The risk of precipitation is therefore reduced. The experiment was performed by Dr. Jia-li Liao in the author's laboratory.

D. Zone Electrophoresis in a pH Gradient

Several additives used to suppress precipitation of focused proteins were discussed in Section VIII. When none of these is sufficiently efficient the isoelectric focusing may be replaced by zone electrophoresis in a preformed pH gradient. The sample is applied at one end of the electrophoresis tube before the pH gradient has reached the steady state. Otherwise, the cathodic drift sets in before the proteins have passed the detector and impairs the resolution (see Section IX). If the pI values of the proteins are outside the pH range of the carrier ampholytes used isoelectric precipitation will not occur, which is the advantage of this method. For instance, if proteins with pI values in the pH range from 5.1 to 8.1 are applied at the cathodic side and the pH gradient runs from 8.5 to 10 (the lower pH at the anode), all proteins will migrate toward the anode. The migration occurs under zone-sharpening, since the front of a protein zone is at a lower pH than the rear and therefore moves slower. The zones will accordingly become very narrow, as in isoelectric focusing. An example is shown in Fig. 12.

Acknowledgment

The author is much indebted to Mrs. Karin Elenbring, who performed the experiments corresponding to Figs. 8, 9, and 11. The work was supported by the Swedish Natural Science Research Council and the Knut and Alice Wallenberg and the Carl Trygger Foundations.

References

Boltz, R. C., Jr., and Miller, T. Y. (1978). *In* "Electrophoresis '78" (N. Catsimpoolas, ed.), pp. 345–355. Elsevier Publishers, Amsterdam.

Chrambach, A., Doerr, P., Finlayson, G. R., Miles, L. E. M., Sherin, R., and Rodbard, D. (1973). *Ann. N.Y. Acad. Sci.* **209**, 44–60.

Foret, F., Deml, M., and Boček, P. (1988). *J. Chromatogr.* 452, 601–613.

Fries, E., and Hjertén, S. (1975). *Anal. Biochem.* 64, 466–476.

Guo, Y.-J., and Bishop, R. (1982). *J. Chromatogr.* **234**, 459–462.

Hjertén, S. (1958). *Arkiv för Kemi* **13**, 151–152.

Hjertén, S. (1967). *Chromatogr. Rev.* 9, 122–219.

Hjertén, S. (1976). *In* "Methods of Protein Separation" (N. Catsimpoolas, ed.), Vol. 2, pp. 219–231. Plenum, New York.

Hjertén, S. (1981). *In* "Methods of Biochemical Analysis" (D. Glick, ed.), Vol. 27, pp. 89–108. Wiley, New York.

Hjertén, S. (1983). *J. Chromatogr.* **270**, 1–6.

Hjertén, S. (1985). *J. Chromatogr.* **347**, 191–198.

Hjertén, S. (1990). *Electrophoresis* **11**, 665–690.

Hjertén, S., and Zhu, M.-D. (1985). *J. Chromatogr.* **346**, 265–270.

Hjertén, S., Pan, H., and Yao, K. (1982). *In* "Protides of the Biological Fluids, Proceedings of the 29th Colloquium, Brussels 1981" (H. Peeters, ed.), pp. 15–27. Pergamon Press, Oxford and New York.

Hjertén, S., Kilár, F., Liao, J.-L., and Zhu, M.-D. (1986). *In* "Electrophoresis '86" (M. J. Dunn, ed.), pp. 451–462. VCH Verlagsgesellschaft, Weinheim, Germany.

Hjertén, S., Liao, J.-L., and Yao, K. (1987). *J. Chromatogr.* **387**, 127–138.

Hjertén, S., Elenbring, K., Sedzik, J., and Valtcheva, L. (1988). *In* "6th International Symposium on Isotachophoresis and Capillary Zone Electrophoresis, Vienna, September 21–23, Abstract Book," p. 5, and in "6th Meeting of the International Electrophoresis Society, Copenhagen, July 4–7, Abstract Book," p. 24.

Johansson, G., Öfverstedt, L.-G., and Hjertén, S. (1987). *Anal. Biochem.* **166,** 267–275.

Jorgenson, J. W., and deArman Lukacs, K. (1981). *J. Chromatogr.* **218,** 209–216.

Kilár, F., and Hjertén, S. (1989). *Electrophoresis* **10,** 23–29.

Kolin, A. (1954). *J. Chem. Phys.* **22,** 1628–1629.

Lederer, M. (1979). *J. Chromatogr.* **171,** 403–406.

McGuire, J. K., Miller, T. Y., Tipps, R. W., and Snyder, R. S. (1980). *J. Chromatogr.* **194,** 323–333.

Mikkers, F. E. P., Everaerts, F. M., and Verheggen, T. P. E. M. (1979). *J. Chromatogr.* **169,** 11–20.

Righetti, P. G. (1984). *J. Chromatogr.* **300,** 165–223.

Righetti, P. G., and Hjertén, S. (1981). *J. Biochem. Biophys. Methods* **5,** 259–272.

Righetti, P. G., Gianazza, E., and Ek, K. (1980). *J. Chromatogr.* **184,** 415–456.

Svensson, H. (1961). *Acta Chem. Scand.* **15,** 325–341.

Svensson, H. (1962). *Acta Chem. Scand.* **16,** 456–466.

Talbot, P. (1975). *In* "Isoelectric Focusing" (J. P. Arbuthnott and J. A. Beeley, eds.), pp. 270–274. Butterworths, London.

Thormann, W., Tsai, A., Michaud, J.-P., Mosher, R. A., and Bier, M. (1987). *J. Chromatogr.* **389,** 75–86.

Tiselius, A. (1930). Thesis. Nova acta regiae societatis scientiarum upsaliensis, Ser. IV, Vol. 7, No. 4, pp. 1–107. Almqvist & Wiksell, Uppsala, Sweden.

Tiselius, A. (1937). *Trans. Faraday Soc.* **33,** 524–531.

Tsuda, T., Nomura, K., and Nakagawa, G. (1982). *J. Chromatogr.* **248,** 241–247.

Tsuda, T., Sweedler, J. V., and Zare, R. N. (1990). *Anal. Chem.* **67,** 2149–2152.

Vesterberg, O. (1969). *Acta Chem. Scand.* **23,** 2653–2666.

Virtanen, R. (1974). *Acta Polytech. Scand. Chem.* **123,** 1–67.

Zhu, M.-D., Hansen, D. L., Burd, S., and Gannon, F. (1989). *J. Chromatogr.* **480,** 311–319.

Öfverstedt, L.-G., Johansson, G., Fröman, G., and Hjertén, S. (1981). *Electrophoresis* **2,** 168–173.

Capillary Electrophoresis in Entangled Polymer Solutions

Paul D. Grossman

I. Introduction

As was pointed out in a previous chapter on free-solution capillary electrophoresis (Chapter 4), for many practically important applications, DNA and sodium dodecyl sulfate (SDS)-denatured proteins in particular, separations based on differences in free-solution electrophoretic mobilities are not possible (Olivera *et al.,* 1964; Hermans, 1953). Therefore, in order to effect electrophoretic separations of mixtures of these molecules, one has had to perform electrophoretic separations in a cross-linked rigid gel matrix, which alters the frictional characteristics of these species in such a way as to introduce a molecular size dependence to their electrophoretic mobility.

However, with the advent of capillary electrophoresis (CE), superior separations of DNA mixtures have been demonstrated without the use of a cross-linked gel matrix (Chin and Colburn, 1989; Zhu *et al.,* 1989). These studies demonstrate that, by using a dilute, low-viscosity polymer solution as the separation medium, high-resolution separations of DNA mixtures can be achieved. Whereas electrophoresis in noncross-linked polymer solutions has been previously demonstrated (Tietz *et al.,* 1986), never before has such high resolution been achieved without the use of rigid gels. The application of entangled polymer solutions as "sieving" media has also been exploited by

Langevin and Rondelez (1978) in the context of centrifugation. This work is a direct analog to electrophoresis in entangled solutions.

In effect, capillary electrophoresis using entangled polymer solutions decouples the two traditional roles of a gel matrix: that of an anticonvective, stabilizing medium and that of a sieving matrix. This is possible because the efficient heat dissipation of the CE format eliminates the need for the anticonvective properties of a rigid gel. Although application of these systems to practically important applications is very recent, this technique promises to combine the advantages of free-solution capillary electrophoresis (system automation, speed, reproducibility, continuously renewable media, and on-line detection) with the range of application and resolving power of gel-based systems.

In this chapter we attempt to show how concepts from polymer physics can be applied to describe the relevant characteristics of the entangled matrix. Furthermore, we demonstrate that the separation mechanism at work in these systems is the same as that of traditional gel-based systems. Much of the work presented in this chapter is taken from a publication by Grossman and Soane (1991).

II. Entangled Polymer Solutions

In this section we introduce the concept of an entangled polymer solution and present some of the simple scaling laws that have been developed to describe their properties.

A. Overlap Threshold

An important difference exists between dilute polymer solutions, in which the polymer chains are hydrodynamically isolated from one another, and more concentrated solutions, in which the chains overlap and interact. The polymer volume fraction at which the polymer chains begin to interact with one another, Φ^*, is called the overlap threshold (the volume fraction is defined as $\Phi = C\rho_p$, where C is the polymer concentration and ρ_p is the density of the polymer). At greater than this concentration, the solution is said to be entangled. A schematic illustration of the entanglement process is given in Fig. 1.

An expression predicting the value of Φ^* as a function of polymer size was first derived by de Gennes (1979). His expression is based on the assumption that when $\Phi = \Phi^*$, the bulk concentration of the solution is the same as the concentration inside a single coil. Thus, according to de Gennes,

$$C^* = \frac{\text{no. of monomers/coil}}{\text{volume/coil}} \approx \frac{N}{R_g^3} \tag{1}$$

where C^* is the polymer number-concentration at the entanglement threshold, N is the number of segments in the polymer chain, and R_g is the radius of gyration of the coil.

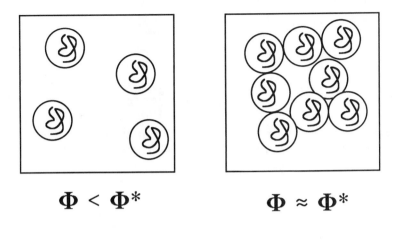

Figure 1   Schematic representation of the entanglement process. Φ is the volume fraction of the polymer and Φ^* is the entanglement threshold.

If the polymer is dissolved in an athermal solvent (i.e., a solvent in which the enthalpy of mixing is zero), then (Flory, 1953)

$$R_g = aN^{0.6} \tag{2}$$

where a is the length of a polymer segment. Therefore, combining Eqs. (1) and (2) gives

$$C^* \approx a^{-3}N^{-0.8} \tag{3}$$

Or, in terms of the volume fraction, Φ, where $\Phi = a^3C$,

$$\Phi^* \approx N^{-0.8}$$

(4)

Note that if N is large, Φ^* can be very small. For example, if $N = 10^4$, Φ^* is on the order of 0.1%. Thus, for a sufficiently large polymer, even seemingly dilute polymer solutions can be in an entangled state.

B. Mesh Size

An entangled solution is characterized by an average mesh size, ξ (see Fig. 2). de Gennes (1979) has derived an approximate relationship relating ξ to the polymer volume fraction, Φ, assuming $\Phi > \Phi^*$. First, it is assumed that when $\Phi > \Phi^*$, ξ depends only on the volume fraction of the polymer, Φ, and not on the size of the polymer chain, N. This simply says that the mesh size is smaller than the overall length of the polymer. Next, it is assumed that when $\Phi \approx \Phi^*$, the mesh size is comparable to the

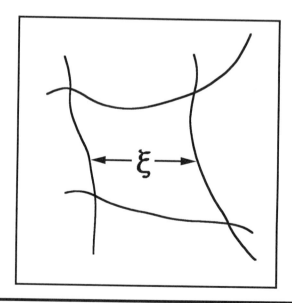

Figure 2   Schematic illustration of the entangled mesh, where ξ indicates the size of the mesh.

size of an individual coil, R_g. These two assumptions lead to the equality

$$\xi \Phi^m = R_g(\Phi^*)^m \tag{5}$$

where a power law relationship between ξ and Φ has been assumed. Substituting Eqs. (2) and (4) for R_g and Φ^* into Eq. (5) leads to the result

$$\xi(\Phi) \approx a\Phi^{-0.75} \tag{6}$$

Again, Eq. (6) assumes that the polymer is dissolved in an athermal solvent. This is an important result, which simply relates the size of the mesh in an entangled solution to the volume fraction of polymer.

In order to apply entangled polymer solutions to the widest range of biopolymer separations, one wants to be able to vary the mesh size in the solution. But, according to Eq. (6), if one wants to form a small mesh, for a given polymer, one must use a high concentration of polymer. However, as the polymer concentration is increased, so is the solution viscosity. Ideally, one would like to maintain the advantages of a low-viscosity solution when going to a smaller mesh. These relationships give an indication of how this might be accomplished. In order to minimize the viscosity of the polymer solution, one wants to operate near Φ^*. But Eq. (6) predicts that in order to achieve a small mesh, one needs a large value of Φ. To satisfy both constraints, one should use a shorter polymer to form a tighter mesh. This can be demonstrated by combining Eqs. (4) and (6) to give the expression

$$\xi(\Phi^*) \approx aN^{0.6} \tag{7}$$

Thus, to create a larger mesh while minimizing the viscosity of the solution, one wants to use a longer polymer, and to create a smaller mesh one wants to use a shorter polymer — in both cases, operating near the entanglement threshold for the given polymer.

C. Experimental Investigation of the Entanglement Threshold and Mesh Size of Hydroxyethyl Cellulose Solutions

1. Entanglement Threshold

Experimentally, the point at which a polymer solution becomes entangled can be determined by plotting the log of the specific viscosity, η_{sp}, as a function of the log of the polymer volume fraction (Hill and Soane, 1989), where η_{sp} is defined as

$$\eta_{sp} = \frac{\eta - \eta_0}{\eta_0} \tag{8}$$

where η is the viscosity of the solution and η_0 is the viscosity of the solvent alone. For independent, noninteracting polymer molecules, i.e., $\Phi < \Phi^*$, dilute solution theories predict that the slope of such a curve should be approximately 1.0 (Allcock and Lampe, 1981). Then, as the polymer coils begin to interact, the slope is expected to increase. As an example of this behavior, experimental results are presented in Fig. 3 for solutions of hydroxyethyl cellulose (HEC) dissolved in a *tris*(hydroxymethyl) aminomethane (Tris)-borate electrophoresis buffer. The solid line in Fig. 3 is the least-squares fit to the first four points, where the slope is 1.07. These data imply that, for this system, $0.29\% < \Phi^*_{HEC} < 0.4\%$. It is significant that the absolute value of the viscosity of this HEC solution at the entanglement threshold is very low — on the order of 1 centipoise (cp). One of the most striking features of these systems is that one can create a network structure in solutions having the viscosity of water.

To check the agreement between the experimental value of Φ^* found in Fig. 3 and that predicted by Eq. (4), we must first determine an approximate value for N for the HEC. This can be done using the Mark–Houwink–Sakurada equation (Brandrup and Immergut, 1989),

$$[\eta] = K^*(MW)^{a_{MHS}}$$ (9)

where $[\eta]$ is the intrinsic viscosity, MW is the polymer molecular weight and K and a_{MHS} are constants characteristic of a given polymer–solvent system at a given temperature. For HEC dissolved in water, $K = 9.53 \cdot 10^{-3}$ ml/g and $a_{MHS} = 0.87$ (Brandrup and Immergut, 1989). The intrinsic viscosity, a measure of the volume oc-

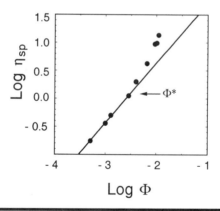

Figure 3   Dependence of the specific viscosity, η_{sp}, of an HEC–buffer solution on HEC concentration. The slope of the line passing through the first four points is 1.07, in good agreement with the value of 1.0 expected from dilute solution theories. Deviation from this line occurs when $0.0029 < \Phi_{HEC} < 0.0040$, indicating the onset of entanglement in this concentration range. Viscosity measurements were performed using an Ostwald viscometer (Allcock and Lampe, 1981) thermostated in a water bath at 30°C ± 0.5°C. Electrophoresis buffer was TBE [89 mM *tris*(hydroxymethyl)aminomethane, 89 mM boric acid, and 5 mM ethylenediaminetetraacetic acid (EDTA)]. Reproduced with permission from Grossman and Soane (1991).

cupied by an individual polymer coil, is defined as

$$[\eta] \equiv \lim_{C \to 0} \frac{\eta_{sp}}{C} \tag{10}$$

where C is the polymer concentration. From viscosity measurements (data not shown), the value of $[\eta]$ for aqueous HEC solutions was found to be 376 ml/g. Finally, given a value for $[\eta]$ and the parameters K and a_{MHS}, we can determine an approximate value for the polymer molecular weight. The resulting value for the molecular weight of the HEC is 191,800. Thus, the value of N is 1026, assuming a monomer molecular weight of 187 for HEC. With this value of N, Eq. (4) predicts $\Phi^*_{HEC} = 0.39\%$; in good agreement with the experimental value of $0.29\% < \Phi^*_{HEC} < 0.40\%$.

2. Mesh Size
In order to apply Eq. (6) to calculate a mesh size, we must first estimate a value for the statistical segment length, a, for HEC. The value of a can be estimated using intrinsic viscosity measurements. From Flory (1953), for a random-coil polymer,

$$[\eta] = \frac{\Phi_c \langle r^2 \rangle^{3/2}}{MW} \tag{11}$$

where Φ_c is a universal constant having a value of $2.1 \cdot 10^{23}$ if $[\eta]$ has the units of ml/g and $\langle r^2 \rangle$ is the root mean squared end-to-end distance between the ends of the polymer chain. Furthermore, for an unperturbed chain,

$$\langle R_g^2 \rangle = \frac{\langle r^2 \rangle}{6} \tag{12}$$

Based on a measured value of 317 ml/g for $[\eta]$ in the electrophoresis buffer (data not shown), Eq. (11) and (12) give a value of 270 Å for R_g.

Next, given the relationship between the segment length, a, and R_g for an unperturbed coil,

$$R_g = aN^{0.6} \tag{13}$$

and given that $N = 1026$, we can see that $a = 4.21$ Å. This is close to the 4.25 Å monomer segment length for HEC (Brandrup and Immergut, 1989).

Using this value for a, it is interesting to compare the approximate mesh sizes of these polymer solutions with those of traditional electrophoresis gels. Righetti et al. (1981) has developed an empirical relationship to correlate mesh size in agarose gels with agarose concentration,

$$\xi = 1407 C^{-0.7} \tag{14}$$

where ξ is the measured mesh size (in Å) and C is the concentration of agarose (in wt%). Note that this empirical expression has the same form and a value of the exponent similar to that predicted by entanglement theory. However, comparing the results of Eqs. (6) and (14), the entangled solution appears to produce a smaller mesh size per

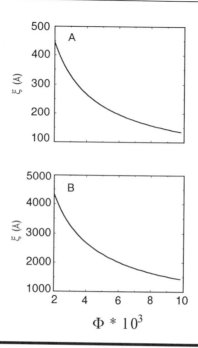

Figure 4   Curves showing the mesh size, ξ, predicted for entangled HEC solutions (A), and agarose gels (B) as a function of mesh-forming polymer volume fraction, Φ. Curve A was generated using Eq. (6), where $a = 4.21$ Å, whereas curve B was produced using Eq. (14).

weight of polymer. This is consistent with the fact that in an agarose gel, the polymer fibers exist as bundles, therefore leaving larger voids. Figure 4 compares the predicted mesh size using Eqs. (6) and (14). Thus, based on these equations, it appears that the mesh sizes achievable using entangled polymers are an order of magnitude smaller than those using agarose gels at the same polymer concentration.

III. Mechanisms of Electrophoretic Migration in a Polymer Network

Once the network structure of the polymer solution has been established, we can address the effects of the polymer–solution network on electrophoresis. As is the case for traditional gel electrophoresis, two main theories describe the migration of a flexible macromolecule through a polymer network: the Ogston sieving model and the reptation model. The applicability of each depends on the size of the migrating molecule relative to the mesh size of the network.

A. The Ogston Model

The Ogston model treats the polymer network as a molecular sieve. It assumes that the matrix consists of a random network of interconnected pores having an average pore size, ξ, and that the migrating solute behaves as an undeformable spherical particle of radius R_g. According to this model, smaller molecules migrate faster because they have access to a larger fraction of the available pores. The mathematical treatment of this problem was first presented by Ogston (Ogston, 1958). Figure 5 is a schematic illustration showing a solute migrating through a polymer network by the Ogston mechanism.

In the Ogston theory, the electrophoretic mobility of the migrating solute through the porous structure is assumed to be equal to its free solution mobility, μ_0, multiplied by the probability that the solute will meet a pore large enough to allow its passage. Thus

$$\mu = \mu_0 P(\xi \geq R_g) \tag{15}$$

where ξ is the radius of the pore in which the coil resides, and $P(\xi \geq R_g)$ is the probability that a given pore has a radius greater than or equal to the radius of the migrating particle. Thus, now the problem simply becomes one of determining the form of $P(\xi \geq R_g)$.

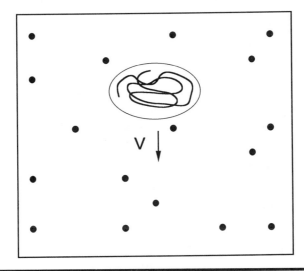

Figure 5   Schematic diagram of a solute migrating through a polymer network by the Ogston mechanism. In this case, because the mesh size of the polymer network is larger than the migrating coil, the coil percolates through the mesh as if the coil were a rigid particle.

The Ogston model of the pore size distribution predicts that, in a random network of linear polymers, the fraction of pores large enough to accommodate a spherical particle of radius R_g is

$$P(\xi \geqslant R_g) = \exp(-\pi n l (r + R_g)^2) \qquad (16)$$

where n is the average number of polymer strands per unit volume, l is the average length of the polymer strands, and r is the thickness of the strands. Furthermore, this model assumes that the product $n \cdot l$ is proportional to the concentration of the gel-forming polymer, C. Thus,

$$P(\xi \geqslant R_g) = \exp(-KC(r + R_g)^2) \qquad (17)$$

where K is a constant of proportionality. In Eq. (17), the term $K(r + R_g)^2$ is known as the retardation coefficient, K_r, and is a characteristic of a given molecular species in a particular polymer system.

Combination of Eqs. (15) and (17) gives the final expression describing the migration of a solute through a polymer network according to the Ogston mechanism

$$\mu = \mu_0 \exp(-KC(r + R_g)^2) \qquad (18)$$

Note that if $r \ll R_g$, one would expect that a plot of log μ versus C would give a straight line with a slope proportional to R_g^2. Such plots are known as Ferguson plots (Ferguson, 1964).

B. Reptation

As stated before, the Ogston model assumes that the migrating solute behaves as an undeformable spherical particle. It does not take into account the fact that the migrating molecule might deform in order to "squeeze" through a pore. Therefore, when $R_g > \xi$, the Ogston model predicts that the electrophoretic mobility of the migrating solute will rapidly approach zero. However, it is well known that large, flexible-chain molecules such as DNA continue to migrate even when $R_g \gg \xi$. This is explained by the second model for migration, the reptation model. The reptation model assumes that instead of migrating as an undeformable particle with radius R_g, the migrating molecule moves "head first" through the porous network. The reptation model is the subject of the following section.

1. Classical Reptation

The basis of the reptation model is the realization that when a long, flexible molecule travels through a polymer network having a mesh size smaller than R_g, it does not necessarily travel as an undeformed particle, but rather "snakes" through the polymer network head first. The migrating solute is assumed to be confined to "tubes" that are formed by the gel matrix. The term *reptation* comes from the reptilelike nature of this motion. Figure 6 is a schematic illustration of a flexible macromolecule undergoing reptation. What follows is a derivation of the size dependence of the translational friction coefficient, f_{rep}, and thus μ, for a chain undergoing reptation. These arguments are scaling arguments rather than complete functional derivations. That is, we are looking for functional dependencies rather than exact equations. The first description of the reptation mechanism was presented by de Gennes (1971, 1979) and Doi and Edwards (1978), whereas the first application of reptation theory to the electrophoresis of biopolymers was presented by Lerman and Frisch (1982).

As described in the previous chapter on free-solution capillary electrophoresis, the definition of the electrophoretic mobility, μ, is

$$\mu = \frac{q}{f} \tag{19}$$

where q is the net charge on the analyte and f is its translational frictional coefficient. In the case of DNA molecules, the structure is such that the total charge is proportional to the size of the fragment,

$$q \sim N \tag{20}$$

where the "$\sim$" symbol indicates proportionality, and N is proportional to the size of the molecule. In order to determine the dependence of μ on molecular size for the case of reptation, we must first derive the dependence of f on N.

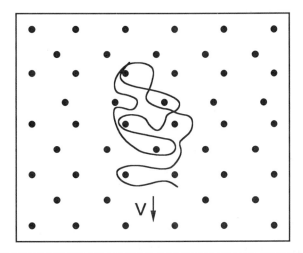

Figure 6 Schematic diagram of a solute migrating by the reptation mechanism. In this case, because the mesh size of the network is smaller than the radius of gyration of the migrating coil, the coil is forced to squeeze "head-first" through the "tubes" formed by the polymer network.

As a starting point, it is recognized that f_{rep} can be related to an apparent diffusion coefficient, D_{rep}, through the Stokes–Einstein equation (Atkins, 1978),

$$D_{rep} = \frac{kT}{f_{rep}} \tag{21}$$

where k is the Boltzmann constant and T is the absolute temperature. Next, in order to derive an expression for D_{rep}, we consider the statistical nature of the diffusion process. If diffusion is considered to be simply the random motion of molecules in a concentration gradient (neglecting any nonidealities of the solution), the diffusion coefficient, D, can be defined as

$$D_{rep} = \frac{l_{rep}^2}{\tau_{rep}} \tag{22}$$

where l_{rep} is a characteristic step length for diffusion and τ_{rep} is the characteristic step time for diffusion [for a clear description of the diffusion process from the statistical point of view, as well as the derivation of Eq. (22), see Berg (1983)]. In order to use Eq. (22) to arrive at an expression for D_{rep}, we must determine the characteristic step time, τ_{rep}, and the characteristic step length, l_{rep}, for the reptation process.

First, we consider the characteristic step length, l_{rep}. Implicit in the statistical description of the diffusion process is the assumption that each step is statistically independent of any previous step. For this to be true, in each step, the molecule must travel a distance at least as large as its own characteristic dimension. If we assume that the solute exists in a random coil conformation, this characteristic dimension would be the radius of gyration of the coil, R_g. For a random coil polymer, the radius of gyration, R_g, is given by (Flory, 1953)

$$R_g^2 = \frac{Na^2}{6} \tag{23}$$

where N is the number of independent segments in the polymer and a is the length of each segment. Therefore,

$$l_{rep} = R_g \sim N^{0.5} \tag{24}$$

Thus, the characteristic length for reptation is proportional to the molecular size to the 0.5 power, assuming the solute behaves as a random coil. Therefore, when the macroscopic displacement of the reptating macromolecule is l_{rep}, its new location is statistically independent of the previous one.

Next, we must consider the characteristic step time in Eq. (22). In order to determine a characteristic time for the reptation process, we will define an *intratube* diffusion coefficient, D_{tube}, where

$$D_{tube} = \frac{kT}{f_{tube}} \tag{25}$$

where f_{tube} is the frictional coefficient of the macromolecule *within an individual tube*. It is important at this point to note the difference between D_{rep} and D_{tube}. Whereas D_{rep} is an apparent "macroscopic" net translational diffusion coefficient, D_{tube} applies to the motion of the DNA along the tube axis. Because within individual tubes, the solute is migrating as if in free solution, it is reasonable to assume that f_{tube} is proportional to N, thus,

$$D_{tube} \sim \frac{1}{N} \tag{26}$$

Furthermore, when considering the motion within an individual tube, the characteristic dimension of the solute now becomes the total extended length, or contour length, of the molecule rather than the radius of gyration. Since the contour length is directly proportional to the number of units in the polymer, N, we obtain

$$l_{tube} \sim N \tag{27}$$

This important conceptual difference between the length scales for macroscopic diffusion and microscopic tube diffusion is the key to providing the molecular size dependence of μ to molecules undergoing reptation. This difference in length scale between macroscopic displacement, l_{rep}, and the displacement within a tube, l_{tube}, is caused by the need for the molecule to move a longer distance within the pore than the overall macroscopic displacement would indicate, because of the tortuous nature of the tube.

Thus, combining Eqs. (22), (26), and (27), we can define a characteristic time for diffusion within a tube as

$$\tau_{rep} = \frac{(l_{tube})^2}{D_{tube}} \sim \frac{N^2}{1/N} \tag{28}$$

or

$$\tau_{rep} \sim N^3 \tag{29}$$

At this point we can determine the dependence of D_{rep}, and thus f_{rep}, on molecular size for molecules undergoing reptation. From Eqs. (22), (24), and (29) we have

$$D_{rep} \sim \frac{(N^{0.5})^2}{N^3} = \frac{1}{N^2} \tag{30}$$

Thus, from Eq. (21) we can see that

$$f_{rep} \sim N^2 \tag{31}$$

This dependence of f_{rep} on N^2 is in contrast to the free-solution case in which f is proportional to N to the first power.

Finally, we are in a position to determine the relationship between electrophoretic mobility and molecular size for a molecule migrating by reptation. From Eqs. (19), (20), and (31) we obtain

$$\mu \sim \frac{N}{N^2} = \frac{1}{N}$$

(32)

Equation (32) is the key result that we have been looking for. It states that for a chain-like molecule undergoing reptile motion under the influence of an electric field, the electrophoretic mobility is inversely proportional to the molecular length. Again, it is important to contrast this result with the free-solution case where μ is independent of N, if the total charge on the molecule is proportional to its length.

2. Biased Reptation

A refinement on the reptation model that takes into account the influence of large electric fields is the *biased*-reptation model. When the electric field becomes large, the assumption that the migrating molecule exists as an unperturbed coil [Eq. (24)], is no longer valid. Because of the induced orientation of the leading segment, as the field strength is increased, the coil becomes more elongated. This is shown schematically in Fig. 7. In the limiting case, the coil becomes a rod. If the migrating molecule becomes a rod, l_{rep} is now proportional to N instead of $N^{1/2}$. If this is substituted into Eq. (30), $f_{rep} \sim N$ and $\mu \sim N^0$. Thus, as the coil becomes more elongated, the $1/N$ size dependence of μ disappears. This effect was first described by Lumpkin *et al.* (1985) who arrived at an expression of the form

$$\mu \approx \left(\frac{1}{N} + bE^2 \right)$$

(33)

where b is a function of the mesh size of the polymer network as well as the charge and segment length of the migrating solute. Note that the first term in Eq. (33) depends on the size of the migrating molecule but does not depend on the electrical field strength, whereas the second term does not depend on molecular size but does depend on electrical field strength. Therefore, as the electrical field or the molecular size increases, the relative importance of the size-dependent term decreases, leading to a decreased dependence of mobility on molecular size. This is the key prediction of the biased-reptation model. The predicted behavior has been observed experimentally (Hervet and Bean, 1987). Because of this effect, the maximal size of DNA that can be separated using traditional electrophoretic techniques is approximately 20,000 base pairs (bp). To go beyond this limit, pulsed-field techniques must be used (Schwartz *et al.,* 1982).

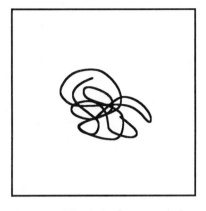

Low Field Strength

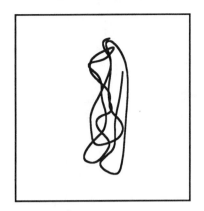

Moderate Field Strength

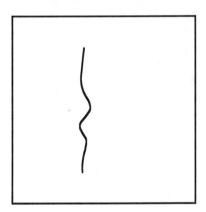

High Field Strength

Figure 7 Schematic illustration showing the elongational influence of the electric field on the size of a molecule migrating by the reptation mechanism. When $E = 0$, $R_g \sim N^{0.5}$, whereas for large E, $R_g \sim N^{1.0}$.

C. Experimental Study of the Migration Mechanism of DNA in Hydroxyethyl Cellulose Solutions

Figure 8 shows a representative electropherogram of DNA fragments ranging in size from 72 to 1353 bp. A description of the capillary electrophoresis apparatus used in these studies as well as the methods used to calculate electrophoretic mobilities is provided elsewhere (Grossman *et al.*, 1989).

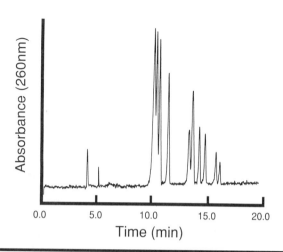

Figure 8 Representative electropherogram showing the separation of 11 DNA restriction fragments ranging in size from 72 to 1353 base pairs. Reading from left to right (not including the first two peaks, which are markers) the species are 1353, 1078, 872, 603, 310, 281 + 271, 234, 194, 118, and 72 base pairs in length, respectively. Electrophoresis conditions: Buffer, 0.25% HEC in TBE [89 mM *tris*(hydroxymethyl)aminomethane, 89 mM boric acid, and 5 mM ethylenediaminetetraacetic acid (EDTA)]; field strength, 301.3 V/cm; UV-absorbance detection at 260 nm; capillary dimensions, 50 cm total length (35 cm to detector) and 50 μm internal diameter; temperature, 30°C ± 0.1°C. Reproduced with permission from Grossman and Soane (1991).

As previously stated, according to the Ogston model, a plot of log μ versus %HEC (a Ferguson plot) should give a linear relationship with a slope equal to K_r and a y-intercept equal to log μ_0. For fragments 118, 194, 234, 281, and 310 in solutions up to $\Phi^*_{HEC} = 0.4\%$, this behavior is indeed observed (Fig. 9). The intercept for these five lines, 0.588 (%RSD = 0.15%), implies a value for μ_0 of 3.87 · 10^{-4} cm^2/V · sec, in complete agreement with the measured electrophoretic mobility of these fragments at $\Phi^*_{HEC} = 0\%$ of 3.86 · 10^{-4} cm^2/V · sec (%RSD = 1.2%, $n = 16$). This is an interesting result. Previously, in rigid-gel systems, the value of μ_0 could only be inferred based on the y-intercept of the Ferguson plot. Here, because μ_0 can be measured directly, the value of μ_0 based on the y-intercept can be compared with the actual measured value. The close agreement of these two values of μ_0 is compelling verification of the Ogston mechanism. For fragments larger than 310 bp, the agreement degrades. This is probably owing to the gradual transition to the reptation regime for these larger fragments. This transition has also been observed in agarose gels (Hervet and Bean, 1987).

According to the Ogston model, a plot of $K_r^{0.5}$ versus R_g should yield a linear relationship, assuming $R_g \gg r$. As seen in Fig. 10, for the smaller fragments, agreement with the prediction of the Ogston model is close, whereas the larger fragments deviate

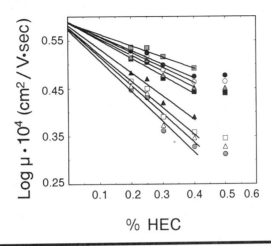

Figure 9 Ferguson plot for sample DNA fragments. (⊞) 118 bp, (●) 194 bp, (○) 234 bp, (▵) 281 bp, (■) 310 bp, (▲) 603 bp, (□) 872 bp, (△) 1078 bp, (⊕) 1353 bp. Reproduced with permission from Grossman and Soane (1991).

significantly. $\langle R_g^2 \rangle^{1/2}$ is calculated for DNA using the Porod-Kratky stiff-chain model, assuming a persistence-length of 450 Å and a contour-length of 3.4 Å per bp (Cantor and Schimmel, 1980). According to Slater and Noolandi (1989), based on experiments and numerical simulations, the transition from the Ogston to the reptation regime takes place when $R_g \cong 1.4\xi$. Therefore, given that according to Fig. 10 the transition occurs when 312 Å $< R_g <$ 490 Å when $\Phi^*_{HEC} = 0.4\%$, Fig. 10 implies that when $\Phi^*_{HEC} = 0.4\%$, 223 Å $< \xi <$ 350 Å. This agrees with the mesh size predicted using Eq. (6) of 264 Å. Thus Fig. 10 provides an independent confirmation of Eq. (6).

Figure 11 shows a plot of log μ as a function of $1/N_{DNA}$. According to Eq. (32), these curves should be linear with slopes of 1.0, if the fragments migrate by the reptation mechanism. As expected from the previous analysis, for the conditions used in these experiments, none of the curves adheres to the reptation behavior. However, as the HEC concentration is increased, the slopes of the curves appear to increase toward 1. Again, this behavior has been observed for low-concentration agarose gels using small DNA fragments (Hervet and Bean, 1987).

It is likely that once $R_g \gg \xi$, i.e., when reptation becomes important, the separation performance of these systems will decrease rapidly. This is because at the high electrical fields typically employed in capillary electrophoresis, the value of the term bE in Eq. (33) is greater than 1, resulting in a saturated, size-independent mobility. The net result will be that the range of DNA sizes that can be successfully separated in a given polymer system will be less than would be the case using low fields. Although this limitation is not unique to the polymer solution system, it does represent a restriction on the ability to exploit the high electric fields and thus enjoy the consequent rapid analysis using CE.

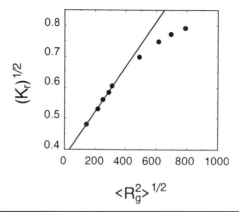

Figure 10   Square root of the retardation coefficient K_r versus the root-mean-square radius of gyration $\langle R_g^2 \rangle^{1/2}$ of the DNA fragments from Fig. 9. K_r values were evaluated from the slopes of the curves in Fig. 9, where $0 \leq \Phi \leq 0.4\%$, and R_g^2 was calculated using the Porod–Kratky stiff-chain model (Cantor and Schimmel, 1980). Reproduced with permission from Grossman and Soane (1991).

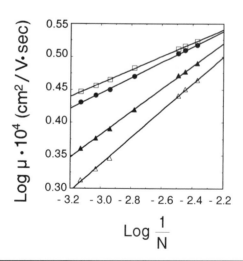

Figure 11   Log of the electrophoretic mobility versus log of inverse molecular size (in base pairs) for different HEC considerations. A slope of 1 would be expected if migration exactly followed the reptation mechanism. ($\square$) 0.20% HEC; slope, 0.091; ($\bullet$) 0.25% HEC; slope, 0.107; ($\blacktriangle$) 0.30% HEC; slope, 0.158; ($\triangle$) 0.50% HEC; slope, 0.201. Reproduced with permission from Grossman and Soane (1991).

IV. Conclusion

In conclusion, it appears that low-viscosity polymer solutions provide a good matrix for electrophoretic separations. Performing polymer-solution electrophoresis in a capillary in effect decouples the two roles of a traditional electrophoresis gel: that of a sieving matrix and that of an anticonvective stabilizing medium.

Because an entangled mesh requires no specific cross-linking or gelation, a wide range of polymers may be easily adapted for these applications. Candidate polymers simply need to be water soluble and preferably uncharged. Whereas this technique is still very new, because of the considerable practical advantages relative to gel-based systems, it promises to be an important alternative to gel-based separations, potentially displacing rigid gels altogether for many applications.

References

Allcock, H. R., and Lampe, F. W. (1981). "Contemporary Polymer Chemistry." Prentice-Hall, Englewood Cliffs, New Jersey.

Atkins, P. W. (1978). "Physical Chemistry." W. H. Freeman, San Francisco, California.

Berg, H. C. (1983). "Random Walks in Biology." Princeton University Press, Princeton, New Jersey.

Brandrup, J., and Immergut, E. H. (1989). "Polymer Handbook," 3rd Ed., Wiley, New York.

Cantor, C. R., and Schimmel, P. R. (1980). "Biophysical Chemistry." W. H. Freeman, New York.

Chin, A. M., and Colburn, J. C. (1989). *Am. Biotech. Lab./News Ed.* **7,** 10A.

de Gennes, P. G. (1971). *J. Chem. Phys.* **55,** 572.

de Gennes, P. G. (1979). "Scaling Concepts in Polymer Physics." Cornell University Press, Ithaca, New York.

Doi, M., and Edwards, S. F. (1978). *JCS Faraday Trans. II* **79,** 1789–1818.

Ferguson, K. A. (1964). *Metabolism* **13,** 985.

Flory, P. J. (1953). "Principles of Polymer Chemistry." Cornell University Press, Ithaca, New York.

Grossman, P. D., and Soane, D. S. (1991). *Biopolymers* **31,** 1221–1228.

Grossman, P. D., Colburn, J. C., and Lauer, H. H. (1989). *Anal. Biochem.* **179,** 28.

Hermans, J. J. (1953). *J. Polymer Sci.* **18,** 257.

Hervet, H., and Bean, C. P. (1987). *Biopolymers* **26,** 727.

Hill, D. A., and Soane, D. S. (1989). *J. Polym. Sci. Poly. Phys.* **B27,** 2295.

Langevin, D., and Rondelez, F. (1978). *Polymer* **19,** 875.

Lerman, L. S., and Frisch, H. L. (1982). *Biopolymers* **21,** 995.

Lumpkin, O. J., Dejardin, P., and Zimm, B. H. (1985). *Biopolymers* **24,** 1573.

Ogston, A. G. (1958). *Trans. Faraday Soc.* **54,** 1754.

Olivera, B. M., Baine, P., and Davidson, N. (1964). *Biopolymers* **2,** 245.

Righetti, P. G., Brost, B. C. W., and Snyder, R. S. (1981). *J. Biochem. Biophys. Methods* **4,** 347.

Schwartz, D. C., Saffran, W., Welsh, J., Haas, R., Goldenberg, M., and Cantor, C. R. (1982). *Cold Spring Harbor Symp. Quant. Biol.* **47,** 189.

Slater, G. W., and Noolandi, J. (1989). *Biopolymers* **28,** 1781.

Slater, G. W., Rousseau, J., Noolandi, J., Turmel, C., and Lalande, M. (1988). *Biopolymers* **27,** 509.

Tietz, D., Gottlieb, M. H., Fawcett, J. S., and Crambach, A. (1986). *Electrophoresis* **7,** 217.

Zhu, M., Hansen, D. L., Burd, S., and Gannon, F. (1989). *J. Chromatgr.* **480,** 311.

APPLICATIONS OF

CAPILLARY

ELECTROPHORESIS

Capillary Electrophoresis Separations of Peptides: Practical Aspects and Applications

Joel C. Colburn

I. Introduction

Peptides play important roles in modern science and technology: as antigens for vaccine production, for determination of unidentified reading frames in molecular biology, for epitope mapping, and in pharmaceutical research. They have significant physiological functions as hormones and neurotransmitters, and there is considerable interest in devising effective analogs of these species. Protein characterization through peptide mapping is ubiquitous.

Although grouped as a single class, peptides exhibit great heterogeneity. They are composed of over 20 amino acid species. Those with biological activity are typically longer than 25 amino acids, whereas immunogens are 8–15 residues, and peptides used for epitope mapping may be only 5–7 residues in length. Since they are amphoteric, all have both acidic and basic properties.

The importance of peptide analysis and purity control cannot be overemphasized. Applications such as biological-activity studies, peptide antigen screening, drug design, and conformational studies [nuclear magnetic resonance (NMR) and X-ray crystallography] may require greater than 98% purity. Purity levels of 80% may be adequate for peptides used for raising antibodies. They have traditionally been analyzed by high-performance liquid chromatography (HPLC) or thin-layer chromatography (TLC), and CE is rapidly becoming an accepted complement to these methods. This chapter demonstrates how capillary electrophoresis (CE) has been used to assess the purity of peptides, extract structural information about peptides, discuss the basis for CE peptide separations, and describe the problems encountered in CE of peptides and their solutions.

II. Modes of Capillary Electrophoresis Separations

A. Charge and Size Effects in Free Solution

Peptides, like any analytes in free-solution capillary electrophoresis, are influenced by various forces. The electrophoretic force is derived from the electric field acting on the charged peptide; the more highly charged the peptide, the stronger the force, and the faster the migration. A drag force due to peptide size and buffer viscosity is also present. Electroosmotic flow (see Chapter 1) is an effect that can act in the same direction as or opposite direction to peptide migration, depending on the pH of the buffer and the charge of the peptide.

In order to explore the effects of various structural elements on mobility and resolution, Grossman *et al.* (1988, 1989b) electrophoresed six 7-amino acid peptides by CE. At pH 2.5 all were resolved, with the peaks eluting in the order predicted by charge calculations: the most positively charged peptide migrating fastest toward the cathode, and eluting first (highest mobility). The peptide pair exhibiting a charge difference of only 0.04 charge units was not completely resolved, but when the pH was adjusted to 4.0, the charge difference increased, and the peptide pair separated (Fig. 1). At pH 4.0 some peptides disappeared from the electropherogram, since they were negatively charged, the electroosmotic velocity being insufficient at this pH to drag the peptides toward the cathode and past the detector. At pH 11.0 electroosmosis is of sufficient velocity for the negatively charged peptides to be swept past the detector. At this pH the most highly (negative) charged species eluted last, since they have the greatest mobility toward the anode (away from the detector). This work shows the tremendous effect that buffer pH has on the resolution of peptides.

The separation of several β-endorphin fragments illustrates the selectivity provided by CE. Since the electrophoretic mobility increases as the number of amino acids decreases (less drag force), as the peptide was reduced from 6-17 β-endorphin peptide to the 7-17 and 8-17 peptides (uncharged amino acids removed), the mobility increased. Reducing the charge should decrease the mobility. When amino acid 8 (negatively charged Glu) was removed from the N-terminus, the mobility increase was greater than that predicted by size considerations alone (Fig. 2). This charge effect is also illustrated by comparing β-endorphin with the *N*-acetyl form; the *N*-acetyl form has one less charge at low pH, and it exhibits a much lower mobility (Grossman *et al.*,

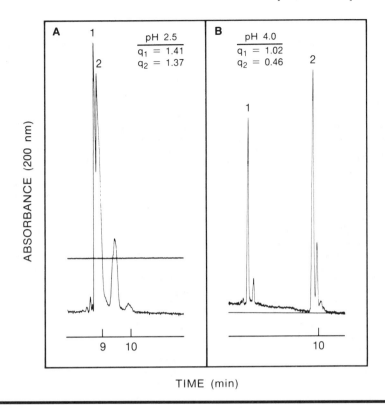

Figure 1 The effect of buffer pH on the selectivity of peptide separations by CE. The sequences of the peptides are AFKAING (1) and AFKADNG (2). Conditions: field, 277 V/cm; current, 24 μA [12 μA for (B)]; buffer, citric acid, 20 mM, pH 2.50 [4.00 for (B)]; capillary length, 65 cm (45 cm to detector); capillary diameter, 50 μm; temperature, 30°C; detection, 200 nm. Reproduced with permission from Grossman *et al.* (1989b).

1989b). It is interesting to note that the separation of these peptides by reverse-phase HPLC (RP-HPLC) required a complex gradient with long separation times. The elution order obtained in RP-HPLC was opposite that obtained by CE, consistent with the different separation mechanisms.

In an early CE paper, Jorgenson and Lukacs (1981a) separated dipeptides as fluorescamine derivatives, using 50 mM phosphate, pH 7.0. Although all peptides had a negative charge at this pH and migrated toward the positive electrode, their net migrations were to the negative electrode (toward the detector) because of electroosmosis. Species are therefore detected as expected, with cations eluting first, then neutrals, then anions. Electroosmosis is considered to be convenient since it allows all ions, whether positive or negative, to be analyzed in a single run. It also decreases run times

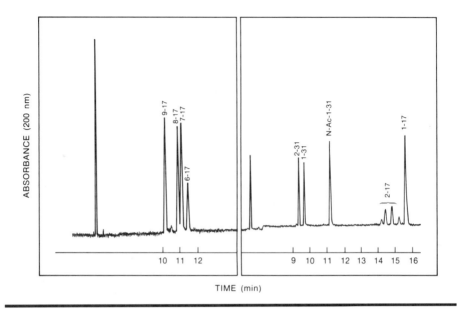

Figure 2 Separation of endorphin fragments by CE. Conditions: as in Fig. 1A. Reproduced with permission from Grossman *et al.* (1989b).

for any weakly charged materials. However, the authors noted that resolution is reduced when the electroosmotic velocity is in the same direction as the peak electrophoretic velocity. Therefore, in general, the best resolution is obtained when the two velocities are nearly equal, although this results in relatively long migration times.

Human insulin is made of two chains (A and B) with a total of 51 amino acids held together with two interchain disulfide bridges. Biosynthetic human insulin (BHI) may be contaminated by desamido-A21, arginyl-A0, diarginyl-B31-B32, and desthreonine-B30, which form during the conversion of proinsulin to insulin. These four species were separated by Grossman *et al.* (1989b) with 10 m*M* tricine, 5.8 m*M* morpholine, 20 m*M* NaCl, at pH 8.0 (Fig. 3). The arginyl peptides are separated from BHI owing to their increased positive charge. The desamido form has a greater negative charge because of the conversion of the amide to a carboxylate group. Desthreonine-B30 insulin did not separate from BHI. The paper by Nielsen *et al.* (1989b) demonstrated CE separations for BHI and its degradation products. This work also investigated the separation of peptides derived from proinsulin: A-chain S-sulfonate, B-chain S-sulfonate, and the C-peptide were resolved in about 16 min.

Peptide hormones have major effects on physiology, so that various synthetic analogs are investigated for enhanced biological activity. Motilin (22 amino acids) is a hormone involved in gut motility. The 1-14 peptide analog of motilin was synthesized, along with the glutamate (for glutamine at positions 11 and/or 14) forms of the peptide (Florance, 1991). Since HPLC was not adequate for monitoring the deamidation of the glutamine peptide, CE separations were attempted. The CE resolved all at pH 9.4,

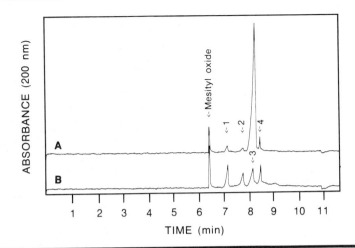

Figure 3 Separation of BHI and its derivatives. (A) BHI plus BHI derivatives; (B) BHI derivatives alone. Peaks: (1) [diarginyl-B31-B32] BHI; (2) [arginyl-A0] BHI; (3) [desthreonine-B30] BHI; (4) [desamido-A21] BHI. Conditions: field, 300 V/cm; current, 20 μA; buffer, 0.01 M tricine, 0.0058 M morpholine, 0.02 M NaCl, pH 8.0; capillary length, 95.5 cm (81.5 cm to detector); capillary diameter, 50 μm; temperature, 24°C, detection, 200 nxm. Reproduced with permission from Grossman *et al.* (1989b).

except the Gln-11, Glu-14 peptide could not be separated from the Glu-11, Gln-14 peptide. Therefore, the separation was the result of a change in charge. As these peptides have alpha-helical structures, it was postulated that increased resolution could arise from modification of the secondary structure with various solvents. The first attempt using hexafluoropropanol did not allow resolution.

Cobb and Novotny (1989) described the use of CE for peptide mapping (see Section IV,A) of phosphorylated and dephosphorylated forms of β-casein (Fig. 4). Any change in net charge should alter the peak profile of the map. Two peaks showed shifts in migration time between the two forms, consistent with the presence of phosphoserine residues in the tryptic peptides. Dephosphorylation decreased the net negative charge of the peptides, which was reflected in their increased mobility toward the cathode. Thus, reduced migration times (greater mobility) of the two peaks were observed in the dephosphorylated peptide map.

Similar studies by this author involved the separation of phosphorylated from dephosphorylated kemptide. This seven-amino acid peptide is a substrate for cyclic adenosine monophosphate (AMP)-dependent protein kinase, and becomes phosphorylated at one serine residue. Based on the relationship developed by Grossman *et al.* (1989a; see Section VI), the difference in mobility between the two forms represented a charge difference of approximately 1.1 charge units, consistent with monophosphorylation. Thus, CE can be used for assaying the extent of phosphorylation, and perhaps other charge-altering modifications of peptides.

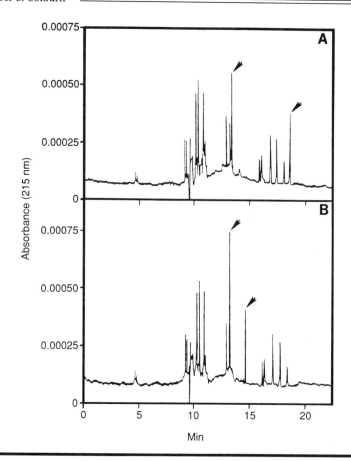

Figure 4 CE separation of tryptic digests from (A) phosphorylated and (B) dephosphoryl-ated forms of β-casein. Arrows denote the two peaks with different migration times between the two forms. Conditions: field, 220 V/cm; buffer, 0.04 M Tris/0.04 M tricine, pH 8.1; capillary length, 100 cm (85 cm to detector); capillary diameter, 50 μm; detection, 215 nm. Reproduced with permission from Cobb and Novotny (1989).

1. Influence of Neutral Amino Acids

In the report by Grossman *et al.* (1988), two peptides were resolved where a neutral amino acid (alanine or isoleucine) was substituted between adjacent lysines or to one side of lysine. Therefore, the peptides had the same charge and size, but could be re-solved (Fig. 5). This phenomenon could be explained by the notion that changing a neutral amino acid next to a charged amino acid affects the local dielectric environ-ment around the charged group, and thus its negative log of ionization constant (pK). As the environment becomes more hydrophobic, the neighboring charged amino acid will lower its pK and become slightly less positively charged at low pH. The data sup-

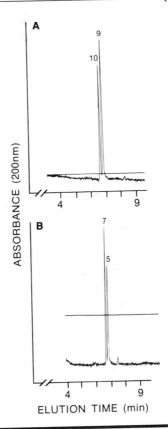

Figure 5 CE separation of peptide mixtures in which two peptides differ by only one uncharged amino acid. Conditions: field, 277 V/cm; current, 24 μA; buffer, citric acid, 20 mM, pH 2.5; capillary length, 65 cm (45 cm to detector); capillary diameter, 50 μm; detection, 200 nm. Peptide sequences are as follows: (5) AFKKING, (7) AFKKANG, (9) AFKIKNG, (10) AFKAKNG. Reproduced with permission from Grossman *et al.* (1988).

port this, since the more hydrophobic isoleucine makes the isoleucine peptide less charged and causes a lower mobility. This effect is, however, minor compared to effects of size and charge.

McCormick (1988) was able to resolve dipeptides differing by one neutral side-chain amino acid at pH 1.5 with 150 mM phosphoric acid. These dipeptides differed by the substitution of a Gly, Ala, Val, or Leu, i.e., only by single methylene group increments. Larger peptides with the same single amino acid change were also resolved. Separation of peptides with Val or Ile substitutions was possible because these substitutions on the side chains alter the pK of the acidic and basic groups in neighboring amino acids. An interesting feature of this paper was that at the beginning of electrophoresis the voltage was kept low, but increased over time to a final 25–30 kV.

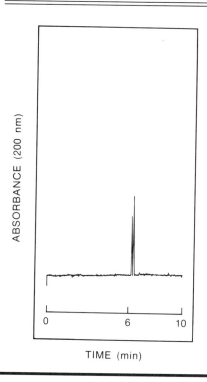

Figure 6   Separation of peptide diastereomers. Conditions: field, 308 V/cm; buffer, 20 m*M* sodium citrate, pH 2.5; capillary length, 65 cm (45 cm to detector); capillary diameter, 50 *μ*m; temperature, 30°C; detection, 200 nm. Samples are (L-Arg 6) dynorphin and (D-Arg 6) dynorphin. Reproduced with permission from Colburn *et al.* (1991).

Studies with proteins indicated that the voltage ramp gives sharper and more symmetrical peaks, but the cause of this improvement in separation efficiency is not understood. If the voltage ramp is too long, band broadening due to diffusion can outweigh any improvements. The author stated that a 100 to 300-sec ramp worked best for this case.

Colburn *et al.* (1991) demonstrated the separation of peptide diastereomers (Fig. 6). Two 13-amino acid peptides, differing only by the substitution of D versus L forms of one amino acid, were resolved at pH 2.5. This resolution is proposed to arise from the local stereochemical effect on neighboring charged amino acid side chains.

The previous examples are given for fairly small peptides. Larger peptides with minor hydrophobic changes may be more difficult to analyze. Hirudin (65 amino acids) is a thrombin inhibitor that gives degradation products where amino acid 65 (Gln) and then amino acid 64 (Leu) are sequentially cleaved from the C-terminus. Although these are neutral, the difference in net charge caused by subtle changes in p*K* of neighboring charged amino acids, or by differences in the p*K* of the C-terminal acid was sufficient to allow separation of the three products (1-65, 1-64, 1-63) (Ludi *et al.,*

1988). In addition, the size differences may contribute to the separation. They used 16.7 mM piperazine-N,N'-bis(2-ethane sulfonic acid) (PIPES), 12 mM borate, 1 mM ethylenediaminetetraacetic acid (EDTA), pH 6.7 at 300 V/cm, and detection at 215 nm for this separation.

2. Additives

A series of histidine-containing heptapeptides with a glycine backbone were studied by Stover *et al.* (1989). These peptides with either one, two, or three histidines were blocked at both termini, so that only the influence of the histidine was observed. The peptides were separated at pH values from pH 6 to 9. At pH 6 the peptides migrate in the order di-histidine peptide (D), mono-histidine peptide (M), and tri-histidine peptide (T), although they were predicted to migrate in the order T, D, M based on charge considerations. Also, T was very broad. When 2 mM putrescine was added to the three peptides at pH 6, T sharpened and eluted first, since putrescine interacts with the wall and thus minimizes any wall sticking of T (Fig. 7; see Section V). Above pH 6, all converged as histidine became deprotonated, and the net charge approached zero.

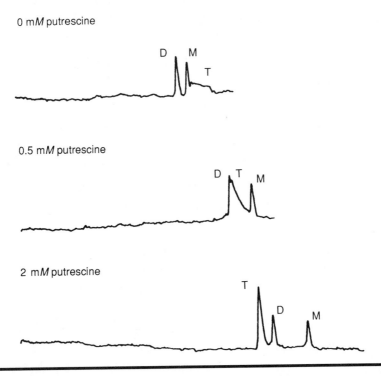

Figure 7   Separation of cationic heptapeptides with putrescine added to the running buffer. Conditions: field, 214 V/cm; current, 17 μA; buffer, 20 mM MES, 10 mM KCl, pH 6; capillary length, 70 cm (40 cm to detector); capillary diameter, 75 μm; detection, 220 nm. Reproduced with permission from Stover *et al.* (1989).

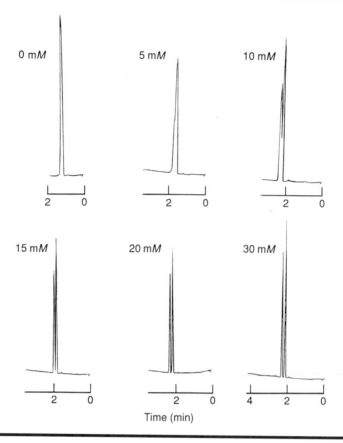

Figure 8 The effect of zinc sulfate on the separation of β-Ala-L-His, which elutes first, and L-Ala-L-His. Conditions: field, 400 V/cm; current, 1 μA; capillary length, 20 cm (18 cm to detector); capillary diameter, 25 μm; detection, 200 nm. Reproduced with permission from Mosher (1990).

At pH values where histidine is neutral, adding Zn (II) facilitated a separation mechanism based on metal binding rather than proton binding. Consequently, on adding 1 mM Zn (II) at pH 7.5, the peptides separated in the order D, M, T. In addition, when longer capillaries were used at constant voltage, the electric field decreased, the current decreased, and the efficiency increased for all peptides. Figure 8 shows how Mosher (1990) used metal ions to effect the separation of histidine-containing peptides. In particular, histidine peptides served as models for metal effects, since histidine is known to interact with Cu (II). Interaction with a metal ion was postulated to affect the mobility of peptides that otherwise may comigrate, and thus cause separation. The dipeptides Ala-His and β-Ala-His were not resolved in acetic acid, but were when 20–30 mM zinc sulfate was added. Also the optical isomers of DL-His-DL-His were resolved with the same metal additive. Cu (II) serves the same purpose with this dipeptide, i.e., increasing migration time and resolving, but Cu (II) gave a much noisier baseline. The data indicated that histidine-containing dipeptides interact with Zn and Cu and alter the electrophoretic mobility. Buffer was not thought to have a major influence on these separations. All work was performed at low pH where all samples were cationic; interaction with metal cations should probably improve at higher pH values.

Cyclodextrins were added to 20 m*M* to 50 m*M* borate, pH 9.50 buffer (Liu *et al.,* 1990). The separation showed very high efficiencies for model peptides and a tryptic digest of cytochrome *c*. The authors postulated a match between the size or shape of the solute and the cyclodextrin hydrophobic cavity as a factor in the separation.

B. Peptide Separations by Micellar Electrokinetic Capillary Chromatography

Separations by Micellar Electrokinetic Capillary Chromatography (MECC) are based on differential interactions with a micelle (see Chapter 6). Since the separation is not based on charge, neutral materials may be resolved. An *Application Note* (#4) from Applied Biosystems discussed the analysis of neutral and charged peptides by MECC. A mixture of six 7-amino acid peptides that are cationic, anionic, and neutral at pH 7.0 was resolved (Fig. 9). The order of elution suggested that the cationic peptides have a high affinity for the negatively charged sodium dodecyl sulfate (SDS) micelle, whereas the anionic peptides may be repelled from the micelle. The electropherogram

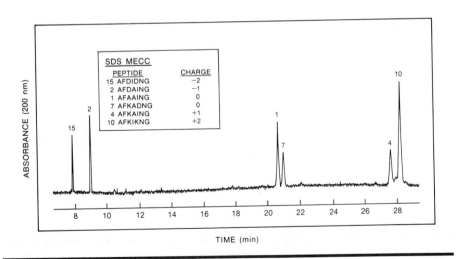

Figure 9   MECC separation of peptides. Conditions: field, 308 V/cm; buffer, 10 m*M* sodium phosphate, 100 m*M* SDS, pH 7.0; capillary length, 65 cm (45 cm to detector); capillary diameter, 50 μm; temperature, 30°C; detection, 200 nm. Samples (charge at pH 7): (15) AFDIDNG [−2]; (2) AFDAING [−1]; (1) AFAAING [0]; (7) AFKADNG [0]; (4) AFKAING [+1]; (10) AFKIKNG [+2].

demonstrated great peak capacity with the MECC method. These data illustrated the ability of MECC to provide another dimension in selectivity for peptides.

Peptides from protease digests and model peptides were separated by CE using MECC and cyclodextrin inclusion-forming compounds (Liu *et al.*, 1990). Sodium dodecyl sulfate and the cationic surfactants, dodecyltrimethylammonium bromide (DTAB) and hexadecyltrimethylammonium bromide (HTAB) were used as micelle-forming reagents. These MECC methods are thought to be very important in the separation of peptides with similar charge. At pH 7.05 each of eight model angiotensin peptides was positively charged and gave poor results with SDS. However, the cationic surfactants HTAB and DTAB did promote resolution (Fig. 10). The eight peptides were separated using 50 m*M* HTAB in a tris(hydroxymethyl)aminomethane

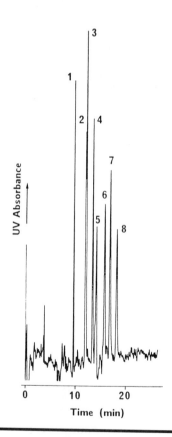

Figure 10   Separation of angiotensin analogs using a cationic detergent in the running buffer. Conditions: field, −250 V/cm; buffer, 25 m*M* Tris–25 m*M* sodium phosphate, pH 7.05, 0.05 *M* HTAB; capillary length, 80 cm (50 cm to detector); capillary diameter, 50 μm; detection, 220 nm. Samples: (1) RVYIHPI; (2) RVYVHPF; (3) NRVYVHPF; (4) RVYIHPF; (5) DRVYVHPF; (6) DRVYIHPF; (7) RVYIHPFHL; (8) DRVYIHPFHL. Reproduced with permission from Liu *et al.* (1990).

(Tris)-phosphate buffer. The experiment required reversed polarity (positive electrode at detector end) since electroendosmosis is reversed as the cationic surfactant is attracted to the negatively charged wall. Angiotensin II and [Val-5]-angiotensin II have the same charge characteristics, since they differ by only one hydrophobic amino acid (valine versus isoleucine), but separate because of the different hydrophobicities of these amino acids. The isoleucine-containing peptide eluted later, as expected, because of its greater hydrophobicity.

At HPCE '90, San Francisco, Moring and Nolan (1990) presented work on the separation of peptides using short-chain ionic surfactants, but below their critical micellar concentrations (CMC). Using model peptides that differ only in degree of hydrophobicity, they manipulated surfactant type, concentration, and pH. Addition of alkyl sulfonates up to C8 increased the resolution and efficiencies of peptides in free solution. The effect was demonstrated on a protein digest by addition of hexane sulfonic acid (HSA) where changes in selectivity were observed. The data supported the idea of a hydrophobic interaction mechanism.

Tran *et al.* (1990) used Marfey's reagent to derivatize dipeptides (Fig. 11) and tripeptide isomers for separation by MECC. Whereas the resolution of peptide enan-

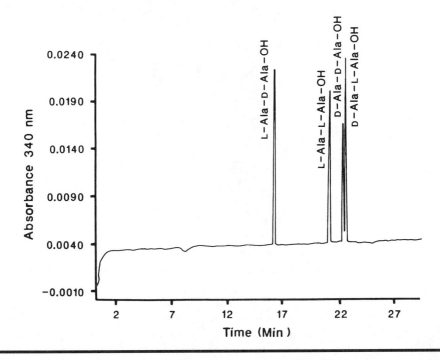

Figure 11 Separation of isomers of dipeptides Ala-Ala-OH derivatized with L-Marfey's reagent. Conditions: field, 210 V/cm; current, 10 μA; buffer, 100 mM sodium borate, pH 8.5; capillary length, 57 cm (50 cm to detector); capillary diameter, 75 μm; temperature, 25°C; detection, 340 nm. Reproduced with permission from Tran *et al.* (1990).

tiomers could not be accomplished, derivatizing the enantiomeric peptides changed them into diastereomeric forms that were separable.

C. SDS-Polyacrylamide Gel Electrophoresis

Slab-gel SDS-polyacrylamide gel electrophoresis (PAGE) is not commonly used for the molecular weight determination of peptides, since peptides typically do not bind SDS in a characteristic, stoichiometric fashion as do proteins. Slab-gel electrophoresis of small peptides can be problematic: special procedures are required, and quantitation and reproducibility are poor. Current methods for the molecular weight determination of peptides include sequencing and amino acid analysis, which are time-consuming procedures. Cohen and Karger (1987) reported on the separation of peptides by SDS-PAGE in a capillary format (Fig. 12). They polymerized acrylamide in 75-μm capillaries, with added urea. Myoglobin fragments, 2500–17,000 daltons, were separated with efficiencies of approximately 40,000 plates. The log molecular weight versus mo-

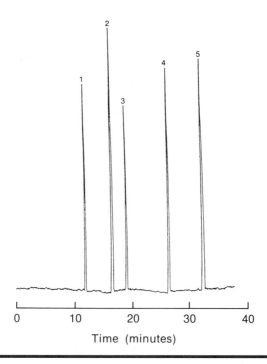

Figure 12 SDS-PAGE separation of myoglobin and myoglobin fragments. Conditions: field, 400 V/cm; current, 34 μA; buffer, 0.1 M Tris-phosphoric acid, pH 6.9, 0.1% SDS, 8 M urea; gel, 12.5% T, 3.3%C; capillary length, 20 cm to detector; capillary diameter, 75 μm; temperature, 25°C. Sample molecular weights: (1) 2510; (2) 6210; (3) 8160; (4) 14,400; (5) myoglobin 17,000. Reproduced with permission from Cohen and Karger (1987).

bility was plotted, and found to give a linear relationship. Ferguson plots indicated that the separations were based on size alone. Insulin A and B chains could be separated in under 10 min using a 10-cm capillary. See Chapter 5 for a complete discussion of capillary gel electrophoresis.

III. Characteristics and Applications

A. Resolution and Comparison with HPLC

A study was undertaken (Vinther *et al.*, 1990) to identify degradation products of bovine aprotinin using a combination of CE, HPLC, and mass spectrometry (MS). Aprotinin is a 58-amino acid peptide with three disulfide bonds, and an isoelectric point (pI) of 10.5. It has been used as a therapy for acute pancreatitis because of its trypsin-inhibitory characteristics. In order to avoid capillary wall sticking (see Section V), the CE experiments were performed with a pH 2.5 citrate buffer. Whereas recombinant aprotinin was found to be 99% pure, bovine aprotinin was only 80% pure. The resolution was observed to be better with CE than with HPLC. Two contaminants were obtained by CE; in order to identify these, the peptide sample was spiked with fractions purified by HPLC, which had been identified by plasma-desorption MS. The contaminants were shown to be the 1-57 and 1-56 forms. These materials were postulated to arise from carboxypeptidases present during the purification.

Impurities in synthetic peptides are usually analyzed by HPLC or TLC. Firestone *et al.* (1987) applied CE to the assessment of purity of the dipeptide L-histidyl-L-phenylalanine. However, they used wide-bore capillaries, with a 0.5-mm internal diameter, and 0.5 μl was injected. Plate heights were high with a value of 100 μm. The authors compared results with HPLC, and observed that CE was able to distinguish some cationic impurities. A purity level of 92% was determined by CE, whereas LC gave 96.4%.

Many different applications of CE to peptide and protein analysis were demonstrated by Grossman *et al.* (1989b); CE was used to assess the heterogeneity of synthetic peptides. Since the mechanism of separation is different from that of RP-HPLC, the two methods may give very different results. In Fig. 13, the HPLC shows a single, symmetrical peak with a purity of 99.2%, whereas CE shows six components, with the major peak only about 50%. The authors concluded that rather than spend time optimizing the HPLC separation, it makes more sense to use a second method with a different separation mechanism to achieve reliable conclusions regarding peptide purity.

In many laboratories CE is used as the final purity check for peptide preparations. However, if mixtures are judged to be heterogeneous, they still cannot be purified since putative preparative capabilities of CE are not universally applicable to all samples, and cannot supply clean samples in millimolar quantities. High-performance ion exchange chromatography (HPIEC) was shown by Guarino and Phillips (1991) to provide selectivity similar to that of CE, although the resolution was far superior with CE. Thus, when CE revealed contaminants in an RP-HPLC-purified pep-

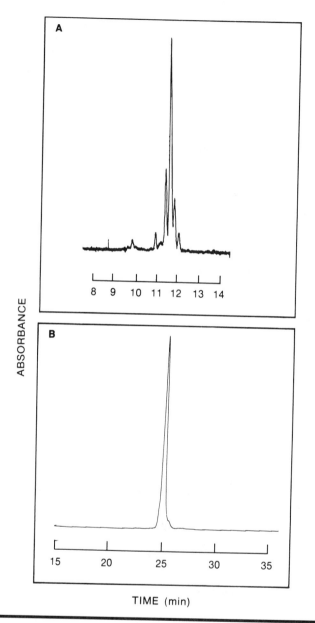

Figure 13 (A) HPLC separation of a synthetic peptide. Conditions: Buffer A: 0.1% TFA/water; B: 0.08% TFA/acetonitrile; gradient, 45 min from 0 to 60% B; column, 220 × 2.1 mm RP-300 (C-8); detection, 220 nm. (B) CE separation of the synthetic peptide. Conditions: field, 277 V/cm; current, 24 μA; buffer, 20 mM sodium citrate, pH 2.5; capillary length, 65 cm (45 cm to detector); capillary diameter, 50 μm; temperature, 30°C; detection, 200 nm. Reproduced with permission from Grossman *et al.* (1989b).

tide, HPIEC gave a similar profile. Collection of fractions from the HPIEC procedure allowed the isolation of the pure peptide material. CE was used because it is a fast screening method that can indicate the potential need for subsequent fractionation by HPIEC.

Instrumentation for two-dimensional (2D) separations combining HPLC and CE were described by Bushey and Jorgenson (1990a,b). The separation power and very high peak capacity of the 2D method was shown to be superior to that of either method alone. Single amino acid changes in proteins would have a greater chance of being detected with this method than with a single dimension.

One potential experimental problem that can be encountered in this 2D method concerns the presence of organic solvents in the HPLC eluant, which then are injected into the CE capillary. They encountered no problems with the CE injections from these organic-solvent-rich HPLC fractions. Though they did not perform it, the authors claim that this apparatus should allow for the formation of buffer gradients during the electrophoresis.

They used the 2D procedure to compare tryptic digest maps of horse heart cytochrome *c* and bovine heart cytochrome *c*. Peptides were labeled with fluorescamine. Selected horse heart cytochrome *c* fragments were commercially available, and spiking experiments were used to identify the various peaks. Bovine heart protein has 16 fragments, and horse heart has 17. They observed minor differences in patterns for the two digests. More than the expected number of peaks were observed, which may be evidence for incomplete or nontrypsin-specific cleavage. Though not identified, species differences were observed in the two maps. The 2D method therefore should be useful for identifying other changes, such as protein modifications.

B. Effects on Separations

A publication from Applied Biosystems, Inc. titled *270A Research News #1*, (Colburn *et al.,* 1990) illustrates various chemical and instrumental effects on peptide separations. This methods development guide shows the effect of pH on resolution and selectivity, and the effect of voltage and temperature on resolution and migration time. Capillary diameter and length also are shown to influence resolution, efficiency, and time. A discussion is given of how HPLC fraction concentrations must be optimized with micro-HPLC or concentrated to be most compatible with CE analysis.

Zhu *et al.* (1989) used CE for the separation of substance P fragments SP(1-4), SP(2-11), and SP(5-11). These materials were loaded at a concentration of 0.5 mg/ml, and separated in 0.1 *M*, pH 2.5, phosphate buffer. When loaded electrophoretically at 8 kV for 8 sec, the bands were thought to have similar widths. When loaded hydrodynamically, the peaks were of different widths, and the peak width was proportional to the migration time of the individual peak. This peak broadening was thought to be caused by the relative velocity differences, and not diffusion. Peaks with a slow velocity will travel past the detector more slowly, and will be integrated for a longer time, and assigned, therefore, a greater area. This is a general phenomenon observed in CE separations, but should not be dependent on the mode of injection. This author questions the similarity of peak widths for the electrophoretically loaded peaks, as stated by Zhu *et al.*

Strickland and Strickland (1990) investigated the conditions for the separation of peptides by CE. Fourteen bioactive peptides with different composition and size were evaluated. The basicity ranged from the acidic tripeptide (Glu-Ala-Glu) to the basic (Lys-Trp-Lys), whereas the size ranged from 3 to 21 amino acids. Migration times gave reproducibilities initially as high as 2.5% relative standard deviation (RSD) when the capillary was pretreated with the running buffer, but decreased to < 1.0% when the capillary was equilibrated with 500 m*M* Na phosphate, pH 2.5, overnight. Also CE was compared to RP-HPLC, and the authors found that CE resolved better, especially those small peptides that eluted very early in RP-HPLC.

C. Determination of Disulfide Bonds

The work of Yuen *et al.* (1989) and Grossman *et al.* (1989b) demonstrated how CE could be used in conjunction with other methods to confirm peptide structure. The peptide STP-3, with unknown structure, was allowed to react with 100 m*M* dithiothreitol, and the reaction was monitored by CE over the course of several hours. Although the reaction volume was only 2 μl, this was sufficient for CE loading purposes. A change in electrophoretic mobility was observed, consistent with the reduction of a disulfide bond (Fig. 14). Based on structural determination methods (e.g., sequencing and specific chemical reactions), a branched structure with a disulfide bond was postu-

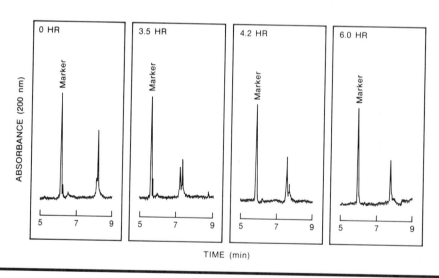

Figure 14   Time course of dithiothreitol (DTT) reduction of STP-3. Conditions: field 308 V/cm; current, 20 μA; buffer, 20 m*M* sodium citrate, pH 2.5; capillary length, 65 cm (45 cm to detector); capillary diameter, 50 μm; temperature, 30°C; detection, 200 nm. First peak is an internal standard, dynorphin 1–13. Reproduced with permission from Grossman *et al.* (1989b).

lated for the unknown. However, the C-terminus could not be definitively assigned to either acid or amide forms. When two proposed structures, with either C-terminal acid or amide, were synthesized and analyzed, CE resolved the amide and acid forms of the peptide. Peptide STP-3 coeluted with the synthetic amide form, thereby leading to the conclusion that the STP-3 was a C-terminal amide form of a branched peptide with a disulfide bond. This assignment was later confirmed.

Insulinlike growth factor (IGF-I) is a peptide with 70 amino acids. A recombinant-IGF byproduct and IGF-I differ by disulfides at 6–47 and 48–52 in the byproduct, but 6–48 and 47–52 in IGF-I. They were separated by Ludi *et al.* (1988) (Fig. 15) using a pH 11.11 buffer [10 m*M* (cyclohexylamino)propanesulfonic acid (CAPS), 5 m*M* sodium tetraborate, 1 m*M* EDTA] at 250 V/cm, and detected at 215 nm.

D. Clinical Applications

The *in vivo* release of luteinizing hormone-releasing hormone (LH-RH) from sheep brains by CE analysis was studied by Advis *et al.* (1989), and it was identified by spik-

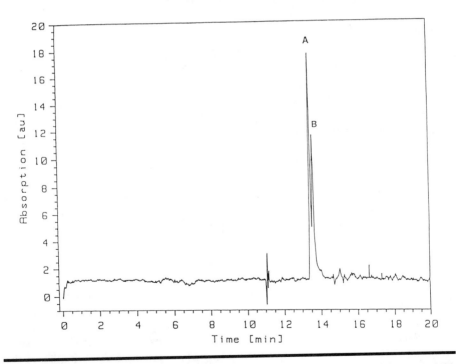

Figure 15 Electropherogram of r-IGF byproduct (A) and R-IGF-I (B). Conditions, field, 250 V/cm; current, 25 µA; buffer, 10 m*M* CAPS, 5 m*M* sodium tetraborate, 1 m*M* EDTA, pH 11.11; capillary length, 120 cm (105 cm to detector); capillary diameter, 75 µm; detection, 215 nm. Reproduced with permission from Ludi *et al.* (1988).

ing experiments. Correlations between CE and radioimmunoassays (RIA) for LH-RH content values proved to be adequate. Although the best sensitivity obtained by CE was at the upper range of the RIA sensitivity (1 pg/μl), CE separated and quantitated several components simultaneously. The authors are currently investigating laser-induced fluorescence (see Section VII,A) to improve the sensitivity of their measurements.

Lal *et al.* (1991) required a method that permits quantitation of histidine-rich polypeptides (HRPs) and degradation products in human saliva when enzyme-linked immunosorbant assay (ELISA) is not suitable. Capillary electrophoresis separated eight components of saliva cationic proteins and other degradation products. Parotid and submandibular sublingual salivas of normals were found to vary in level of HRPs, histatin 6, and lysozyme. The authors intend to use CE to quantitate and correlate these findings with the antimicrobial activity of the salivas. This may prove to have broad ramifications, since the antifungal activity of salivas of acquired immunodeficiency syndrome (AIDS) patients can be significantly depressed.

IV. Protein Digests

A. Peptide Mapping

Traditionally, the confirmation of protein identity involves the fragmentation of the protein into peptides, which can then be analyzed. Fragmentation is most often accomplished by digesting the protein enzymatically or chemically. The digestion results in a complex mixture of peptides, which is resolved to form a fingerprint or map characteristic for that protein. Reverse-phase HPLC is the most widely used method for analysis of the peptide mixture, but the resolving power of HPLC may not prove sufficient. This may lead to multicomponent fractions that cannot be separated. With its high resolving power and alternative selectivity, CE can provide a complementary method for generating the map.

The ability of CE to serve as a tool for peptide mapping was recognized early in the history of CE. In 1981, Jorgenson and Lukacs (1981b) reported using CE to resolve peptides obtained from the tryptic cleavages of egg-white lysozyme (Fig. 16). The separation was complete in 20 min using a simple 50 mM phosphate buffer, pH 7.0. The peptide fragments were treated with fluorescamine, and the peptides were observed by a fluorescence detector. Green and Jorgenson (1984) also used CE to analyze labeled peptide fragments obtained from the tryptic digest of chicken ovalbumin. The authors concluded that the number of peaks resolved represented a significant improvement over the traditional electrophoresis methods.

Frenz *et al.* (1989) correlated the elution pattern observed for the CE separation of tryptic peptides from recombinant human growth hormone (hGH) with that of the HPLC map: as expected, because of the different separation mechanisms, no correlation was noted (Fig. 17). This clearly demonstrates the utility of CE as a complementary method that provides the extra level of confidence necessary for confirmation of protein identity. Karger (1989) also found no correlation between HPLC and CE data

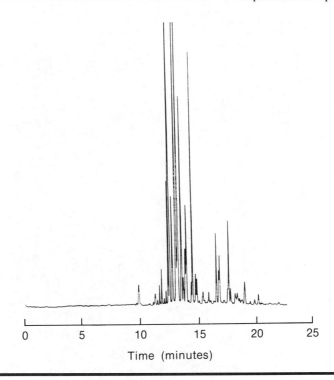

Time (minutes)

Figure 16   Separation of fluorescamine-labeled peptides obtained from a tryptic digest of lysozyme. Conditions: field, 300 V/cm; buffer, 50 mM phosphate, pH 7; capillary length, 1 m; capillary diameter, 80 μm. Reproduced with permission from Jorgenson and Lukacs (1981b).

for a tryptic digest. Steuer *et al.* (1990) also compared the CE and HPLC analyses of digests and concluded that CE is particularly suitable because of its tremendous peak capacity, very high efficiencies, and potential for rapid methods development. However, they did point out issues with concentration sensitivity, which could present a possible limitation of CE.

The issue of concentration sensitivity was discussed by Cobb and Novotny (1989). Because of the nanoliter sample injection volume, only 2 ng (80 fmol) of β-casein tryptic digest could be introduced into the capillary from 50 ng of total sample; the small sample mass and the short path length led to sensitivity problems. Whereas microcolumn HPLC is capable of accepting the entire digest sample, it was impossible to use the entire digest sample for CE analysis. Injection volumes greater than 20 nl were found to result in inferior electropherograms with peak broadening, apparently due to electric field perturbations. In this paper, peptide maps from both phosphorylated and dephosphorylated forms of β-casein were distinguishable. Phosphorylation at serine effectively increases the net negative charge of peptides, which is reflected in their decreased mobility toward the cathodic electrode at the detector.

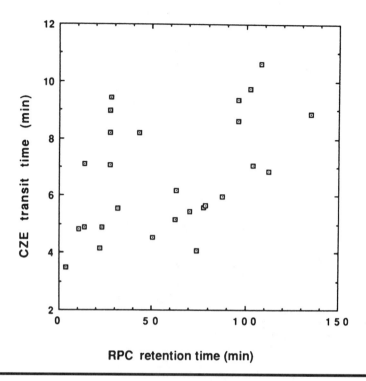

Figure 17 Correlation of RP-HPLC retention times of peptides obtained from the tryptic digestion of recombinant hGH with their migration times by CE. Reproduced with permission from Frenz *et al.* (1989).

The authors used a Tris-tricine running buffer, pH 8.1. They noted that at this pH, the complex mixture should contain both positively and negatively charged peptides, which can be separated because of the strong electroosmosis. Positively charged peptides did not appear to be sticking to the negatively charged wall, as evidenced by the absence of peak tailing or broadening. Thus, resolution of a complex mixture of peptides was accomplished by CE without the use of any buffer additives or capillary coatings, thereby demonstrating the simple nature of the method.

The same study also described migration-time reproducibility studies of 14 major peaks from the β-casein maps. Relative standard deviation values were less than 1% for all peaks, better than that obtained with microcolumn HPLC. This improvement was attributed to the lack of a gradient and the simplicity of the CE method. The small changes that were observed by CE were probably the result of changes in electroosmosis. In addition, the time of analysis by microcolumn HPLC was reported to be 70 min, as compared to only 20 min for CE.

In a series of papers, Nielsen *et al.* (1989a,b), Nielsen and Rickard (1990), and Grossman *et al.* (1989b) discussed the use of CE to generate a tryptic map of hGH.

They used 10 m*M* tricine, 20 m*M* NaCl, 4.5 m*M* morpholine, pH 8.0, as the CE running buffer. Human growth hormone is cleaved into 21 peptides, but only 19 peptides are obtained without prior reduction of the disulfides (6–16 and 2–21). Fragments range from a single lysine to a 35 amino-acid peptide, from highly basic (pI = 10.4) to highly acidic (pI = 3.5), and some are hydrophobic, whereas others are hydrophilic. Even with a 1- to 2-hr complex gradient, RP-HPLC did not give complete resolution.

The map was correlated between CE and RP-HPLC after fragment identity had been assigned (Fig. 18). Capillary electrophoresis migration times of the HPLC-iso-

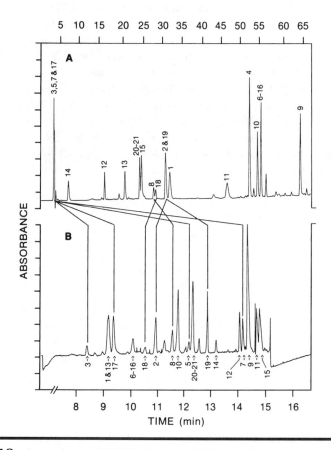

Figure 18 Comparison of RP-HPLC (A) and CE (B) tryptic maps. Peak assignments and correlations for selected fragments are noted. CE conditions: field, 316 V/cm; current, 20 µA; buffer, 0.01 *M* tricine, 0.045 *M* morpholine, 0.02 *M* NaCl, pH 8.0; detection, 200 nm. Chromatographic conditions: column, Aquapore RP-300 (4.6 × 250 mm, Brownlee Labs, San Jose, CA); flow rate, 1 ml/min; detector, 214 nm; solvents: A, 0.1% TFA in water; B, 0.1% TFA in acetonitrile; gradient, 0–20% B in 20 min, 20–25% B in 20 min, 25–50% B in 25 min. Reproduced with permission from Grossman *et al.* (1989b).

lated fractions could not be directly compared to the times obtained for the mixture since the matrix perturbed the electrophoretic environment; therefore, spiking experiments were used to identify peaks. Fragments 3, 5, 7, and 17 were reported to be small, very hydrophilic peptides that do not separate by RP-HPLC, but rather elute in the void volume. Because of the differences in size and charge of these peptides, they separate well by CE. Fragments 8 and 18, and 2 and 19 have similar hydrophobicities and were not separated by RP-HPLC, but were by CE. Fragments 1 and 13 and 4 and 9, which are of similar size and charge, were not separated by CE, but could be by the RP-HPLC method. The elution orders do not correlate between the two methods. These results clearly stress the importance of analysis by both CE and HPLC, which, in combination, produce a rapid, reproducible, verification of hGH identity.

Nielsen and Rickard (1990) showed the effect of pH (2.4, 6.1, 8.1, 10.4) on selectivity since pH determines charge, a prime contributor to selectivity. Buffer, ionic strength, and mobile-phase additive concentrations were varied at pH 8.1; all resulted in effects on electroosmosis and current but little on selectivity. Optimal separation conditions were 0.1 M tricine, 0.02 M morpholine, and no salt (salt was found to give poorer resolution, presumably because of increased heating) at pH 8.1. Modifiers may be added to reduce wall interactions or maintain solubility. High buffer concentrations were chosen in order to maintain a constant pH and conductivity within the analyte zone, but high concentrations of ionic buffers can give high conductivities (and high temperature gradients), so zwitterionic buffers with relatively low conductivities were used. Increasing the tricine concentration to 0.1 M gave improved efficiency and resolution; excess buffer can prevent perturbations in field and pH. Additives, such as morpholine, are thought to coat wall-surface silanols, thereby accounting for the lower electroosmosis. Although the additive gave higher currents, it did lead to better results. Some early eluting fragments were better resolved at pH 2.4, so in general, complete resolution may not be possible by a single elution condition. The authors concluded that the use of two buffers with different pH values can give improvements in resolution without changing other operating parameters.

A commercial CE instrument described by Albin *et al.* (1991) has the ability to thermostat samples from 0 to 60°C. This system was used to incubate a tryptic digest of ribonuclease T1 and inject from this reaction mixture under software control. In this fashion the progress of the tryptic reaction was monitored automatically, and the data are shown in Fig. 19. Although not shown, it can be inferred that information concerning the kinetics of generation of specific fragments may be obtained by this method. Results showed that at 37°C the reaction took about 10 hr whereas at 50°C, the digestion was complete in 2 hr.

Zhu *et al.* (1990) used polyacrylamide-coated capillaries to separate tryptic fragments of bovine serum albumin (BSA). When an uncoated capillary was used, results were very poor. However, it should be noted that the buffer contained methylcellulose, which may contribute to the reduction of electroosmosis. This may result in the absence of peaks, since the fragments (negative at pH 8.5) may be traveling away from the detector, and there is little electroosmosis to carry them through. Frenz *et al.* used these same coated capillaries to separate fragments from the tryptic digestion of hGH.

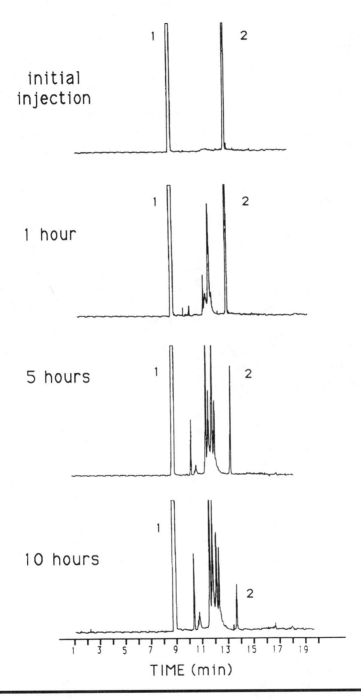

Figure 19 Electropherogram illustrating the time dependence of tryptic digestion at elevated sample temperature. Conditions; field, 139 V/cm; buffer, 20 mM sodium phosphate, pH 7.0; capillary length, 72 cm (50 cm to detector); capillary diameter, 50 μm; capillary temperature, 30°C; sample temperature, 37°C; detection, 200 nm. Peak 1 is a neutral marker (β-mercaptoethanol); peak 2 is ribonuclease T1. Reproduced with permission from Albin *et al.* (1991).

Although they reported advantages such as speed, resolution, and small sample requirements, they also saw poor migration-time reproducibility with these capillaries. Low-pH buffer (100 mM phosphate, pH 2.56) was used. Fragments were identified by spiking with HPLC fractions of known sequence. Reproducibility problems were attributed to material possibly adsorbed to the capillary wall. Whereas coatings may mask wall charges, peptides may still interact with the coating in a hydrophobic fashion. Poor instrument temperature control may also be responsible for reproducibility performance problems.

B. Screening HPLC Fractions

Not only is CE useful for the generation of peptide maps, but it also has been successfully used to screen the individual fractions derived from the HPLC purification of the digest. The most widely used method for peptide purification is RP-HPLC, and ultimately these peptide fragments are sequenced. However, the resolving power of the HPLC method may not always prove sufficient, which can lead to contaminated fractions and confusing sequence information. With its high resolving power and alternative selectivity, CE can provide a rapid method for determining the purity of individual HPLC fractions before sequencing. Since CE uses only nanoliter quantities, the overwhelming majority of sample is still available for sequencing or further purification. Schlabach *et al.* used this concept when sequencing carboxypeptidase P. For the initial purification of the fragments, RP-HPLC was inadequate, so ion exchange chromatography followed by RP-HPLC was used. After this 2D purification, CE was used to analyze the fractions. Heterogeneity was observed and confirmed by actually sequencing these fractions. This method is said to be an important application of CE, since it can dramatically increase sequencer productivity. Frenz *et al.* (1989) also used this approach and did observe that a fraction thought to be clean by HPLC actually contained two fragments as defined by CE (Fig. 20). Grossman *et al.* (1989b) utilized this same approach. A sample of β-lactoglobulin A was digested with trypsin, fractions were isolated by RP-HPLC, and CE was used to screen the fractions. Fraction A gave one peak by RP-HPLC but three by CE; subsequent sequence analysis showed three sequences. Samples from RP-HPLC may be eluting in low-ionic- strength solvents, an excellent CE sample buffer system due to potential stacking effects for detectability improvements (see Section VII,C).

V. Wall Interactions

The most general and simplest method for separating peptides involves the use of low-pH buffers. At low pH the capillary wall is virtually neutral, so that ionic interactions between peptides and the capillary wall are essentially eliminated. Acidic buffers are also thought to provide more reproducible results, since electroosmosis is almost eliminated at low pH (McCormick, 1988; see Chapter 1 for a thorough treatment of electroosmosis). Separations that do not require electroosmosis are not subject to any flow

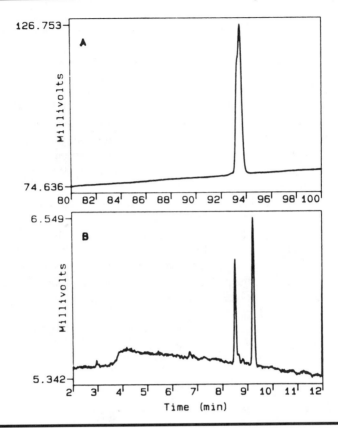

Figure 20 Analysis of the "T11, T10c2" fraction from the RP-HPLC tryptic map by (A) HPLC and (B) CE. Conditions: (A) column, 150 × 4.6 mm Nucleosil C-18; temperature, 35°C; detection, 214 nm; solvents: A, 0.1% TFA in water; B, acetonitrile; gradient, 0–38% B from 5–120 min, 38–57% B in 10 min; flow rate, 1 ml/min. Conditions (B) field, 400 V/cm; buffer, 100 mM sodium phosphate, pH 2.56; capillary length, 20 cm; capillary diameter, 25 μm; detection, 200 nm. Reproduced with permission from Frenz et al. (1989).

variability that might result from subtle changes in electroosmosis. McCormick proposed that the reduction in electroosmotic velocity for phosphate buffers could arise from the stabilization of the silica via the formation of a complex of silanol and phosphate groups on the capillary surface. Experiments showed that phosphates binds to the wall at the rate of 23 phosphates/silanol. Phosphate on the wall may also serve to reduce the adsorption of peptides to the wall. In addition, capillaries were modified with poly(vinylpyrrolidinone) (PVP), acrylamide, and acrylic acid to decrease electroendosmosis, although these yielded low-efficiency separations.

At high pH (above the pI of the peptide analyte), peptide–wall interactions are also reduced, since peptides acquire a negative charge, and are repelled from the negatively charged wall. However, this method is not universal. Aprotinin is a 58-amino acid peptide with a pI of 10.5. Because of capillary-wall interaction considerations and the high pI, CE was performed at pH 2.5 (Vinther *et al.*, 1990). Using a high-pH buffer below the pI led to peptide sticking to the wall, and the peptide is not stable at the very high pH values required for efficient separation.

The use of high or low pH to reduce interactions between the capillary wall and tryptic digest fragments was noted by Nielsen and Rickard (1990), but they suggested that these buffer conditions may not be advantageous for optimizing selectivity. They therefore recommended minimizing wall sticking by other means, such as adjusting buffer composition or ionic strength, or using additives, so that the pH is chosen for the best separation.

Wiktorowicz and Colburn (1990) described a reagent (for treating the capillary surface) that effectively reverses the surface to a positive charge. In this manner, separations can be performed at pH values below the pI. Thus, separation conditions can be selected in which the native states of basic proteins can be maintained, at pH values where the analyte is positively charged. Whereas the paper discussed the application to protein separations, the treatment also permits peptide analysis under these pH conditions. Even at low pH values where wall interactions are minimal, the reagent can offer improvements in peptide peak efficiencies.

The work of Stover *et al.* (1989) demonstrates the use of additives, including metals, to reduce wall interactions (see Section II,A,2).

VI. Models of Peptide Mobility

Various researchers have developed models to describe the electrophoretic mobility of peptides. A report by Grossman *et al.* (1989a) (see Chapter 4) derived a semiempirical relationship that expressed the electrophoretic mobility as a function of charge and size. Electrophoretic mobility is the only parameter measured by CE that has any physical significance with respect to solute structure, and was therefore chosen as the parameter to best describe the characteristics of CE separations.

In its simplest form, mobility is a function of charge (q) and an inverse function of size. For equal-sized peptides, with each added charge, the incremental increase is a smaller percentage of the total charge, so that mobility should be a logarithmic function of the charge. Peptides were treated as classical polymers in solution, so that the size term is defined as function of number of amino acids (n) to a power. The relationship

$$\mu = D \ln(q + 1)/n^{0.43}$$

was derived. This equation was tested with 40 different peptides that ranged in size from 3 to 39 amino acids, whereas the charge varied from 0.33 to 14. The results are shown in Fig. 21. For peptides with known sequence, a computer program incorporat-

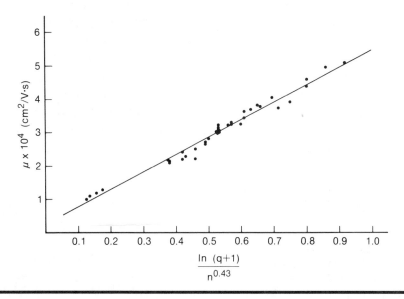

Figure 21 Electrophoretic mobility as a function of charge and size for 40 different peptides. Y intercept $= 2.47 \times 10^{-5}$, slope $= 5.23 \times 10^{-4}$, $r = 0.989$. Reproduced with permission from Grossman *et al.* (1989a).

ing the Grossman model was described (Pennino, 1989). The program calculates theoretical charge, electrophoretic mobility, and the expected elution times of the peptides.

This model has been used to assign the C-terminal structure of a native peptide. When native and synthetic forms of a tridecapeptide were electrophoresed by CE at pH 2.5, instead of coeluting, they separated. The mobility difference corresponded to a difference in charge of about 0.5 units. This charge difference was consistent with the native peptide being a C-terminal amide: the synthetic material is known to be a C-terminal acid, with a 50% ionization (0.5 charge units) of the acid at this pH. The native peptide structure was confirmed by chemical methods and fast atom bombardment (FAB)-MS.

Deyl *et al.* (1989a,b) investigated the relationship between peptide relative migration time and peptide character. The relative migration times of the cyanogen bromide (CNBr) cleavage products of collagen alpha chains correlated linearly with Offord's relationship $M^{0.67}/Z$, where M is the molecular weight and Z is the charge. However, there was some anomalous behavior with one fragment, since it migrated 2 min faster than predicted by the Offord relationship. CNBr-cleaved peptides of collagen type I and III were plotted versus $M^{0.67}/Z$, and a linear relationship was obtained. Masses of fragments were 13,000–25,000 and $Z = 18$–30. Frenz *et al.* (1989) used the Offord relationship for peptide fragments: they claim the correlation is good, but this author observed appreciable scatter in the data and no r value was given (Fig. 22).

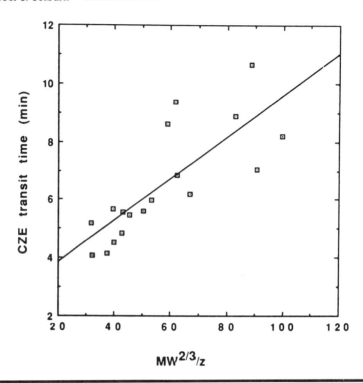

Figure 22 Correlation of CE migration times of tryptic peptides with molecular weight and charge. Peptides were assumed to carry charges only at the N-terminus and on lysine, arginine, and histidine side chains for charge calculations. Reproduced with permission from Frenz *et al.* (1989).

VII. Detection Methods for Capillary Electrophoresis of Peptides

A. Direct and Indirect Fluorescence

Ultraviolet-visible (UV) absorbance detection is the most common method for analyte detection, primarily because it is general, i.e., many substances absorb in the UV. Also, UV detectors are reasonably robust and inexpensive. Whereas UV detectors are most often used for CE peptide applications, other detectors have been used because of the need for increased sensitivity. Since CE instruments typically inject only nanoliter quantities, extremely small masses of analytes are loaded. Thus, some minimal concentration must be maintained in order to load a detectable mass. In addition, the very short optical path lengths used in CE (50–100 μm) limit absorbance and therefore sensitivity. See Chapter 2 for a complete discussion of detection methods for CE.

Fluorescence detection is capable of providing increased sensitivity, but most materials are not natively fluorescent so must be derivatized with a fluorescent "tag." Fluorescence detectors were used by Jorgenson and Lukacs (1981a,b) for observing fluorescamine-labeled tryptic peptides derived from lysozyme and for labeled dipeptides. Novotny *et al.* (1990) and Liu *et al.* (1990) observed peptides in their native forms (UV), and as fluorescently labeled compounds derivatized with orthophthalaldehyde (OPA), fluorescamine, and 3-(4-carboxybenzoyl)-2-quinolinecarboxaldehyde (CBQCA). The relative fluorescence intensity was found to increase 10-fold for CBQCA-Des-Asp 1-angiotensin I when β-cyclodextrin was added to the buffer in the range from 0 to 20 m*M*; a smaller increase in intensity was observed for CBQCA-Gly-Gly-Tyr-Arg.

Fluorescence detection does present some problems: since the incorporation of label depends on the rate at which reactive moieties can be modified and the total number of such moieties, different peptides will be modified to different degrees regardless of concentration, thus generating extra species. Therefore, quantitation is difficult. In addition, the reaction alters the structure of the peptide permanently, which means that one observes the adduct and not the actual peptide.

Indirect fluorescence detection permits high sensitivity typical of fluorescence detection for nonfluorescent analytes without the need for derivatization. It is thus a nearly universal method and should be quantitative. Detection is based on the displacement of a fluorescent buffer ion by the nonabsorbing analyte ion, and not on absorption or emission properties of the analyte. Hogan and Yeung used indirect fluorescence CE to observe the separation of subfemtomole quantities of β-casein digest mixtures in 3 min. The quantitative and qualitative data were reproducible, and the mass limit was 180 times lower than that obtained with UV-absorbance detection. An excellent review of indirect detection methods for CE was presented by Yeung and Kuhr (1991).

Low concentrations (1 m*M*) of sodium salicylate were added to the buffer; because of its native fluorescence, it produced a high level of laser-induced fluorescence as it passed the detector. The migration of a negatively charged analyte requires the displacement of a salicylate ion in order to maintain charge balance. In this way, the fluorescence signal decreases, indicating a migrating zone of analyte.

Gross and Yeung (1990) also used indirect fluorescence methods, but used low-pH buffers (0.58 m*M* sulfuric acid, pH 3.7) to analyze cationic peptides derived from the tryptic digest of β-casein and BSA. In this case a positively charged fluorescent ion had to be used, so 0.38 m*M* quinine sulfate was added to the buffer.

B. Capillary Electrophoresis-Mass Spectrometry

One of the chief disadvantages of CE is that, because of the small sample size, identification of unknown peaks becomes very difficult. For this reason, many researchers have attempted the interfacing of CE with MS (see Chapter 2). Reinhoud *et al.* (1990) used CE with continuous flow fast atom bombardment (CF-FAB) MS and scanning array detection to analyze 5 pmol of β-endorphin fragments. Moseley *et al.* (1989) used FAB-MS for acquisition of both MS and MS–MS spectra from femtomoles of peptides while maintaining $>10^5$ plates, and as low as 500 amol with a narrow mass range scan. The same authors (Moseley *et al.,* 1991) reported the use of CE-MS for bioactive peptides and neuropeptides. Analysis of basic peptides was performed in aminopropyl-silylated CE capillaries to minimize adsorption effects. A coaxial interface was used in much of this work, with the advantage that it permits optimization of the CE buffer for separations and the FAB buffer composition and flow rate independently. Caprioli *et al.* (1989) also used CE with coaxial CF-FAB and showed tryptic digests of horse heart cytochrome *c* and hGH.

Smith *et al.* (1988, 1989) described the application of electrospray ionization (ESI) in CE-MS. Examples include leucine, enkephalin, and vasotocin. Smith *et al.* (1989) discussed the potential for on-line sequencing by CE-MS-MS. The same group (Loo *et al.,* 1989) used ESI CE-MS where the buffers had pH values greater and significantly less than the pI in order to reduce adsorption onto capillaries.

Takigiku *et al.* (1990) described a method for peak identification using CE. They devised an elegant method for fraction collection and then, in a separate procedure, subjected the fraction to plasma desorption (PD) MS. Though not so convenient as on-line methods, this offers the possibility of independently optimizing both the CE and the MS, since the optimal pH for MS may differ from the pH of the CE buffer.

By employing a porous glass joint at the detector end of the capillary, coupled to an electrically isolated capillary, analytes and buffer migrate by electroosmosis, not electrophoretically, and can then be collected. Radioactivity measurements showed that 99.7% of the analyte is transferred through the joint to the end of the capillary. Plasma desorption spectra of 10 pmol of bradykinin and bovine insulin were shown. The PD–MS response was observed at the 500 fmol level.

C. Stacking

Since CE is limited in sensitivity, methods have been devised to increase sample loading (total mass) without sacrificing peak efficiency (see Chapter 3). These so-called "stacking" methods usually require that the sample be dissolved in a buffer of significantly lower conductivity than that of the running buffer. Satow compared dynorphin diluted in 0.01% trifluoroacetic acid (TFA) to the peptide dissolved in run buffer. The sample in dilute TFA exhibited a linear peak height increase with injection time up to 7 sec of hydrostatic injection, while the peak width remained constant due to stacking. The sample in buffer, however, increased peak height linearly up to 4 sec only, and then the width began increasing. Thus, maintaining a low-ionic-strength sample gave high-efficiency results with a relatively large mass injected.

For bioactive peptides, 30 mM salt in the sample buffer allowed resolution of three peaks at both 10- and 20-sec injection times. With 100 mM salt, the 10-sec injection led to peak broadening, whereas 20 sec of loading resulted in peak distortions and poor resolution. Whereas the authors concluded that 20 sec was the maximal injection time for salt-containing samples, one must consider the total capillary volume and the percentage of that volume that represents the low-conductivity sample zone; greater volumes may be injected if the total capillary volume is large, and also perhaps if the salt is a low-conductivity species. For samples containing salt, broad peaks are expected, so one should dilute the samples or increase the ionic strength of run buffer to equalize the ionic strengths of both run buffer and sample. These conditions would, however, disallow any stacking from occurring.

Aebersold and Morrison (1990) demonstrated a stacking technique that may allow even greater sample volumes to be applied to the capillary, although the method requires working at pH extremes only. Under the conditions used, the authors were able to inject for 60 sec, although base-line problems were noted. Isotachophoretic concentration methods have been described showing the effect on loading of CNBr fragments (Dolnik *et al.*, 1990).

D. Other Detection Methods

Olefirowicz and Ewing (1990) used indirect amperometric detection for the analysis of dipeptides—3,4-dihydroxybenzylamine (DHBA) served as a continuously eluting electrophore. As DHBA was oxidized, a constant background current was produced. Cationic DHBA was displaced by cationic peptides, and as the peptide zone passed the detector, the current decreased, and a negative peak was obtained. Dipeptides were separated by CE where 0.01 mM DHBA in 25 mM 2[N-morpholino]ethanesulfonic acid (MES), pH 5.65 was used as the running buffer. Only 6–8 fmol of dipeptide were loaded in the extraordinarily minute injection volume of 52 pl.

The very short optical path length of the capillaries used in CE severely limit absorbance. Chervet *et al.* (1991) altered the geometry of detection by using a Z-shaped flow cell to increase the path length. With this design, the signal obtained for Val-5-angiotensin II was improved 14-fold over standard detection schemes, although this was considerably less than the predicted increase. However, the configuration did lead to small losses of efficiency because of the larger cell volume.

References

Advis, J. P., Hernandez, L., and Guzman, N. A. (1989). *Peptide Res.* **2**, 389–394.

Aebersold, R., and Morrison, H. D. (1990). *J. Chromatogr.* **516**, 79–88.

Albin, M., Wiktorowicz, J. E., Black, B., and Moring, S. E. (1991). *Am. Lab.* October, 27–35.

Bushey, M. M., and Jorgenson, J. W. (1990a). *Anal. Chem.* **62**, 978–984.

Bushey, M. M., and Jorgenson, J. W. (1990b). *J. Microcol. Sep.* **2**, 293–299.

Caprioli, R. M., Moore, W. T., Martin, M., DaGue, B. B., Wilson, K., and Moring, S. (1989). *J. Chromatogr.* **480**, 247–257.

Chervet, J. P., Van Soest, R. E. J., and Ursem, M. (1991). *J. Chromatogr.* **543**, 439–449.

Cobb, K. A., and Novotny, M. (1989). *Anal. Chem.* **61**, 2226–2231.

Cohen, A. S., and Karger, B. L. (1987). *J. Chromatogr.* **397**, 409–417.

Colburn, J., Black, B., Chen, S.-M., Demarest, D., Wiktorowicz, J., and Wilson, K. (1990). Model 270A Research News, Issue 1, Winter.

Colburn, J. C., Grossman, P. D., Moring, S. E., and Lauer, H. H. (1991). *In* "HPLC of Peptides and

Proteins: Separation, Analysis, and Conformation" (C. T. Mant and R. S. Hodges, eds.), pp. 895–901. CRC Press, Boca Raton, Florida.

Deyl, Z., Rohlicek, V., and Adam, M. (1989a). *J. Chromatogr.* **480**, 371–378.

Deyl, Z., Rohlicek, V., and Struzinsky, R. (1989b). *J. Liq. Chromatogr.* **12**(13), 2515–2526.

Dolnik, V., Cobb, K. A., and Novotny, M. (1990). *J. Microcol. Sep.* **2**, 127–131.

Firestone, M. A., Michaud, J., Carter, R. H., and Thormann, W. (1987). *J. Chromatogr.* **407**, 363–368.

Florance, J. (1991). *Am. Lab.* May, 32L–32O.

Frenz, J., Wu, S.-L., and Hancock, W. S. (1989). *J. Chromatogr.* **480**, 379–391.

Green, J. S., and Jorgenson, J. W. (1984). *J. High Res. Chromatogr.* **7**, 529–531.

Gross, L., and Yeung, E. S. (1990). *Anal. Chem.* **62**, 427–431.

Grossman, P. D., Wilson, K. J., Petrie, G., and Lauer, H. H. (1988). *Anal. Biochem.* **173**, 265–270.

Grossman, P. D., Colburn, J. C., and Lauer, H. H. (1989a). *Anal. Biochem.* **179**, 28–33.

Grossman, P. D., Colburn, J. C., Lauer, H. H., Nielsen, R. G., Riggin, R. M., Sittampalam, G. S., and Rickard, E. C. (1989b). *Anal. Chem.* **61**, 1186–1194.

Guarino, B. C., and Phillips, D. (1991). *Am. Lab.* March, 68–69.

Hogan, B. L., and Yeung, E. S. (1990). *J. Chromatogr. Sci.* **28**, 15–18.

Jorgenson, J. W., and Lukacs, K. D. (1981a). *Anal. Chem.* **53**, 1298–1302.

Jorgenson, J. W., and Lukacs, K. D. (1981b). *J. High Res. Chromatogr.* **4**, 230–231.

Karger, B. L. (1989). *Nature* **339**, 641–642.

Lal, K., Xu, L., Colburn, J., Hong, A. L., and Pollock, J. J. Accepted for publication, *Archives of Oral Biology.*

Liu, J., Cobb, K. A., and Novotny, M. (1990). *J. Chromatogr.* **519**, 189–197.

Loo, J. A., Jones, H. K., Udseth, H. R., and Smith, R. D. (1989). *J. Microcol. Sep.* **1**, 223–229.

Ludi, H., Gassman, E., Grossenbacher, H., and Marki, W. (1988). *Anal. Chim. Acta* **213**, 215–219.

McCormick, R. M. (1988). *Anal. Chem.* **60**, 2322–2328.

Moring, S. E. and Nolan, J. A. (1990). *In* "270A Research News Issue 2, Summer 1990." Applied Biosystems, Foster City, California.

Moseley, M. A., Deterding, L. J., Tomer, K. B., and Jorgenson, J. W. (1991). *Anal. Chem.* **63**, 109–114.

Moseley, M. A., Deterding, L. J., Tomer, K. B., and Jorgenson, J. W. (1989). *J. Chromatogr.* **480**, 197–209.

Mosher, R. A. (1990). *Electrophoresis* **11**, 765–769.

Nielsen, R. G., and Rickard, E. C. (1990). *J. Chromatogr.* **516**, 99–114.

Nielsen, R. G., Riggin, R. M., and Rickard, E. C. (1989a). *J. Chromatogr.* **480**, 393–401.

Nielsen, R. G., Sittampalam, G. S., and Rickard, E. C. (1989b). *Anal. Biochem.* **177**, 20–26.

Novotny, M. V., Cobb, K. A., and Liu, J. (1990). *Electrophoresis* **11**, 735–749.

Offord, R. E. (1966). *Nature* **211**, 591.

Olefirowicz, T. M., and Ewing, A. G. (1990). *J. Chromatogr.* **499**, 713–719.

Pennino, D. J. (1989). *BioPharm* September, 41–44.

Reinhoud, N. J., Schroder, E., Tjaden, U. R., Niessen, W. M. A., Ten Noever De Brauw, M. C., and Van Der Greef, J. (1990). *J. Chromatogr.* **516**, 147–155.

Satow, T., Machida, A., Funakushi, K., and Palmieri, R. (1990). Presented at the 12th International Symposium on Capillary Chromatography, Kobe, Japan, September 11–14, 1990.

Schlabach, T. D., Colburn, J. C., Mattaliano, R. J., and Yuen, S. (1989). *In* "Techniques in Protein Chemistry" (T. Hugli, ed.), pp. 497–505. Academic Press, San Diego, California.

Smith, R. D., Olivares, J. A., Nguyen, N. T., and Udseth, H. R. (1988). *Anal. Chem.* **60**, 436–441.

Smith, R. D., Loo, J. A., Barinaga, C. J., Edmonds, C. G., and Udseth, H. R. (1989). *J. Chromatogr.* **480**, 211–232.

Steuer, W., Grant, I., and Erni, F. (1990). *J. Chromatogr.* **507**, 125–140.

Stover, F. S., Haymore, B. L., and McBeath, R. J. (1989). *J. Chromatogr.* **470**, 241–250.

Strickland, M., and Strickland, N. (1990). *Am. Lab.* November, 60–65.

Takigiku, R., Keough, T., Lacey, M. P., and Schneider, R. E. (1990). *Rapid Commun. Mass Spectr.* **4**, 24–29.

Tran, A. D., Blanc, T., and Leopold, E. J. (1990). *J. Chromatogr.* **516**, 241–249.

Vinther, A., Bjorn, S. E., Sorensen, H. H., and Soeberg, H. (1990). *J. Chromatogr.* **516**, 175–184.

Wiktorowicz, J. E., and Colburn, J. C. (1990). *Electrophoresis* **11,** 769–773.

Yeung, E. S., and Kuhr, W. G. (1991). *Anal. Chem.* **63,** 275A–282A.

Yuen, S. W., Otteson, K. M., Colburn, J. C., Moore, W. T., Schlabach, T. D., Dupont, D. R., and Mattaliano, R. J. (1989). *In* "Techniques in Protein Chemistry" (T. Hugli, ed.), pp. 589–597. Academic Press, San Diego, California.

Zhu, M. D., Hansen, D. L., Burd, S., and Gannon, F. (1989). *J. Chromatogr.* **480,** 311–319.

Zhu, M. D., Rodriguez, R., Hansen, D. L., and Wehr, T. (1990). *J. Chromatogr.* **516,** 123–131.

Protein Analyses by Capillary Electrophoresis

John E. Wiktorowicz and Joel C. Colburn

I. Introduction

Because of their complexity of structure, heterogeneity of amino acid compositions, and diverse solution properties, proteins present a challenge to those who wish to separate them or gain insights into their structure. Their complexity has dictated that their separation and analysis require the application of many analytical principles, including spectroscopy, chromatography, electrophoresis, and ultracentrifugation. Each of these techniques exploits specific properties or combinations of properties of proteins, yielding structural information; some techniques also serve as separation tools. To a large degree, capillary electrophoresis (CE) is a hybrid, utilizing many of the principles that underlie these techniques, as well as contributing some unique analytical features of its own. We are just beginning to apply CE to the analysis of protein structure, and with the knowledge gained, better exploit its separation potential for proteins. This chapter will focus on the fundamental properties of proteins and peptides that influence electrophoretic separations, describe the important considerations in devising separation strategies using CE, and present examples of separation techniques and applications.

A. Charge, Size, and Shape

In the earliest forms of electrophoresis, charged molecules were separated in free solution by their different mobilities under the influence of an electrical field (Tiselius, 1937). Any effort designed to increase resolution by increasing field strength was plagued by the internal heating that occurred when moderate field strengths were applied. This heating generated cross-sectional temperature gradients that created ruinous convection currents, destroying the resolution obtained by these early attempts. In order to overcome these problems, anticonvective agents, such as starch gels, paper, or cellulose acetate, were used. These agents exerted essentially no effect on the migration of proteins. With the invention of polymers whose polymerization permitted rudimentary control over pore size, however, molecular sieving became possible. Thus, in addition to their historical role in limiting convection, gels effected separations by exerting molecular weight-dependent resistance to migration. In short, electrophoretic separations in a gel format are dependent on the charge and frictional coefficient of the analytes (see Chapter 5).

Because of the narrow bore of the capillary and therefore efficient heat dissipation, cross-sectional thermal gradients are considerably less problematic in CE. There is no need for anticonvective agents, other than for their sieving properties. Separations in free solution are once again possible, and the advantages of this format have been rediscovered. The basic principles of electrophoresis are nevertheless the same. Peptides, whose charge is strictly related to the ionization of their constituent amino acid side-chains and termini, can be modeled empirically by determining their mobility as a function of their *calculated* charge and chain length, which can be correlated to a frictional coefficient (Grossman *et al.*, 1989b). Thus, for simple analytes (such as short peptides), mobility is a predictable function of size and charge (Fig. 1).

For more complex peptides and proteins, i.e., those that exhibit significant three-dimensional folding, charge cannot be simply calculated from their sequence by assuming that their amino acid side-chain ionization constants (pK) are equivalent to those of free amino acids. The environment of the side-chains in the folded protein often causes shifts in pK values, depending on whether or not the residues are buried, engaged in hydrogen bonding, or associated in salt bridges with nearby residues. Charge, therefore, is a function of the influence of local shape, rather than simply free amino acid side-chain pK values. In addition, since the frictional coefficient no longer correlates to the molecular size, mobility of complex proteins and peptides is truly a function of *shape,* or tertiary structure, rather than simply of charge and molecular size.

Clearly, unless tertiary structure is taken into account, calculated charges will be in gross error. Unfortunately, detailed information regarding tertiary structure is lacking in all but a few selected proteins. For these, high-resolution X-ray structure reveals the local environment of specific residues in a hydrated crystal, and two-dimensional nuclear magnetic resonance (NMR) reveals the pK values of important residues under near-native conditions. Since deviation from the predicted behavior (derived from the amino acid sequence) may reflect this structure, analysis of electropherograms, generated under closely controlled conditions in free solution, may yield rules governing the

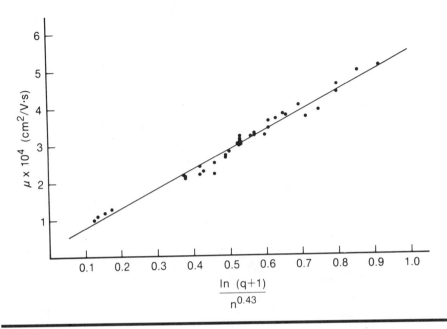

Figure 1 Relationship between the mobilities of 39 peptides in 20 mM sodium citrate, pH 2.5, calculated charge, and molecular size. Reprinted with permission from Grossman *et al.* (1989b).

folding behavior of proteins in solution without the need for highly pure preparations and may supplement X-ray data (Wiktorowicz *et al.*, 1991). Thus capillary electrophoresis is capable of yielding information of importance relating to the higher-order structure of proteins, in addition to generating separations of high efficiency and speed.

B. Problems Encountered in Capillary Electrophoresis: Detection, Control of Electroendosmosis, and Wall Interactions

Capillary electrophoresis also exhibits some unique difficulties (with respect to detection) that must be overcome before its full analytical potential in the study and separation of proteins and peptides is realized. The advantage of on-column monitoring is obvious, but it requires high-sensitivity detector strategies, since the short path length of the capillary makes detection of low concentrations difficult. Laser detection (see Kuhr, 1990) is clearly the choice for fluorescent molecules; however, most laboratories do not have the luxury of such equipment, nor the expertise to incorporate these devices in a home-built CE device. For most protein analyses, ultraviolet (UV) detection may be sufficient if precautions designed to optimize sensitivity are taken. For example, UV spectrophotometric detection of proteins and peptides at the highest sensitivity requires short wavelengths. Thus the material from which the capillary is fabricated must be compatible with this requirement, and so is critical to the sensitivity of detection.

Although many materials have been used (Mikkers *et al.*, 1979; Jorgenson and Lukacs, 1983), fused silica has achieved widespread use in many laboratories because of its reproducible physical dimensions, transparency to short wavelengths, high thermal conductivity and the vast experience of the analytical community with its chemical and physical manipulation. Whereas fused silica is UV transparent and therefore provides flexibility in the detection strategy, the ionization of silanols, created on hydration of the silica, presents complications for many CE separation strategies.

Two phenomena are associated with the negative surface charge acquired by the ionization of the silanols on hydration. The first, known as electroendosmosis, is the bulk movement of liquid through the capillary due to an applied electrical field (see Chapter 1 for a thorough discussion of electroendosmosis). The ionization of silanols results from a complex cascade of chemical equilibria that exhibit an average pK, or isoelectric point (pI), of around 1.5 to 2.0 (Wiese *et al.*, 1971). Thus, at pHs above this value, the fused-silica capillary wall is unprotonated and negatively charged. This immobile surface of negative charge density attracts cations from the aqueous mobile phase, establishing a gradient of positive charge, which is at its greatest level near the capillary surface. When voltage is applied, these cations move toward the cathode, entraining water and other neutral molecules. Electroendosmotic flow in this case is cathodic. If the capillary surface were positively charged, endosmotic flow would be anodic, since anions would be attracted to the surface and migrate toward the anode on application of voltage. Endosmotic flow, therefore, is a vectorial entity in which the magnitude and direction of the flow is dependent on the total charge on the capillary surface.

Since capillary surface charge is an ionic phenomenon, its magnitude is dependent on those parameters that control the degree of ionization. Thus above pH 1.5, endosmotic flow increases until ionization is complete, and flow reaches a maximal value; increasing temperature increases endosmotic flow by altering the pK of silanols as well as decreasing the viscosity of the mobile phase; increasing the ionic strength of the separation buffer shifts the silanol ionization equilibrium toward the unionized state, thereby decreasing endosmotic flow; finally, decreasing the voltage quite obviously decreases endosmotic flow. Endosmotic flow can also be controlled by a slow, continuous process in which ionized silanols are selectively neutralized by a polymeric solution, while endosmotic flow is monitored. When the desired value is reached, the process is stopped, the nonbound polymer is removed, and the endosmotic flow remains quite stable for up to 6 to 7 hr of continuous running (Wiktorowicz and Colburn, 1989). The control of electroendosmosis is discussed further in Chapter 1.

The second phenomenon associated with the negative capillary surface charge is the attraction of positively charged ions, including analytes. Indeed, it is this attraction that engenders endosmotic flow above pH 2. However, when the ions attracted are the analytes themselves, this "wall interaction" interferes significantly with the separation: peak broadening will occur, and at worst, complete inhibition of migration prohibits analysis. Because of this phenomenon, separation conditions must be chosen that minimize these coulombic wall interactions. Thus, efficient separations of proteins and peptides require buffer pH values above their pI, where their main- and side-chain carboxyls are largely deprotonated and are therefore repelled from the capillary wall (Lauer and McManigill, 1986; Fig. 2) (although regions exhibiting a local positive charge density may be attracted). Alternatively, separations at low pH are permitted because silanols are protonated, and thus less available for ionic interaction with cationic species. Figure 3 shows the separation of three batches of recombinant interleukin-2. These materials were run under low-pH conditions, in which wall interactions are minimized. Each batch gave a single, symmetrical peak by reverse-phase high-performance liquid chromatography (RP-HPLC); CE, however, demonstrated degrees of heterogeneity in each batch.

Although the number of options available for protein and peptide analysis in free-solution capillary electrophoresis is limited, this should be considered to be a temporary condition. High- and low-pH separation conditions may render proteins and peptides denatured, eliminating structural uniqueness that may facilitate separation. In addition to causing denaturation, low pH abolishes endosmotic flow, thereby eliminating an important separation component unique to CE. In addition, the requirement for pH extremes in the analysis of proteins precludes detailed higher-order structural anal-

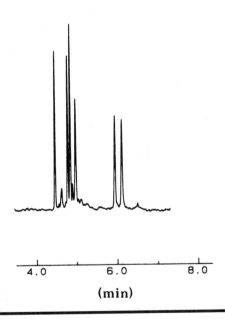

(min)

Figure 2   Analysis of proteins. Conditions: field, 297 V/cm; current 38 μA; buffer, 20 mM borate, pH 8.25; capillary length, 101 cm (55 cm to detector); capillary diameter, 52 μm. Protein migration (increasing times): whale skeletal myoglobin, horse heart myoglobin, carbonic anhydrase B, carbonic anhydrase A, β-lactoglobulin B, β-lactoglobulin A. Reprinted with permission from Lauer and McManigill (1986).

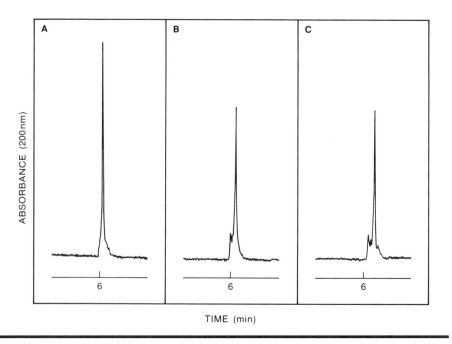

Figure 3 Electropherogram of three batches of recombinant interleukin-2. Conditions: field, 308 V/cm; current, 25 μA; buffer, 20 mM sodium citrate, pH 2.5; capillary length, 65 cm (45 cm to detector); capillary diameter, 50 μm; temperature, 30°C; detection, 200 nm. Reprinted with permission from Colburn *et al.* (1991).

yses of proteins, separations such as those of multimeric proteins held together by noncovalent interactions, the study of intermolecular protein : protein interactions such as antigen : antibody or receptor : ligand complexes, and postseparation, on-column functional assays of enzymes.

Another approach available for reducing wall interactions (which relies on masking the negative charge of the capillary surface by covalent modification with a neutral compound) is also effective; however, this strategy requires special expertise and equipment in order to achieve effective coatings, abolishes endosmotic flow, and suffers from an inherent instability of the coating at precisely the pH range generally desired for separation (Iler, 1979; Bruin *et al.*, 1989a). Coating agents such as methylcellulose (Hjertén, 1967, 1985), carbohydrates (Bruin *et al.*, 1989b; Fig. 4), arylpentafluoro moieties (Swedberg, 1990; Fig. 5), polyethylene glycol (Bruin *et al.*, 1989a; Fig. 6), and polyethyleneimine (Towns and Regnier, 1990) have been investigated; although these approaches have met with some success, the poor stability of the siloxane bond severely restricts their application. The polyethylene coating described by Bruin *et al.* (1989a) was stable only at pH values under 5, and wall interactions were observed. The aryl pentafluoro-coated capillary (Swedberg, 1990) did minimize wall interactions, and gave reasonable migration-time reproducibility values.

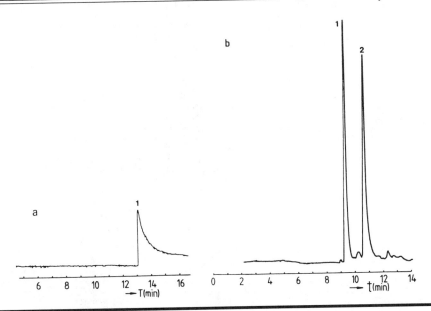

Figure 4 Separation of (a) lysozyme (1) on an aminopropyl-modified capillary and (b) of lysozyme (1) and cytochrome c (2) on a maltose-modified fused-silica capillary. Conditions: field, 253 V/cm; buffer, 50 mM phosphate, pH 6; capillary length, 39.5 cm (20 cm to detector); capillary diameter, 50 μm; temperature, 20–22°C; detection, 205 nm. Reprinted with permission from Bruin *et al.* (1989b).

Recently a coating strategy has appeared (Cobb *et al.*, 1990; Fig. 7) that suffers less from the instability of the -Si-O-Si-C- based coatings described above. This method achieves covalent bonding of polyacrylamide through the more stable -Si-C-bonds. Whereas all inhibit wall interactions of some proteins, coatings of neutral character do not prohibit proteins that contain uncharged surface domains from adsorbing to the capillary wall by virtue of other weak interactions. Only those that actively repel proteins have the greatest likelihood of generally permitting unretarded migration of proteins.

A simpler approach utilizes "dynamic" coating agents, which interact weakly with the capillary surface and or analytes in order to minimize adsorption (Lauer and McManigill, 1986; Bullock and Yuan, 1991). Various agents have been utilized to achieve this effect, including surfactants with various ionic properties (Towns and Regnier, 1991; Fig. 8). The common feature of all the agents is that the nature of their interaction with the wall requires that they be included in the separation buffer. Since the heart of this strategy is that these agents out-compete or displace the analytes from the reactive sites on the capillary wall, they must therefore be present at high-enough concentrations for them to be effective. In addition, this method suffers from the potential interaction of coating agent with analytes, altering the selectivity of the system and thus limiting the usefulness of the strategy. A subset of this approach utilizes rela-

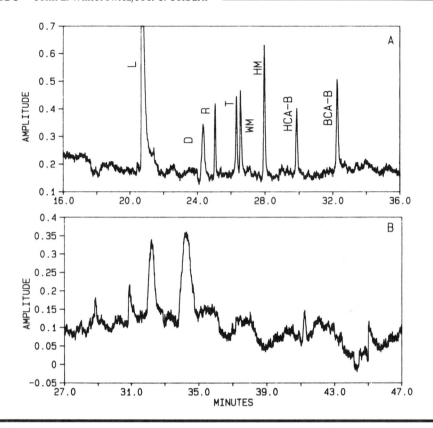

Figure 5   Electropherogram of seven proteins and dimethylsulfoxide (DMSO; D) on an aryl pentafluoro-treated capillary (A) and uncoated fused silica (B). Conditions: field, 250 V/cm; buffer, 200 mM phosphate, 100 mM KCl, pH 7; capillary length, 100 cm to detector; capillary diameter, 20 μm; detection, 219 nm. Proteins: (L) lysozyme, (R) ribonuclease, (T) trypsinogen, (WM) whale myoglobin, (HM) horse myoglobin, (HCA-B) human carbonic anhydrase B, (BCA-B) bovine carbonic anhydrase B. Reprinted with permission from Swedberg (1990).

tively high levels of salt and zwitterionic detergents to prohibit ionic interaction (Green and Jorgenson, 1989; Bushey and Jorgenson, 1989) and has been commercialized by the Waters Corporation. Whereas this strategy is effective in minimizing some wall interactions, the use of up to 1 M concentrations of detergents or salt in the separation buffer alters the selectivity of the system and increases the potential for denaturation of some proteins. In addition, this method results in higher currents and higher temperatures, requiring lower separation voltages and consequently longer separation times.

    A final approach, which combines aspects of both covalent and dynamic coating

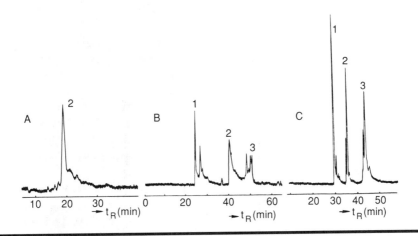

Figure 6 Separation of proteins (A) in untreated and (B and C) in polyethylene glycol-coated capillaries. Conditions: buffer, (A and B) 10 mM potassium phosphate, pH 6.8; (C) 50 mM potassium phosphate, pH 4.1. Reprinted with permission from Bruin *et al.* (1989a).

strategies, is called surface-charge reversal (Wiktorowicz and Colburn, 1989). This technique achieves separation of proteins and peptides below their pI values by stably reversing the charge on the capillary surface with noncovalent coating agents, which are not required to be present in the separation buffer (Wiktorowicz and Colburn, 1989; Wiktorowicz and Colburn, 1990). Although these agents are similar to the dynamic coating agents in that their interaction with the surface is mediated by a combination of ionic and hydrophobic interactions, this approach yields a coating sufficiently strong to permit analysis of proteins and peptides in the pH range of 2 to 10 without including the coating agent in the separation buffer. In addition, this strategy permits the removal of coating agent from the capillary surface, when desired, and reapplication, when desired.

Under the conditions of coating, at pH values below their pI values, proteins and peptides are repelled by the capillary wall, and separations are permitted with characteristically high efficiencies (see Fig. 9). The coating agent is applied by the user and is absent in the separation buffer; therefore, the requirement for special expertise and equipment — an inherent disadvantage of covalent coatings, and the potential for analyte interactions with free coating agent (an inherent disadvantage of the "dynamic" coating approach) — are avoided. This method allows the analysis of basic proteins below their pI values, e.g., at physiological pH. Although the method of Bullock and Yuan (1991) is also designed for basic proteins, that method uses a dynamic coating approach, which may suffer from the disadvantages previously discussed.

With capillary wall–protein interactions minimized, it is possible to address separation strategies in order to optimize protein separations. Buffer selection is lim-

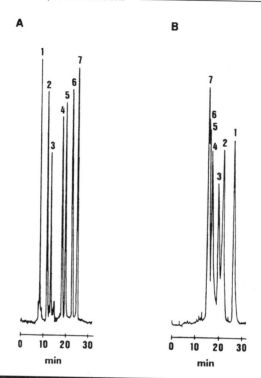

Figure 7 Electropherograms of proteins using (A) coated and (B) uncoated fused-silica capillaries. Conditions: field, (A) 333 V/cm, (B) 167 V/cm; current, (A) 15 μA, (B) 7 μA; buffer, 50 mM glutamine, triethylamine, pH 9.5; capillary length, 60 cm (45 cm to detector); capillary diameter, 50 μm; detection, 214 nm. Proteins: (1) porcine insulin chain A, (2) bovine serum albumin, (3) chicken egg ovalbumin, (4) porcine insulin, (5) bovine α-lactalbumin, (6) bovine β-casein, (7) porcine insulin chain B. Reprinted with permission from Cobb *et al.* (1990).

Figure 8 Electropherograms demonstrating the separation of five basic proteins on surfactant-coated capillaries. (A) Tween 20 (top) and (B) BRIJ 35 (bottom) alkylsilane capillaries. Conditions: field, 300 V/cm; buffer, 10 mM phosphate, pH 7; capillary length, 50 cm; capillary diameter, 75 μm; detection, 200 nm. Proteins: (1) lysozyme, (2) cytochrome *c*, (3) ribonuclease A, (4) α-chymotrypsinogen, (5) myoglobin. Reprinted with permission from Towns and Regnier (1991).

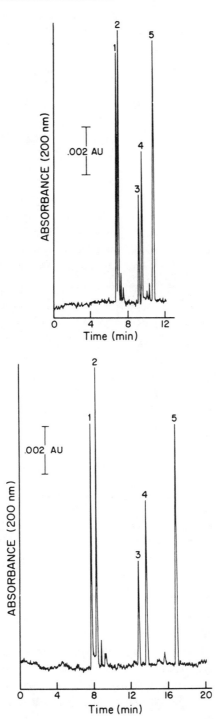

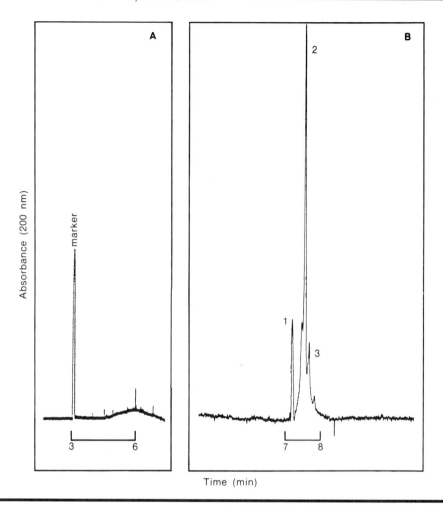

Figure 9 (A) Separation of lactate dehydrogenase (LDH) isoenzymes in an uncoated capillary. (B) Separation of LDH isoenzymes by capillary charge reversal. Conditions: field, 400 V/cm (reversed polarity in B); current, 4 μA; buffer, 5 mM sodium phosphate, pH 7.0; capillary length, 90 cm (68 cm to detector); capillary diameter, 50 μm; temperature, 30°C; detection, 200 nm. Peaks 1, 2, and 3 have pI values of 8.3, 8.4, and 8.55, respectively. Reprinted with permission from Wiktorowicz and Colburn (1990).

ited by the solubility and stability properties of the proteins to be analyzed, and the sensitivity of the detection system. Clearly, aromatic buffers may yield high backgrounds at short UV wavelengths, as well as organic buffers containing carbonyls or unsaturated carbon–carbon bonds. Thus buffers can only be selected on a case-by-case basis, dependent on the special needs of the separation strategy, detection method, and protein properties.

Thus, the interference due to wall interactions will plague those who pursue protein and peptide separations for the foreseeable future. Many strategies have been devised to minimize this difficulty, some more effective than others, but all at the mercy of the very property of proteins and peptides that makes them interesting: their complexity and variability.

II. Protein Analysis

A. Molecular-Weight Determination

Traditional electrophoretic analyses of proteins considered whether information sought could best be obtained under native or denaturing conditions. Analysis by denaturation examines the protein in its most elemental state: in the absence of native tertiary or quaternary structure, i.e., minus natural shape or subunit complexity. Thus basic properties, such as molecular weight or subunit composition, can be deduced by reference or by subtractive analysis, respectively. This is accomplished through denaturation by sodium dodecyl sulfate (SDS) in the presence of urea and a disulfide reducing agent, and molecular sieving *via* polyacrylamide gel electrophoresis (SDS-PAGE). The SDS achieves denaturation by disruption of the intramolecular hydrophobic interactions that govern three-dimensional structures, at the same time imparting the same unit charge density to proteins regardless of their primary structure. By equalizing the mass : charge ratio of all proteins, separation in a restrictive medium provided by the polyacrylamide is dependent on mass only.

Polyacrylamide of differing cross-linker concentrations has been polymerized and covalently bonded to the inside of a fused-silica capillary for standard SDS-PAGE analyses (Cohen and Karger, 1987; Hjertén *et al.*, 1987; see Chapter 5; see Fig. 10). In a capillary, SDS-PAGE takes advantage of the speed and mass sensitivity of the separation format, but suffers from the difficulty in obtaining reproducibly uniform gels. When acrylamide degrades, carboxylate anions are incorporated into the gel matrix. When voltage is applied to such a gel, these regions attempt to migrate toward the anode, pulling against their neutral neighbors, thus tearing the gel apart. This is accompanied by bubble formation, and breakdown of current. Thus polyacrylamide gels must be made from ultrapure acrylamide, and the polymerization process must minimize oxidative deamidation — a difficult goal, since the polymerization process itself relies on the generation of free radicals. Attractive alternatives may exist, but must await further development and significant scrutiny before they become truly viable. Until then, SDS-PAGE will remain the method of choice for molecular-weight determination of proteins.

B. Isoelectric Focusing

Isoelectric focusing (IEF) (see Chapter 7) achieves separation by forcing proteins to migrate through a stable pH gradient until they exhibit no net charge. At this point,

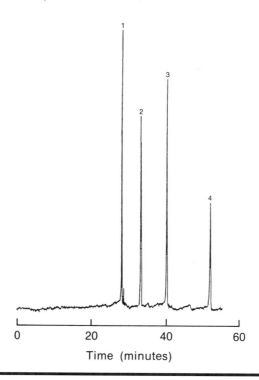

Figure 10   Separation of proteins by capillary SDS-PAGE. Conditions: field, 400 V/cm; current, 24 μA; buffer, 90 mM Tris-phosphate, pH 8.6, 8 M urea, 0.1% SDS; capillary diameter, 75 μm; temperature, 27°C; gel, 10% T, 3.3% C. Proteins and (molecular weight): (1) α-lactalbumin (14,200); (2) β-lactoglobulin; (3) trypsinogen; (4) pepsin (34,700). Reprinted with permission from Cohen and Karger (1987).

their isoelectric point, their mobility is zero. Any diffusion causes them to reacquire charge and move back into their isoelectric zone. Thus proteins are dynamically maintained, or "focused," into sharp bands. Originally performed in vertical devices through a density gradient in free solution (Vesterberg and Svensson, 1966), IEF has achieved wide acceptance as a separation alternative for complex mixtures. In combination with SDS-PAGE, IEF in gels forms the first dimension of a powerful two-dimensional separation strategy (O'Farrell, 1975).

Adaptation of IEF to the capillary format presents unique challenges. In the first place, because of on-line detection, endosmotic flow must be greatly diminished if not abolished. Once focusing is complete, some means must then be utilized to mobilize the gradient, without loss of resolution, so that focused peaks move past the detector. Hydrodynamic mobilization induces a parabolic cross-sectional flow profile, resulting in poor peak efficiencies and resolution. Thus, some means of electrokinetic mobilization must be used. Once these challenges are overcome, the persistent difficulty with unequal ampholine distribution over the analysis pH range must be considered. This problem not only limits the resolution of separation range, but also, since ampholines exhibit slight absorbance at 280 nm, high-sensitivity detection may result in unstable base lines.

Covalently coated capillaries effectively abolish endosmotic flow to permit focusing without loss of peaks. High-resolution mobilization schemes permit detection of peaks after focusing. These may consist of increasing the salt concentration of the anolyte or catholyte buffer (depending on the direction of the mobilization desired) after focusing (Hjertén et al., 1987). Similar results can be achieved by decreasing the pH of the catholyte buffer or increasing the pH of the anolyte after focusing (Hjertén and Zhu, 1985). No significant difference was observed as a result of either mobilization scheme. Because IEF is performed in the entire capillary, larger sample volumes may be analyzed than in free solution; however, because many proteins precipitate at their pIs, precautions should be taken (such as the addition of nonabsorptive, nonionic detergents, or urea to the ampholyte mixture) to minimize precipitation.

C. Purity Determination and Methods Development

Besides permitting the determination of structural properties of proteins, most separation techniques are typically utilized to gauge purity. The efficacy of these procedures revolves around their inherent resolution, selectivity, and sensitivity. Electrophoresis and chromatography have been traditionally used as complementary strategies in order to establish purity. Capillary electrophoresis adds to the electrophoretic armamentarium by contributing increased resolution, sensitivity, and speed to the analysis.

The single most effective strategy in demonstrating purity is one that examines the protein preparation in multiple "dimensions." Thus homogeneity as demonstrated by RP-HPLC is strengthened by homogeneity of the same preparation by ion-exchange (IEC) HPLC. Similarly, purity demonstrated on IEF is strengthened by purity demonstrated by SDS-PAGE. Alternatively, homogeneity may be demonstrated in native PAGE at several different pH values of overloaded gels. Each strategy has its own strengths and weaknesses. For example, RP-HPLC must be performed using different solvents for elution in order for homogeneity to be established, and IEC must achieve single peaks with different salt or pH gradients. At each of these steps, considerable investment with precious sample must be made, since recovery is never 100%. For the same reasons, multiple dimensions in slab gels is unattractive, in addition to the time-consuming nature of these analyses. In contrast to these techniques, multiple dimensions in CE can be performed simply, and quickly, without the loss of significant amounts of sample.

In the free-solution format, multiple dimensions may consist of different buffer pH values, different buffer compositions, or different additives that change the selectivity of the separations. All of these alternatives can be performed in a single capillary by simply flushing the appropriate solution through. Variously coated capillaries discussed previously permit the analysis of a protein preparation at several widely different pH values, preferably above and below the isoelectric point of the protein.

Additives that may alter the selectivity of the separation or alter the properties of the proteins include denaturing and nondenaturing surfactants, such as SDS and Triton X-100 respectively, salt, urea, etc. These treatments would normally be unacceptable, since the entire sample to be analyzed would be exposed to the agents. Capillary electrophoresis requires only miniscule amounts of material (pg by mass, nl by volume), and therefore is effectively nondestructive. Since the length of time from sample injection to a final result is often less than 30 min, appropriate buffers and additives may be screened quickly for efficacy. Clearly, whereas CE as a purity evaluation strategy may meet with initial resistance from those who fear greater scrutiny, the greater prestige that comes from having purified a protein judged to be homogeneous by CE, and the certainty that no other method short of sequencing is quite so convincing, will soon overcome the initial resistance.

D. Peptide Mapping

Also known as protein fingerprinting, peptide mapping (also see Chapter 9) has traditionally been used as a technique to verify the identity of a protein. Since each protein is unique in structure, if proteins vary in the number of proteolytic cleavage sites, the spacing of those sites, and the amino acid composition of the peptides generated by the cleavage, then the separation of the resultant peptides may serve as a signature, or fingerprint, of the intact protein. Clearly, the more "dimensions" that can be brought to bear in the separation scheme, the more accurate the pronouncement of identity. Historically, peptide mapping consisted of two-dimensional analysis: electrophoresis (mass : charge ratio) in the first, followed by thin-layer chromatography (TLC) (partitioning coefficient) in the second. Currently, peptide mapping is loosely defined but most often consists of proteolytic digestion followed by RP-HPLC.

As electrophoresis (on paper or microcrystalline cellulose) served as the first dimension of traditional mapping, it can also accomplish mapping in the capillary format. Since separations by CE are characterized by high-efficiency peaks with high peak capacities, mapping by CE is particularly appropriate. An added feature of on-line detection, low sample utilization, reproducible migration time, and rapid analysis, is the ability to rerun digests and monitor at different wavelengths. Should certain peptides exhibit characteristic absorption ratios (e.g., 200 : 280) owing to the presence of tryptophan, tyrosine, or phenylalanine, identification of peak(s) may be made on this basis. In addition, modification of specific residues with reagents that impart unique absorption spectra to peptides containing modifiable residues [such as the modification of sulfhydryls with 4-vinylpyridine (Cavins and Friedman, 1970), which absorbs at 254 nm] may permit peak identification. In general, however, since the chemical nature of the peptides that are generated may in large part be unknown, peptide mapping in CE is normally done at the pH extremes in order to avoid wall interactions. The exquisite sensitivity of CE to the overall charge characteristics of the analytes makes peptide mapping by CE also the method of choice for the determination of artifactual or postsynthetic modifications of protein structure, such as deamidation (Nielsen *et al.,* 1989; Wu *et al.,* 1990), sulfhydryl oxidation/reduction (Grossman *et al.,* 1989a), acetylation (Wiktorowicz and Colburn, 1990), and phosphorylation (Meyer *et al.*, 1990). The rapidity of analysis also suggests that CE can provide valuable insights into the monitoring of processes in the manufacture and purification of biologicals.

E. Native Structural Analysis

Proteins exhibit complex higher-order structures that derive from the interaction of their constituent amino acids with each other in aqueous solution. The structures they generate play a major role in the expression of protein function. The great challenge to protein chemists is comprehension of the interplay among the protein structural hierarchies, which will permit an understanding of the relationship between protein structure and function. An appreciation of this relationship is the first step in controlling function by creating or manipulating structure, and predicting structure from function.

In addition to the many tools available for the analysis of protein structure, CE contributes speed, accuracy, and high resolving power. Beyond the ability to simply separate closely migrating proteins, capillary electrophoresis is capable of generating data from which structural information can be extracted.

An example of this is shown in Fig. 11. Three site-directed mutants and the wild-type form of *Aspergillus oryzae* ribonuclease T1 are shown separated at pH 7.0. The difference in charge predicted from the primary sequence between mutants I (Gln-25 $\rightarrow$ Lys) and II (Glu-58 $\rightarrow$ Ala), and the wild-type enzyme (WT) is $+1$, indicating identical charge : mass ratios of the two single mutants. Because these are acidic proteins (pI $= 3-4$), at pH values above 4, WT RNase T1 exhibits electrophoretic mobility away from the detector (toward the positive electrode, the anode). At pH 7, the endosmotic flow is toward the detector (the negative electrode, the cathode) and overwhelms its countercurrent migration, sweeping it past the detector. Since the Lys-25 substitution in mutant I changes the net charge by $+1$ (Fig. 11 inset), its mobility toward the positive electrode (anode) is lower, and is therefore separated from the WT enzyme. Since the Ala-58 mutation in mutant II similarly changes the net charge by $+1$, its mobility will also be less than that of the WT. Mutant III, containing both mutations, results in a net charge difference from WT of $+2$, exhibits the least mobility of all the isoforms, and is the first to appear in the electropherogram.

Despite the same net charges predicted from their amino acid sequences, mutants I and II are completely resolved in their native conformations by CE. This analysis suggests that the substituted residues in the isoforms exist in different ionic environments. Indeed, Gln-25 in the wild-type enzyme exhibits near 70% solvent accessibility, whereas Glu-58 exhibits less than 5% (Sugio *et al.*, 1988). These mutations induce few if any changes in the tertiary structure of the isoenzymes since studies have shown that (1) the kinetic properties of mutant I do not significantly depart from those of the wild type, and (2) although the specific activities of mutant II and III are drastically reduced ($\sim$ 7% of WT), this loss is largely the result of the involvement of Glu-58 in catalysis (k_{cat}) but not binding (K_m) of substrate (Shirley *et al.*, 1989). The solvation of the substitutions in the mutants, therefore, reflect those of the parent molecule (WT); that is, the Lys-25 mutation in mutants I and III exhibits greater sol-

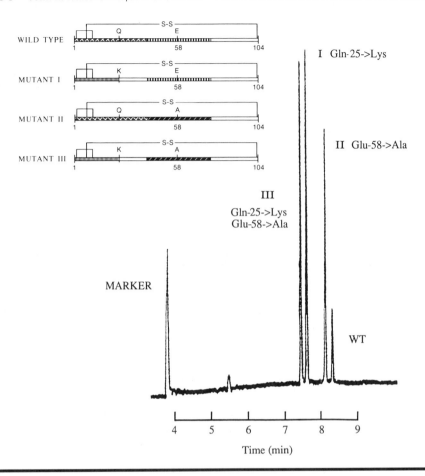

Figure 11 Electropherogram of four isoforms of RNase T1. Inset: Schematic of isoforms with substitutions indicated. Conditions: field, 430 V/cm; current, 32 μA; buffer, 25 mM sodium phosphate, pH 7.0; capillary length, 90 cm (68 cm to detector); capillary diameter, 50 μm; temperature, 30°C; detection, 200 nm. Reprinted with permission from Wiktorowicz *et al.* (1991).

vation than the Ala-58 mutation in mutants II and III (70 versus 5%, respectively). In addition, His-40 in WT exhibits a pK of about 7.8, elevated from the norm (p$K \sim 6.6$) by the proximity of Glu-58. Since mutant I contains Glu-58, it experiences the near full charge contribution of His-40 as well as Lys-25 (net charge difference $\sim +1.8$). In the absence of Glu-58, however, the pK of His-40 is depressed to about 6.5 to 7.1 (McNutt *et al.*, 1990; Steyaert *et al.*, 1990). This results in a net loss in charge contribution of +0.3 (from +0.8 to +0.5) in mutant II, in addition to the loss of the charge imparted by the substitution of Glu-58 (net charge difference $\sim +1.5$). Qualitatively, the mutants behave consistent with this analysis. However, the large difference in mobility between mutants I and II indicates that other factors are operative. One possibility is that whereas Glu-58 contributes a full -1 charge to WT at pH 7, its contribution is minimized by the shielding from the field it experiences by its hydrophobic environment ($<5\%$ solvent accessibility). Thus, substitution by Ala makes little difference in its electrophoretic behavior. Proof of the involvement of the tertiary structure of RNase T1 in the separation shown in Fig. 11 is presented in Fig. 12. Thermal denatu-

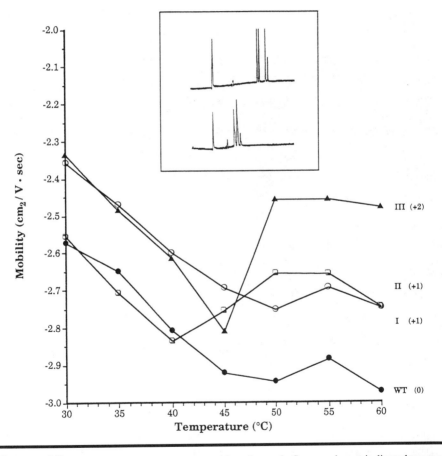

Figure 12 Electropherograms were run with endosmotic flow marker at indicated temperatures—in 20 mM TES/5 mM NaCl, pH 7.0—and mobilities were calculated and plotted. At denaturation, single mutant mobilities were identical, as predicted from primary sequence. Differences between WT, single-mutant, and double-mutant mobilities at 60°C are identical, also as predicted from primary sequence and comparison of charges at pH 7.0. Reprinted with permission from Wiktorowicz *et al.* (1991).

ration abolishes the difference in mobility of the single mutants, since in the absence of tertiary structure, they exhibit identical mass : charge ratios. In addition, since in their denatured states WT, mutants I and II, and mutant III differ by $+1$, their mobility differences are identical.

Thus without prior knowledge of the three-dimensional structure, but knowing primary structure, amino acid substitutions, and kinetic effects, it should be possible to predict the structural differences of the mutants. Continued study of well-characterized protein systems should expand our understanding of the general rules that relate protein behavior in high electrical fields (in free solution, at nondenaturing pH values) to higher-order structure.

III. Applications

Selected applications of CE for protein analysis will be discussed. Although many such applications have been reported, this section is not meant to represent a comprehensive review, but rather, the examples are chosen to illustrate some particular feature or capability of CE.

A. Human Growth Hormone

Human growth hormone (hGH) is a single polypeptide chain of 191 amino acids, with two intrachain disulfide bonds, a molecular mass of 22,250 daltons (d), and a pI of 5.2. The biosynthetic form and its desamido derivatives were analyzed using CE by Grossman *et al.* (1989a) and Nielsen *et al.* (1989). The CE separation buffer conditions used were 0.01 M tricine, 0.0058 M morpholine, and 0.02 M NaCl, adjusted to pH 8. The pH chosen was above the pI of the protein (5.2) in order to minimize wall interactions. Figure 13 shows the separation of hGH, desamido-149 hGH, and didesamido-149–152 hGH. With respect to the parent hGH, the net charge is -1 for desamido-149 and -2 for didesamido-149–152, owing to the replacement of the neutral amide group with a charged carboxylate. The derivative sulfoxide-14 hGH could not be separated from hGH under these conditions, but this was expected because of the lack of any charge difference between species.

Whereas CE has most often been used as a tool in basic research, its application as a rapid, automated method for quality control of proteins was reported by Nielsen and Rickard (1990). Any quality control method must provide excellent performance characteristics to correctly determine purity. Method validation schemes include the assessment of linearity, precision, and sensitivity. For hGH and the desamido-149 derivative, CE was concluded to be comparable to RP-HPLC for linearity and precision, and the sensitivity was adequate for monitoring minor impurities.

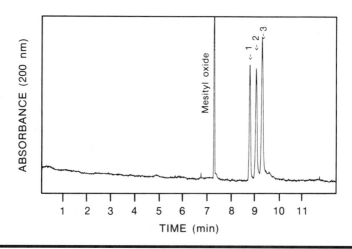

Figure 13 Separation of hGH and its derivatives. Conditions: field, 300 V/cm; current, 20 μA; buffer, 10 mM Tricine, 5.8 mM morpholine, 20 mM NaCl, pH 8.0; capillary length, 105 cm (81.5 cm to detector); capillary diameter, 50 μm; temperature, 24°C; detection, 200 nm. Proteins: (1) hGH; (2) [desamido-149] and [desamido-152] hGH; (3) [didesamido-149-152] hGH. Reprinted with permission from Grossman *et al.* (1989a).

B. Antibody Complexes

Nielsen *et al.* (1991) studied monoclonal antibody–antigen complexation with CE with hGH as the antigen. An IgG monoclonal antibody specific for hGH was combined with hGH in various ratios. Two types of complexes were expected: IgG–hGH and IgG–(hGH)$_2$, corresponding to the two antigen-binding sites on the monoclonal antibody. Capillary electrophoresis was performed in a 0.1 M tricine, pH 8 buffer, and hGH and the IgG antibody are well resolved under these conditions (Fig. 14). The IgG (pI = 7–8) has a lower negative charge at this pH than hGH (pI = 5.2) and therefore migrates more rapidly. The peak at 531 sec in electropherograms E3 and E4 was a new species assigned to IgG–(hGH)$_n$ complexes. In electropherogram E5, two peaks designated as complexes were found at 513 and 531 sec. The peak at 531 sec decreased as the molar ratio of IgG : hGH changed from 1 : 1 to 3 : 1. The peak at 531 sec was probably the IgG–(hGH)$_2$ complex, whereas the peak at 513 was the IgG–hGH complex. The complexes migrate more slowly than the IgG and closer to hGH, and it is on the basis of stoichiometry and this migration pattern that the two peaks representing complexes were assigned. The causes of the relatively broad peaks observed for the antibody and the complexes were presumed to be multiply-charged IgG species with different electrophoretic mobilities. However, all species were better resolved by CE than by high-performance size-exclusion chromatography, and the analysis was considerably faster. Another advantage of CE the authors noted was the absence of

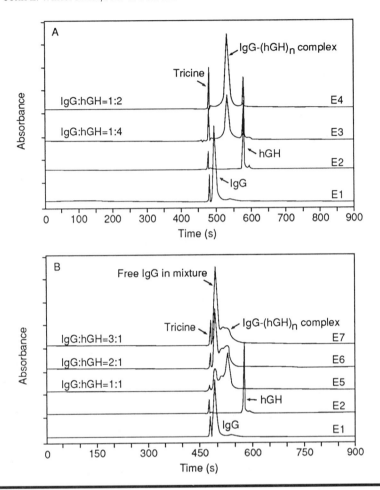

Figure 14   Separation of IgG, hGH, and IgG–(hGH)$_n$ complexes: (A) complexes formed from excess hGH, (B) complexes formed from excess IgG. Conditions: field, 300 V/cm; current, 19 μA; buffer, 100 mM tricine, pH 8; capillary length, 100 cm (80 cm to detector); capillary diameter, 50 μm; temperature, 30°C; detection, 200 nm. Reprinted with permission from Nielsen *et al.* (1991).

any supporting media, which might have enhanced the dissociation of the noncovalent complexes during the separation process.

C. Erythropoietin

Recombinant human erythropoietin (r-HuEPO) is a 34,000–38,000 dalton protein with 165 amino acids and a pI of 4.5 to 5.0. Its extensive carbohydrate structure makes up about 40% of its weight. Tran *et al.* (1991) used free-solution CE to separate the various glycoforms of r-HuEPO. The authors proposed that resolution could be improved by increasing the charge differences between the glycoforms, or by decreasing the electroosmotic flow. Thus, for those analyses at pH values from 6 to 9, the best results were obtained at pH 6, where electroosmosis is the lowest among these conditions (see Fig. 15).

In order to reduce electroosmosis, organic modifiers were used, since they can increase buffer viscosity and reduce buffer dielectric constant. When methanol or ethylene glycol was used, the separation of glycoforms improved (Fig. 16). Low pH should also reduce electroosmosis, so that the optimal conditions utilized a 100 mM acetate-phosphate buffer, pH 4.0, with 10 hr of preequilibration time. Phosphate buffers worked very well, presumably because of the ability of phosphate to bind to silanol groups on the capillary surface to reduce charge and the resulting electroosmosis.

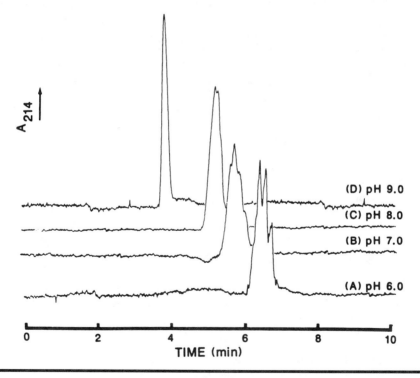

Figure 15 Analysis of r-HuEPO. Conditions: field, 439 V/cm; current, 44 μA (A), 15 μA (B), 70 μA (C), 85 μA (D); buffer, 50 mM MES, pH 6.0 (A), 50 mM Bis-Tris, pH 7.0 (B), 50 mM tricine, pH 8.0 (C), 50 mM tricine, pH 9.0 (D); capillary length, 57 cm (50 cm to detector); capillary diameter, 75 μm; temperature, 25°C; detection, 214 nm. Reprinted with permission from Tran *et al.* (1991).

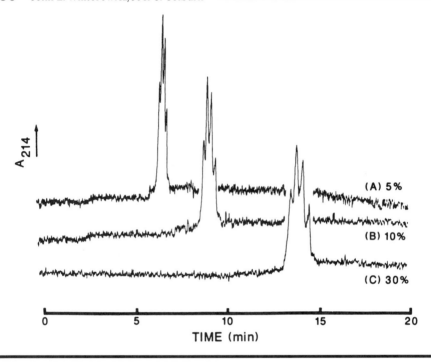

Figure 16   Analysis of r-HuEPO. Conditions: field, 439 V/cm; current, 27 μA (A), 21 μA (B), 10 μA (C); buffer, 50 mM tricine, pH 8.0 with 5% (A), 10% (B), and 30% (C) ethylene glycol; capillary length, 57 cm (50 cm to detector); capillary diameter, 75 μm; temperature, 25°C; detection, 214 nm. Reprinted with permission from Tran *et al.* (1991).

D. Membrane Proteins

Hydrophobic membrane proteins represent a class of proteins that often presents handling difficulties because of their tendency to precipitate in aqueous solutions; detergents are typically used to prevent this problem. However, in CE experiments, Josic *et al.* (1990) showed that the detergent in the sample solution was actually separated during electrophoresis, leading to protein precipitation. Addition of 7 M urea to the separation buffer prevented precipitation and resulted in reproducible migration-time and peak-height values.

E. Fraction Collection

Although the very small sample requirement of CE is usually thought to be an advantage, this can be a drawback if new peaks are obtained from the electropherogram without the ability to identify those peaks. Hecht *et al.* (1989a,b) analyzed recombinant interleukin-3 by CE, and demonstrated the collection of sufficient material from individual peaks for sequence analysis. This collection procedure, although not at present completely universal for all CE peaks, is being used by many researchers as a micropreparative method for identification by sequencing or mass analysis.

An instrument for fraction collection was described by Rose and Jorgenson (1988). These collection experiments allowed the determination of electrophoresis effects on activity. Heating may affect the activity of proteins, and wall interactions may denature the protein, also resulting in loss of activity. When the activity of the collected protein was compared to that of the control, the activities were found to be nearly equal, indicating the gentle nature of the CE electrophoresis process.

F. Physical Parameters

Capillary electrophoresis has been used to determine the effective size and charge of proteins. Walbroehl and Jorgenson (1989) calculated the electrophoretic mobility and diffusion coefficient for sperm whale myoglobin. Diffusion coefficients can be measured from peak widths, assuming that diffusion is the only source of band broadening; knowing the diffusion coefficient allows the calculation of effective size, and knowing both the electrophoretic mobility and diffusion permits the calculation of the effective charge of the protein. The authors used KCl in the separation buffer to minimize wall interactions, but the voltage was kept very low (2.5 kV) to prevent band broadening due to thermal effects. The results indicated that CE gave values for the diffusion coefficient identical to those obtained by other methods. The subsequent calculation of size agreed with crystallographic data.

Myoglobin was also studied with respect to column temperature by Rush *et al.* (1991). Whereas the protein gave one peak at 20°C, another peak appeared at higher temperatures until 50°C, where only the second peak was present. A conformational change was shown not to be the cause of the new peak. The data were demonstrated to be consistent with the reduction of Fe^{3+} to Fe^{2+} in the heme portion of the protein. Based on charge differences, the more positive ferric form migrated faster (Fig. 17). Column temperature was concluded to affect the rate of the reduction, and first-order kinetics were observed.

IV. Future Directions

It is entirely likely that all analytical procedures that rely on electrophoresis for separation will eventually find their counterpart in CE. When the difficulties brought about by wall interactions are overcome, the advantages of protein analysis by CE will become compelling. This is not to say that these competitors will no longer find use; on the contrary, these separation formats will become complementary, and as such, they will always provide useful information.

The future of protein analysis by CE will depend on its wide acceptance by the protein analytical community. Assuming such acceptance, CE technology will be eventually used for every aspect of structural analysis, ranging from amino acid composition and sequence to subunit composition and structure, i.e., from primary to qua-

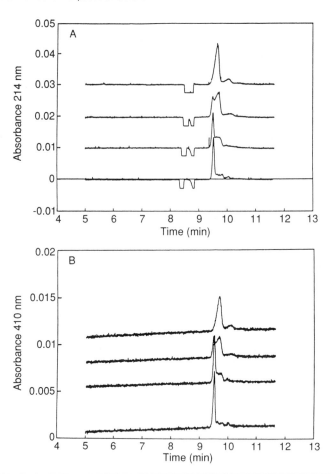

Figure 17 (A) Effect of temperature on the analysis of horse heart myoglobin. Conditions: field, 262, 213, 178, and 153 V/cm (ascending direction); current, 9.8 μA; buffer, 100 mM Tris, 25 mM boric acid, pH 8.6; capillary length, 57 cm (50 cm to detector); capillary diameter, 75 μm; temperature, 20°C, 30°C, 40°C, 50°C (ascending direction); detection, 214 nm. (B) Same conditions as in (A) but detection at 410 nm. Reprinted with permission from Rush *et al.,* 1991.

ternary structural analysis. Peak identification may pose some problems, although fraction collection and CE-MS hyphenation methods (for a review, see Kuhr, 1990) will continue to improve. With the marriage of high detection sensitivity and postseparation reactors (Pentoney *et al.,* 1988; Nickerson and Jorgenson, 1989), selectivity in detection will permit identification of peaks.

From the perspective of chemistry, functionalized capillary coatings may permit exquisite control of selectivity, completing the circle with chromatography. Efforts in this arena, called capillary electrochromatography, have been ongoing for years and hold a bright future. By permitting analysis of proteins at any pH, capillary coatings — covalent, dynamic, or charge reversal — foreshadow functional assays of proteins. Immunoassays, enzymatic assays, or native structure analysis require the ability to separate at native, nondenaturing pH values. By expanding the range of conditions for analysis, capillary coatings and buffer chemistries hold the future for capillary electrophoresis.

References

Bruin, G., Chang, J., Kuhlman, R., Zegers, K., Kraak, J., and Poppe, H. J. (1989a). *J. Chromatogr.* **471**, 429–436.

Bruin, G., Huisden, R., Kraak, J., and Poppe, H. J. (1989b). *J. Chromatogr.* **480**, 339–350.

Bullock, J. A., and Yuan, L-C. (1991). *J. Microcol. Sep.* **3**, 241–248.

Bushey, M. M., and Jorgenson, J. W. (1989). *J. Chromatogr.* **480**, 301–310.

Cavins, J. F., and Friedman, M. (1970). *Anal. Biochem.* **35**, 489–493.

Cobb, K. A., Dolnik, V., and Novotny, M. (1990). *Anal. Chem.* **62**, 2478–2483.

Cohen, A. S., and Karger, B. L. (1987). *Chromatographia* **397**, 409–417.

Colburn, J. C., Grossman, P. D., Moring, S. E., and Lauer, H. H. (1991). In "HPLC of Peptides and Proteins: Separation, Analysis, and Conformation" (C. T. Meat and R. S. Hodges, eds.), pp. 895–901. CRC Press, Boca Raton, FL.

Green, J. S., and Jorgenson, J. W. (1989). *J. Chromatogr.* **478**, 63–70.

Grossman, P. D., Colburn, J. C., Lauer, H. H., Nielsen, R. G., Riggin, R. M., Sittampalam, G. S., and Rickard, E. C. (1989a). *Anal. Chem.* **61**, 1186–1194.

Grossman, P. D., Colburn, J. C., and Lauer, H. H. (1989b). *Anal. Biochem.* **179**, 28–33.

Hecht, R. I., Morris, J. C., Stover, F. S., Fossey, L., and Demarest, C. (1989a). *Prep. Biochem.* **19**, 201–207.

Hecht, R. I., Coleman, J. F., Morris, J. C., Stover, F. S., and Demarest, C. (1989b). *Prep. Biochem.* **19**, 363–366.

Hjertén, S. (1967). *Chromatog. Rev.* **9**, 122–219.

Hjertén, S. (1985). *Chromatographia* **347**, 191–198.

Hjertén, S., and Zhu, M. (1985). *J. Chromatogr.* **346**, 265–270.

Hjertén, S., Elenbring, K., Kilar, F., and Liao, J-L. (1987). *J. Chromatogr.* **403**, 47–61.

Iler, R. K. (1979). *In* "The Chemistry of Silica." Wiley, New York.

Jorgenson, J. W., and Lukacs, K. D. (1983). *Science* **222**, 266–272.

Josic, D., Zeilinger, K., Reutter, W., Bottcher, A., and Schmitz, G. (1990). *J. Chromatogr.* **516**, 89–98.

Kuhr, W. G. (1990). *Anal. Chem.* **62**, 403R–414R.

Lauer, H. H., and McManigill, D. (1986). *Anal. Chem.* **58**, 166–170.

McNutt, M., Mullins, L. S., Raushel, F. M., and Pace, C. N. (1990). *Biochemistry* **29**, 7572–7576.

Meyer, H. E., Hoffmann-Posorske, E., Korte, H., Donella-Deana, A., Brunati, A. M., Pinna, L. A., Coull, J., Perich, J., Valerio, R. M., and Johns, R. B. (1990). *Chromatographia* **30**, 691–695.

Mikkers, F., Everaerts, F., and Verheggen, T. J. (1979). *Chromatographia* **169**, 11–20.

Nickerson, B., and Jorgenson, J. (1989). *J. Chromatogr.* **480**, 157–168.

Nielsen, R. G., Sittampalam, G. S., and Rickard, E. C. (1989). *Anal. Biochem.* **177**, 20–26.

Nielsen, R. G., Rickard, E. C., Santa, P. F., Sharknas, D. A., and Sittampalam, G. S. (1991). *J. Chromatogr.* **539**, 177–185.

Nielsen, R. G., and Rickard, E. C. (1990). *ACS Symp. Ser.* **434**, 36–49.

O'Farrell, P. H. (1975). *J. Biol. Chem.* **250**, 4007.

Pentoney, S., Huang, X., Burgi, D., and Zare, R. (1988). *Anal. Chem.* **60**, 2625–2630.

Rose, D. J., and Jorgenson, J. W. (1988). *J. Chromatogr.* **438**, 23–34.

Rush, R. S., Cohen, A. S., and Karger, B. L. (1991). *Anal. Chem.* **63,** 1346–1350.

Shirley, B. A., Stanssens, P., Steyaert, J., and Pace, C. N. (1989). *J. Biol. Chem.* **264,** 11621–11625.

Steyaert, J., Hallenga, K., Wyns, L., and Stanssens, P. (1990). *Biochemistry* **29,** 9064–9072.

Sugio, S., Amisaki, T., Ohishi, H., and Tomita, K. (1988). *J. Biochem. (Tokyo)* **103,** 354–366.

Swedberg, S. A. (1990). *Anal. Biochem.* **185,** 51–56.

Tiselius, A. (1937). *Trans. Faraday Soc.* **33,** 524.

Towns, J. K., and Regnier, F. E. (1990). *J. Chromatogr.* **516,** 69–78.

Towns, J. K., and Regnier, F. E. (1991). *Anal. Chem.* **63,** 1126–1132.

Tran, A. D., Park, S., Lisi, P. J., Huynh, O. T., Ryall, R. R., and Lane, P. A. (1991). *J. Chromatogr.* **542,** 459–471.

Vesterberg, O., and Svensson, H. (1966). *Acta Chem. Scand.* **20,** 820.

Walbroehl, Y., and Jorgenson, J. W. (1989). *J. Microcol. Sep.* **1,** 41–45.

Wiese, G. R., James, R. O., and Healy, T. W. (1971). *Disc. Faraday Soc.* **52,** 302–305.

Wiktorowicz, J. E., and Colburn, J. C. (1989). Presented at the *Ninth International Symposium of Proteins, Peptides, and Polynucleotides,* Paper #704.

Wiktorowicz, J. E., and Colburn, J. C. (1990). *Electrophoresis* **11,** 769–773.

Wiktorowicz, J. E., Wilson, K. J., and Shirley, B. A. (1991). *In* "Techniques in Protein Chemistry II" (J. Villafranca, ed.), pp. 325–333. Academic Press, New York.

Wu, S-L., Teshima, G., Cacia, J., and Hancock, W. S. (1990). *J. Chromatogr.* **516,** 115–122.

Separation of Small Molecules by High-Performance Capillary Electrophoresis

Charles W. Demarest,
Elizabeth A. Monnot-Chase,
James Jiu,
Robert Weinberger

I. Introduction

High-performance capillary electrophoresis (HPCE) consists of a family of related techniques that can be applied to the separation of both large and small molecules. For the separation of small ions, small molecules, proteins, DNA, bacteria and nano-particle viruses, HPCE has perhaps the greatest molecular weight dynamic range of all the known separation techniques.

Small-molecule separations are a specialized subclass of HPCE. Among the techniques suitable for these separations are capillary zone electrophoresis (CZE), electrokinetic capillary chromatography (ECC), and isotachophoresis (ITP). By proper selection of the mode of electrophoresis, anionic, cationic, and neutral small molecules can be separated, often during the course of one run. Some of the applications that have been reported include inorganic cations (Gross and Yeung, 1990), amino acids (Otsuka *et al.*, 1985), carbohydrates (Garner and Yeung, 1990), catecholamines (Wallingford and Ewing, 1988), nucleic acids (Row *et al.*, 1987), pharmaceuticals (Nishi *et al.*, 1990d), and vitamins (Fujiwara *et al.*, 1988). Quantitative analysis of small molecules from various matrices, such as aqueous solutions, pharmaceutical formulations (Nishi *et al.*, 1990d), plasma (Nishi *et al.*, 1990b), and urine (Weinberger *et al.*, 1990) have been reported as well.

This chapter discusses how the HPCE separation modes of CZE, ECC, and ITP are utilized for the analysis of small molecules. In addition, some of the quantitative aspects of HPCE, including method validation, along with selected applications for small molecules, will be described.

In the large molecule arena, the advantages of HPCE are compelling. In particular, the low surface area of the capillary wall (as opposed to the high surface area of a column packing) permits facile separations with high recoveries. Whereas gel techniques such as sodium dodecyl sulfate-polyacrylamide gel electrophoresis (SDS-PAGE) and isoelectric focusing work well in the slab-gel format, the automation aspects of HPCE are an attractive alternative. In addition, the ability to perform CZE provides numerous separation mechanisms that cannot be achieved in the slab-gel format.

The fundamental equation for the number of theoretical plates in capillary electrophoresis is

$$N = \frac{\mu_{ep}V}{2D_m} \tag{1}$$

where N = the number of theoretical plates, μ_{ep} is the electrophoretic mobility, V is the voltage and D_m is the diffusion coefficient. This equation approximates the plate count when molecular diffusion is the limiting factor. Refer to Chapter 1 for more details.

The relationship between the diffusion coefficient and molecular weight in liquids is given in Table I. Actual values of D_m for large and small molecules are given in Table II. For the most part, the difference in the diffusion coefficients between large and small molecules is about 10- to 100-fold. Plugging some numbers into the plate-count equation for horse heart myoglobin, where μ_{ep} = 0.65 × 10^{-4} cm²/V · sec and D_m = 1 × 10^{-6} cm²/sec at 30 kV, yields 975,000 theoretical plates.

For small molecules, μ_{ep} is relatively large. This can compensate in part for the high D_m, since the separations can be performed rapidly. The mobility of quinine sulfate is 4 × 10^{-4} cm²/V · sec (Altria and Simpson, 1987). If D_m is 0.7 × 10^{-5} cm²/sec (a reasonable assumption), then at 30 kV, Eq. (1) solves at 857,000 theoretical plates. From this simple analysis, capillary electrophoresis, at least theoretically, can yield very high efficiencies for both small and large molecules.

Whereas this explanation establishes a framework for employing HPCE for small molecules, mature techniques like high-performance liquid chromatography (HPLC) have provided adequate and sometimes spectacular solutions to complex separations problems for years. Because HPCE is based on different physicochemical

Table I Diffusion Constants in Liquids and Molecular Diameters of Nonelectrolytes[a]

Molecular weight	Diffusivity (cm²/sec × 10⁵)	Molecular diameter (Å)
10	2.20	2.9
100	0.70	6.2
1,000	0.25	13.2
10,000	0.11	28.5
100,000	0.05	62.0
1,000,000	0.025	132.0

[a] Data from Karger *et al.* (1973).

principles from those of HPLC, it is expected that HPCE will address areas in which HPLC has shortcomings. One such area is time of separation. With its ultrahigh efficiency, HPCE is expected to give substantial improvement in speed. It may be best utilized as an alternative to HPLC for small molecules when (1) the separation is complex and the peak capacity of the HPLC separation is limiting; (2) complex mobile phases or gradient elution is required to obtain the separation; (3) the column medium is detrimental to the separation (i.e., adsorptive effects need to be minimized); or (4) the cost of the modifiers and solvents (and/or the disposal of the materials) is substantial.

Table II Diffusion Coefficients in Liquids of Large and Small Molecules

Compound	Molecular weight	Diffusivity (cm²/sec × 10⁵)
β-Alanine	89	0.933[a]
Phenol	94	0.84[b]
o-Amino-benzoic acid	137	0.840[a]
Glucose	180	0.671[a]
Citric acid	192	0.661[a]
Cytochrome c	13,370	0.114[c]
β-Lactoglobulin	37,100	0.075[c]
Catalase	247,500	0.041[c]
Myosin	480,000	0.011[c]
Tobacco mosaic virus	40,590,000	0.0046[c]

[a] Data from CRC (1965).
[b] Data from Karger *et al.* (1973).
[c] Data from Lehninger (1970).

Some comparisons between HPLC and HPCE (Weinberger and Lurie, 1991) are presented, since the features that support the capillary format with small molecules are less clear cut compared to those of large-molecule separations. The HPLC separation of impurities found in a heroin seizure sample is illustrated in Fig. 1A. Employing a quaternary gradient elution system, about 35 peaks are resolved in 20 min. Using the same sample, HPCE resolved twice as many peaks in 50 min, as illustrated in Fig. 1B. The mode of separation here is micellar electrokinetic capillary chromatography (MECC). Whereas the time of separation is longer, a comparison of the actual run times yields very compatible numbers because of the gradient reequilibration in HPLC.

More convincing evidence is shown in the following comparisons between HPLC and HPCE (Weinberger and Lurie, 1991) for heroin, heroin impurities and adulterants (Fig. 2A,B), and cocaine, benzoylecgonine, and cocaine degradation products (Fig. 3A,B). The most important factor seen in the heroin separation by MECC is its speed. Even without optimization, the separation is about three-fold faster than a quaternary gradient-optimized HPLC separation. Clearly, the resolution is sufficient that the MECC separation could be run considerably faster by increasing the field strength.

The separation of cocaine illustrates another salient feature of MECC. The excellent peak symmetry found for cocaine compared to HPLC is clear evidence for the lessening of interactions with, in all probability, free silanol groups (See Chapter 1). In addition, benzoylecgonine does not even appear in the HPLC run, owing to adsorption on the stationary phase surface. A sharp, clearly resolved peak is seen in the MECC separation.

The preceding examples present overwhelming evidence that HPCE will play an important role in the determination of small molecules. The major limitation of HPCE is the sensitivity of the current detectors. The concentration limit of detection (CLOD) of HPCE is approximately 50 times poorer than that of HPLC. This is owing in part to the path-length differential (50 μm versus 10 mm), but this factor of 200 is partially compensated by the lesser dilution factor in HPCE. There is increasing evidence that this detection gap will be narrowed by improving instrumental designs for detection

---

Figure 1   (A) HPLC separation of acidic and neutral impurities in an illicit heroin sample. Conditions: injection size: 50 μl; column 11.0 cm $\times$ 4.7 mm Partisil 5-ODS-3; phosphate buffer (23 mM hexylamine, pH 2.2). Initial conditions: 13% methanol, 8.9% acetonitrile, 6.7% tetrahydrofuran (THF), 71.1% phosphate buffer. Final conditions: 21.7% methanol, 14.5% acetonitrile, 10.8% THF, 53% phosphate buffer; gradient: 15-min linear gradient, hold for 5 min at final conditions; flow rate 1.5 ml/min; detector wavelength, 210 nm. Reprinted with permission from Weinberger and Lurie (1991). (B) MECC separation of acidic and neutral heroin impurities in an illicit heroin sample. Conditions: capillary, 50 cm (length to detector) $\times$ 50 μm i.d.; voltage, 30 kV; temperature, 50°C; buffer, 85 mM SDS/8.5 mM phosphate/8.5 mM borate/15% acetonitrile, pH 8.5; detector wavelength, 210 nm. Reprinted with permission from Weinberger and Lurie (1991), American Chemical Society.

and developing trace enrichment techniques that can be used in CE. For example, stacking, which is covered in Chapter 3 of this volume, can yield gains of 5 to 200 times in CLOD, depending on the specific chemical compositions of the run and injection buffers. Hyphenated techniques like LC/CE (Merion *et al.*, 1991) or ITP/CE (Dolnik *et al.*, 1990; Foret *et al.*, 1990) present intriguing possibilities. Furthermore, low-ultraviolet (UV) absorption at 185 nm (Fuchs *et al.*, 1991) is possible in HPCE,

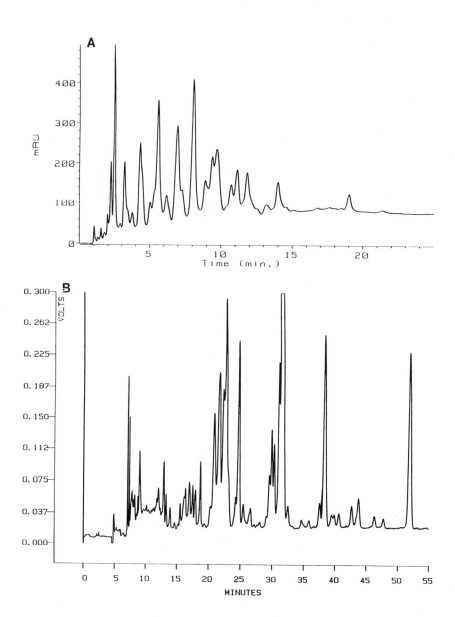

since the short path length lowers the UV cutoff of most buffer systems. Many solutes have higher molar absorptivity at 185 nm than at other wavelengths. The capability to perform on-capillary detection with a "Z"-cell (Chervet *et al.*, 1991) can yield an optical path length of 3 mm (5.88 nl in a 50 μm capillary). Whereas such a "large" volume might be detrimental to some separations, it provides the opportunity to trade some theoretical plates for sensitivity.

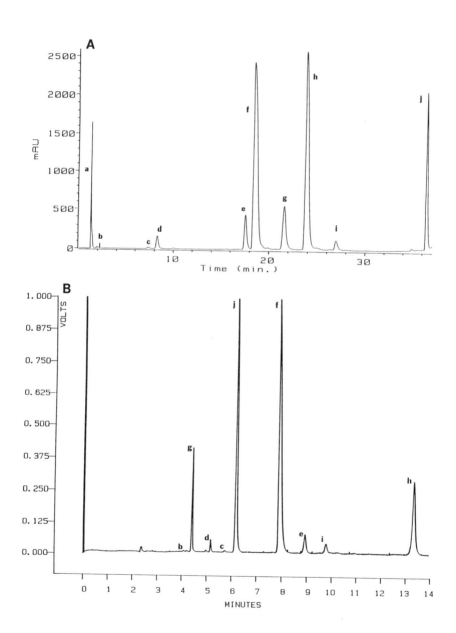

II. Capillary Zone Electrophoresis

The fundamentals of CZE have been covered in another chapter of this volume; therefore, only a brief description of the separation principle will be given. This is a high-efficiency analytical technique that may be used to separate a series of charged solutes. The most important characteristic of CZE, relative to the other HPCE techniques, is the use of a homogeneous buffer system. Because of this feature, the voltage drop is uniform across the entire length of the capillary. The uniformity of the buffer composition and voltage drop leads to simplified methods development compared to that of a more complex technique such as ITP.

When the voltage is applied across the capillary, cationic solutes will migrate to the negative electrode (cathode). Likewise, anionic solutes will migrate to the positive electrode (anode). Under conditions of zero electroosmotic flow (EOF), neutral solutes will not migrate at all since they are not affected by the electric field. In most systems that employ bare silica capillaries, at buffer pH values above 6.0, the EOF is substantially greater than the electrophoretic mobility for most solutes. Under these conditions most solutes, regardless of charge, will migrate toward the negative electrode. At low pH, most ionizable solutes will have a positive charge and migrate as well toward the negative electrode.

Electrophoretic mobility (μ), the fundamental separation parameter of CZE, is a function of a solute's charge : mass ratio. When properly corrected for EOF, the mobility is independent of capillary length and field strength. On the other hand, mobility is profoundly influenced by the buffer composition, pH, and temperature. When the sample-side electrode is positive, cations are always detected first, since both the electrophoretic mobility and EOF are in the same direction. Within a class of cationic species, the solute with the greatest mobility will be detected first. Since the migration of the neutral solutes is not affected by the electric field, only the EOF mediates their migration through the capillary; thus, they are detected as a single band. Consequently, CZE cannot be utilized to separate neutral solutes. The mobility of anions is directed toward the positive electrode, but since the EOF is generally greater than the elec-

---

Figure 2 (A) HPLC of bulk heroin, heroin impurities, degradation products, and adulterants. Conditions: column, 11.0 cm × 4.7 mm Partisil 5-ODS-3; mobile phase A, phosphate buffer (23 mM hexylamine, pH 2.2); mobile phase B, methanol; gradient, 5–30% B over 20 min, 30% B for 6 min, 30–80% B over 10 min, 80% B for 4 min, then reequilibrate over 5 min; flow rate, 1.5 ml/min; detector wavelength, 210 nm. (a) acetic acid, (b) morphine, (c) O^3-monoacetylmorphine, (d) O^6-monoacetylmorphine, (e) acetylcodeine, (f) heroin, (g) phenobarbital, (h) noscapine, (i) papaverine, (j) methaqualone. Reprinted with permission from Weinberger and Lurie (1991). (B) MECC of bulk heroin, heroin impurities, degradation products, and adulterants. Conditions: capillary, 25 cm (length to detector) × 50 μm i.d.; voltage, 20 kV; temperature, 40°C; buffer, 85 mM SDS/8.5 mM phosphate/8.5 mM borate/15% acetonitrile, pH 8.5; detector wavelength, 210 nm. (a) through (j) as per (A). Reprinted with permission from Weinberger and Lurie (1991), American Chemical Society.

trophoretic mobility, the anions migrate toward the negative electrode. In this instance, the anion with the highest mobility will be detected last.

Surprisingly, a broad range of small molecules can be separated by CZE. (Wainright, 1990) reported on the separation of nonsteroidal antiinflammatories, sulfonamides, cephalosporins, penicillins, antibacterials, and pharmaceuticals in

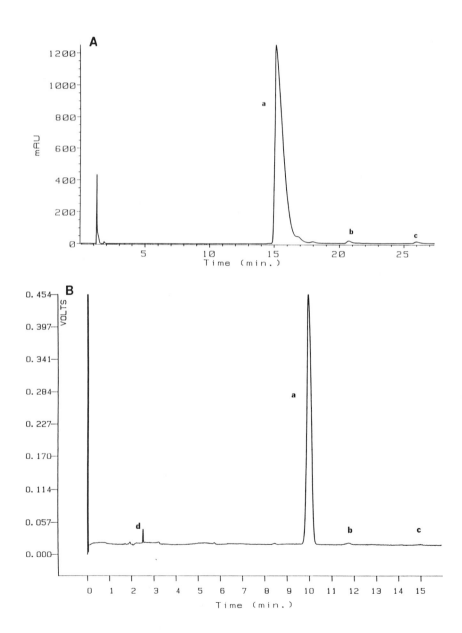

cough/cold preparations. Classes of molecules are amenable to CZE whenever their molecular structural features provide for changes in the charge : mass ratio.

A. Buffers

The most important consideration in the selection of the buffer is choosing a pH where the analytes are ionized. If a pH is selected that is above the negative log of the ionization constant (pK_a) for bases or below the pK_a for acids, separation will not occur, since there is no charge developed on the analyte. For molecules that differ only in hydrophobicity, MECC (described later in this chapter) must be employed to permit the separation to occur.

A wide variety of buffers can be employed in CZE. A buffer is most effective within 1 pH unit of its isoelectric point (pI). For example, citrate is used around pH 2.5; phosphate, around pH 7; and borate, around pH 9. A series of buffers, along with their useful pH range, is given in Table III. Included in Table III is a series of zwitterionic buffers that are commonly used for large-molecule separations. The advantage of the zwitterionic buffer is the low conductivity of the buffer system, particularly when the pH is adjusted to be close to the pI of the zwitterion.

The typical buffer concentrations used for small-molecule separations range from 15 to 25 mM. These buffer concentrations have sufficient buffering capacity yet minimal conductivity, that Joule heating is usually not a problem. Higher buffer concentrations can be utilized, but often the electroosmotic flow is suppressed. This feature can be used advantageously to adjust the selectivity of the separation. The higher-concentration buffers tend to give greater linear dynamic range and peak symmetry at the expense of high conductivity. For a 75-μm capillary, the power generated should not exceed 2.5 W/m (Atamna *et al.*, 1990a). Peak asymmetry occurs owing to field inhomogeneity. The sample, if ionized, can contribute to the conductivity of the sample zone (Thormann, 1983; Mikkers *et al.*, 1979), which results in the migration velocity, v_{ep} being different in the run buffer from that in the sample zone. The result of differing migration velocities in the two buffers is peak fronting or tailing, which is solute concentration dependent.

Figure 3 (A) HPLC of cocaine. Conditions: column, 11.0 cm × 4.7 mm Partisil 5-ODS-3; mobile phase A, phosphate buffer (23 mM hexylamine, pH 2.2); mobile phase B, methanol; gradient, 5–30% B over 20 min, 30% B for 6 min, 30–80% B over 10 min, 80% B for 4 min, then reequilibrate over 5 min; flow rate, 1.5 ml/min; detector wavelength, 228 nm. (a) cocaine, (b) *cis*-cinnamoylcocaine, (c) *trans*-cinnamoylcocaine. Reprinted with permission from Weinberger and Lurie (1991). (B) MECC of cocaine. Conditions: capillary, 25 cm (length to detector) × 50 μm i.d.; voltage, 20 kV; temperature, 40°C; buffer, 85 mM SDS/8.5 mM phosphate/8.5 mM borate/15% acetonitrile, pH 8.5; detector wavelength, 210 nm. (a) cocaine, (b) *cis*-cinnamoylcocaine, (c) *trans*-cinnamoylcocaine, (d) benzoylecgonine. Reprinted with permission from Weinberger and Lurie (1991), American Chemical Society.

Table III Buffers for CZE

Buffers	Useful pH range
Chemical	
Citrate	2.08–5.74
Acetate	3.76–5.76
Phosphate	1.14–3.14
	6.20–8.20
Borate	8.14–10.14
Biological	
MES	5.15–7.15
ACES	5.75–7.75
PIPES	5.80–7.80
BES	6.0–8.10
HEPES	6.55–8.55
EPPS	6.90–9.25
Tricine	7.15–9.15
Bicine	7.25–9.25
Tris	7.30–9.30
TAPS	7.40–9.40

Atamna *et al.* (1990a) studied the role of the cation in CZE buffers. The lower atomic weight Group IA metals, e.g., lithium, sodium, have lower mobilities, since their hydrated radii are actually larger, resulting in a lowered charge density. Some of the physical attributes of the alkali metal cations are included in Table IV. The zeta potential, and thus the EOF, is directly related to the charge density; consequently, cations with smaller Van der Waals radii produce a stronger EOF. Thus equimolar preparations of sodium- or potassium-based buffers will give different conductivities, EOF, and migration times. The selectivity does not appear to be affected. It appears that standardizing on sodium-based buffers is adequate, with the EOF adjusted by varying the buffer concentration or using additives in the electrophoretic buffer.

Table IV Physical Properties of Group 1A Alkali Metals

Element	Hydrated radii (Å)	μ_{ep} (cation) $(10^5 \, cm^2/V \cdot sec)$	μ_{eo} $(10^5 \, cm^2/V \cdot sec)^a$
Li	3.40	40.2	49.20
Na	2.76	51.9	46.80
K	2.32	76.1	45.70
Rb	2.28	86.5	40.45

[a] Mesityl oxide in 50 mM, pH 6.5 acetate buffer; voltage: 20 kV; temperature: 20°C; capillary; 50 cm (length to detector) × 75 μm i.d.

Atamna *et al.* (1990b) studied the role of the anion in CZE buffers. The data reported for anions are more complex since controlled experimental conditions (e.g., pH) result in working outside of the buffer's useful pH range. Based on our present knowledge, it appears that selecting an anion from those presented in Table III will give adequate results.

Numerous papers reported buffers comprising phosphate–borate combinations, particularly in the MECC field. The major advantage with this mixed-buffer system is that the pH of the buffer can be varied between 6 and 10 without changing the specific anion content of the solution.

When doing methods development, analysts frequently change buffer solutions simply by flushing the new solution through the capillary. This procedure is generally adequate when using a common anion. When switching from a phosphate buffer to a different buffer system, anomalous results (migration-time variation) have been reported, presumably as a result of binding of phosphate ions to the capillary wall (McCormick, 1988). In some instances, phosphate ions may help to prevent interaction between the capillary wall and a solute.

B. Buffer Additives

A wide variety of buffer additives are useful in CZE of small molecules. A listing of some of these additives is given in Table V. Micelle-forming surfactants and cyclodextrins are covered later in this chapter. In general, additives can be divided into two types: those that interact with the analyte, and those that influence only the EOF.

1. Organic Modifiers

Organic solvents, such as methanol, 2-propanol, acetonitrile, and tetrahydrofuran (THF), can be utilized as modifiers in either CZE (Griest *et al.*, 1988; Liu *et al.*, 1988; Wallingford and Ewing, 1988, 1989; Wallingford *et al.*, 1989), or MECC (Balchunas *et al.*, 1988; Bushey and Jorgenson, 1989a; Fujiwara and Honda, 1987a; Liu *et al.*, 1988; Nishi *et al.*, 1990c; Snopek *et al.*, 1988). These organic modifiers are added to

Table V   Buffer Additives for CZE

Additive	Function
Organic solvents	Modify EOF, solubilize analyte or buffer
Urea	Solubilize analyte or buffer additive
Ethylene glycol	Reduce wall interactions
Polymer ions	Ion exchange
Hydroxypropylcellulose	Suppress or control EOF
Amines	Cover free silanol groups

the electrophoretic buffer to enhance the selectivity of capillary electrophoresis (Balchunas *et al.*, 1988; Bushey and Jorgenson, 1989a; Fujiwara and Honda, 1987a; Gorse *et al.*, 1988; Griest *et al.*, 1988; Nishi *et al.*, 1990c; Snopek *et al.*, 1988; Wallingford and Ewing, 1989; Wallingford *et al.*, 1989). In most situations, increasing the organic concentration results in increased resolution; however, the migration times of the components also increase. The power of organic modifiers in HPCE was clearly demonstrated by Bushey and Jorgenson (1989a). Two extremely similar compounds, dansylated-NHCH$_3$ and dansylated-NHCD$_3$, were resolved utilizing MECC and an electrophoretic buffer containing 20% methanol.

The impact of the organic modifier in MECC differs from that in CZE. The most important role of the organic modifier is to alter the solute's micelle–water partition coefficient. This feature has been covered in other chapters. For CZE, increasing the proportion of methanol in the electrophoretic buffer causes a decrease in the EOF; whereas increased acetonitrile gives the reverse effect (Fujiwara and Honda, 1987a). However, a number of parameters influence the velocity of the EOF (v_{eo})

$$v_{eo} = \left(\frac{\varepsilon \zeta}{4\pi\eta}\right) E \qquad (2)$$

where ε is the dielectric constant; ζ is the zeta potential at the capillary wall; η is the viscosity of the medium; and E is the field strength. The inclusion of an organic solvent in an aqueous electrophoretic buffer alters not only the dielectric constant (Balchunas *et al.*, 1988; Wallingford and Ewing, 1989; Wallingford *et al.*, 1989) and viscosity of the medium, but also the zeta potential at the capillary wall.

2. Urea

Urea is generally used to solubilize large molecules such as proteins and DNA. Fresh urea is neutral and nonionic, so it does not contribute to the conductivity of the buffer. Concentrations as high as 7 M have been utilized. For small-molecule separations, urea has been employed to better solubilize cyclodextrins, which can be used for chiral recognition. This feature will be discussed in more detail later in the chapter.

3. Polymer Ions

Polymer ions such as poly(diallyldimethylammonium chloride) (PDDAC) or (diethyl-amino)ethyldextran (DEAE-dextran) are useful in separating analytes that have identical mobilities by CZE (Terabe and Isemura, 1990b). For example, DEAE-dextran can ion-pair with anions at pH 7. The positively charged ion-polymer migrates toward the cathode, whereas the anions migrate to the anode. The degree of ion interaction between the analyte and the polymer influences the overall migration velocity. This technique has been applied to separations of isomeric naphthalene sulfonates. The separation mechanism has some similarities with MECC, except that the ion-pairing agent is a solubilized polymer rather than a micellar aggregate.

4. Hydroxypropylcellulose

Cellulose additives are useful in suppressing the EOF. Whereas this procedure is usually necessary for techniques such as isotachophoresis and required for isoelectric focusing (see Chapter 7), in certain circumstances cellulose additives are very useful in CZE. Suppose one wants to separate only the anions present in a mixture of anions and cations. In a bare silica capillary at alkaline pH, cations and neutrals will always elute before the anions. On addition of 0.1% hydroxypropylcellulose, the EOF is almost completely suppressed (Terabe and Isemura, 1990a). Consequently, by reversing the polarity of the applied voltage, only anions will be electrophoretically transported toward the positive detector end and be detected.

III. Electrokinetic Capillary Chromatography

Electrokinetic capillary chromatography separation modes were developed to help resolve neutral solutes that were not separable by CZE. Using these techniques, solutes could differentially partition between the aqueous component of the buffer and an additive that has the property of creating a pseudostationary phase. The key property of the pseudophase is its own electrophoretic migration in a direction opposite that of the EOF. A solute, when located or partitioned within the pseudophase, migrates more slowly than when found in the bulk solution. Solutes with differing affinities for the pseudophase can be separated in this electrochromatographic technique. Not only can neutral solutes be resolved, but the selectivity toward anionic and cationic species may also be dramatically affected. Unlike IEF, CGE, ITP, and CZE, electrokinetic separations are devoted exclusively to the separation of small molecules.

A number of additives have properties that permit the formation of stable reproducible pseudophases. Among these additives are cyclodextrins and surfactants. The use of surfactants forms the basis of MECC, the most versatile and studied of the electrokinetic techniques. The theory and fundamentals of MECC have been covered in Chapter 6, so this section will be devoted primarily to surfactant chemistry, methods development, and the utilization of cyclodextrins as buffer additives.

Because of strong binding of surfactant molecules to the protein molecules, MECC is not useful for protein separations. In the presence of sodium dodecyl sulfate (SDS), all proteins have the same charge : mass ratio. Approximately 1.4 g of SDS is adsorbed onto each gram of protein. Small peptides like angiotensins are not strongly bound by SDS and can be separated based on native charge. Uncharged peptides are separated as well, based on hydrophobic effects. Nucleotides, nucleosides, bases, and single-stranded oligonucleotides (Cohen *et al.*, 1987) up to 21 base pairs (bp) are separable. Despite these often important applications for biotechnology, MECC has only modest utility here, since proteins and DNA cannot be separated. However, for small molecules, MECC has proven to be a very useful and robust separation mode, which is applicable to a wide variety of analyses. An applications table is given elsewhere in this volume and will not be repeated here.

A. Micelles

Micelles are amphophilic aggregates of molecules known as surfactants. A surfactant is a long-chain molecule (10–50 carbon units) and is characterized as possessing a long hydrophobic tail and a hydrophilic head group. Normally, micelles are formed in aqueous solution with the hydrophobic tail pointing inward and the hydrophilic head pointing outward into the aqueous solution. Micelles form as a consequence of the hydrophobic effect. They form to reduce the free energy of the system. The hydrophobic tail of the surfactant cannot be solvated in aqueous solution. Above a surfactant concentration known as the critical micelle concentration (CMC), the aggregate is fully formed. Physical changes such as surface tension, viscosity, and the ability to scatter light accompany micellization. The CMC of a surfactant depends on many parameters, including the nature of the surfactant and its physicochemical environment. Organic modifiers affect the CMC, the aggregation number, and the micellar ionization (rate of exchange of surfactant moiety and micelle). Through salting processes, the CMC can be lowered by increasing the ionic strength of the buffer (Burton *et al.*, 1987; Snopek *et al.*, 1988). Reverse micelles, which form in organic solvents, have not been studied in MECC.

There are four major classes of surfactants: anionic, cationic, zwitterionic, and nonionic, examples of which are given in Table VI. Of these four classes, the first two are most useful for resolving neutral, as well as charged solutes, in MECC. Both synthetic and naturally occurring surfactants have been employed for MECC. The synthetic varieties include anionic SDS, cationic cetyltrimethylammonium bromide (CETAB). The naturally occurring surfactants consist of the bile salts, such as sodium taurocholate, sodium deoxycholate, sodium cholate, and taurodeoxycholate.

Micelles have the ability to organize analytes at the molecular level, based on hydrophobic and electrostatic interactions. Even neutral molecules can bind to micelles since the hydrophobic core has very strong solubilizing power. Surfactant solutions have been employed in spectroscopy (Skrilec and Cline Love, 1980) and

Table VI   Surfactant Classes

Surfactant[a]	Type	CMC in water (M)	Aggregation No.
SDS	Anionic	8.1×10^{-3}	62
CTAB	Cationic	9.2×10^{-4}	61
Brij-35	Nonionic	1.0×10^{-4}	40
Sulfobetaine	Zwitterionic	3.3×10^{-3}	55

[a] SDS, sodium dodecyl sulfate; CTAB, cetyltrimethylammonium bromide; Brij-35, polyoxyethylene-23-lauryl ether; sulfobetaine, *N*-dodecyl-*N*,*N*-dimethyl-ammonium-3-propane-1-sulfonic acid.

chromatography (Armstrong and Nome, 1981) to take advantage of these unique micellar properties. For example, room temperature phosphorescence is readily observable in micellar media since the micellar environment prevents many of the normal quenching mechanisms from operating. More significantly with regard to MECC, these same surfactant solutions can serve as chromatographic mobile-phase modifiers. Micellar chromatography can mimic reverse-phase LC, in that increasing the surfactant concentration increases the eluting power of the mobile phase. The analyte can partition between the micelle and the bulk solution, the micelle and the stationary phase, or the bulk solution and stationary phase. Thus "pseudophase" or micellar LC has more complex equilibria than conventional LC. This three-phase equilibrium can be likened to ion-pair chromatography in many instances. In certain aspects, the mechanism of MECC, because of a two-phase equilibrium, is simpler than micellar chromatography. The complicating factor in MECC is that analyte electrophoretic mobility often contributes to the overall separation. This complexity is illustrated in Fig. 4A,B, comparing CZE and MECC for the separation of some nonsteroidal antiinflammatory agents (Weinberger and Albin, 1991). The problem therein is not the comparison between CZE and MECC; the selectivity is expected to change between these two modes of HPCE. By reverse-phase HPLC (chromatogram not shown), an elution order of peak numbers 3, 1, 5, 2, 4 is obtained (refer to the figure captions for peak identification). If a hydrophobic mechanism were solely responsible for MECC, the elution order would be identical to that of HPLC. Because of the complicated mechanism in MECC, the prediction of the migration times for charged species is more difficult when compared to CZE or HPLC.

B. Surfactants

1. Anionic Surfactants

Surfactants that possess a net negative charge on dissolution in an aqueous medium make up this subclass. The most widely used anionic surfactant is SDS (Burton *et al.*, 1986, 1987; Fujiwara *et al.*, 1988; Gorse *et al.*, 1988; Nakagawa *et al.*, 1988, 1989; Nishi *et al.*, 1989a,b,c,d, 1990a,c,d; Otsuka and Terabe, 1989; Rasmussen and McNair, 1989; Snopek *et al.*, 1988; Terabe *et al.*, 1984, 1985a,b, 1986, 1989a; Wallingford *et al.*, 1989; Wallingford and Ewing, 1988). Other ionic surfactants that have been utilized include sodium pentanesulfonate, sodium octanesulfonate, N-lauroyl-N-methyltaurate (Nishi *et al.*, 1989b,c), sodium octyl sulfate (Wallingford *et al.*, 1989) and sodium decyl sulfate (Burton *et al.*, 1987). Generally, surfactants with alkyl tails of C8 or less do not form micelles under conditions found in capillary electrophoresis and are added as ion-pairing reagents (Nishi *et al.*, 1989b,c). Alternatively, surfactants with alkyl tails greater than C14 have poor aqueous solubility (Burton *et al.*, 1987).

The selectivity with regard to anionic micelles for neutral molecules is dependent on the hydrophobicity of the solutes. The most hydrophobic solute will interact with the lipophilic interior of the micelle to the greatest extent and will elute last. If the solutes are completely solubilized by the micelles, the components will elute with the

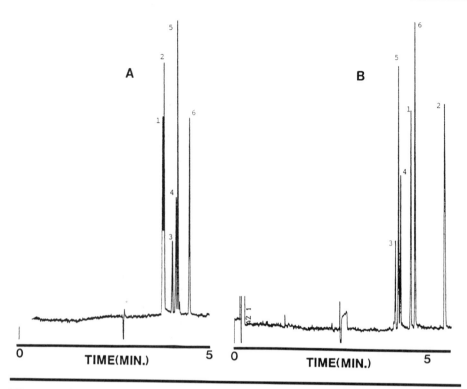

Figure 4 (A) Capillary zone electrophoresis of nonsteroidal antiinflammatory drugs. Conditions: capillary, 42.5 cm (length to detector) × 25 μA; buffer, 20 m*M* phosphate, pH 7.0; temperature, 30°C; voltage, 25 kV; current, 5 μA; injection, vacuum, 2 sec; detection, UV 230 nm. (1) sulindac, 100 μg/ml; (2) indomethacin, 100 μg/ml; (3) tolmetin, 100 μg/ml; (4) ibuprofen, 100 μg/ml; (5) naproxen, 10 μg/ml; (6) diflunisal, 50 μg/ml. Reprinted with permission from Weinberger and Albin (1991). (B) MECC of nonsteroidal antiinflammatory drugs. Buffer, 20 m*M* phosphate, pH 7.0, 25 m*M* SDS. Other conditions as per (A). Reprinted with permission from Weinberger and Albin (1991), Marcel Dekker, Inc.

micelles, and no separation will take place. The separation of ionic solutes in MECC with anionic surfactants is controlled in part by the electrophoretic mobility of the compound, the interaction between the compound and the surfactant, and the degree of solubilization (Nishi *et al.*, 1989b,c). For example, cationic analytes participate in an ion-exchange mechanism with the surface of the micelle, which can alter the selectivity (Wallingford and Ewing, 1988, 1989; Wallingford *et al.*, 1989). For anionic solutes, the ability of the negatively charged analytes to partition into the micelle is suppressed, if not eliminated, owing to repulsion of the analyte by the negatively charged surface of the micelle (Nishi *et al.*, 1989b,c).

2. Cationic Surfactants

This subclass of surfactants contains those that, on dissolution in an aqueous medium, have a net positive charge. The most popular type of cationic surfactant is the tetraalkylammonium salt. Some of the tetraalkylammonium salts most commonly utilized are cetyltrimethylammonium chloride (CTAC) (Burton *et al.,* 1987), cetyltrimethylammonium bromide (CTAB) (Altria and Simpson, 1987, 1988a; Liu *et al.,* 1989; Tsuda, 1987), tetradecyltrimethylammonium bromide (TTAB) (Huang *et al.,* 1989), dodecyltrimethylammonium chloride (DTAC) (Burton *et al.,* 1987), dodecyltrimethylammonium bromide (DTAB) (Liu *et al.,* 1989; Altria and Simpson, 1988a), hexyltrimethylammonium bromide (Altria and Simpson, 1988a), and propyltrimethylammonium bromide (Altria and Simpson, 1988a). Again, surfactants with alkyl tails of C8 or less do not form micelles under conditions found in capillary electrophoresis, whereas surfactants with alkyl tails greater than C14 have poor aqueous solubility (Burton *et al.,* 1987). The chloride salts tend to be more soluble than the corresponding bromide salts, and the heavy atom effect of bromide ion can quench fluorescence (Burton *et al.,* 1987).

The selectivity using cationic micelles with respect to neutral molecules is similar to that of anionic surfactants (Burton *et al.,* 1987). Hydrophobic analytes partition preferentially into the core of the micelle and tend to elute in the order of increasing hydrophobicity. Cationic surfactants are often added to the electrophoretic buffer to enhance the separation of organic acids, nucleosides, and nucleotides (Huang *et al.,* 1989; Liu *et al.,* 1989). The electrostatic effects are similar to those of anionic surfactants, except in reverse. Cations tend to be repelled from the micellar aggregate, whereas anions are attracted either as native species or as ion pairs (Liu *et al.,* 1989).

3. Nonionic Surfactants

As the name suggests, this subclass of surfactants consists of those surfactants that, when dissolved in water, possess no net charge. Nonionic surfactants that have been utilized in capillary electrophoresis include octyl glucoside (Hjertén *et al.,* 1989; Swedberg, 1990a). In addition, ethoxylated surfactants such as polyoxyethylene-23-lauryl ether (Brij 35) have been employed as modifiers in SDS-based MECC separations to adjust the selectivity (Rasmussen *et al.,* 1990).

Since nonionic surfactants possess no net charge, their micelles have a net charge of zero and migrate with the electroosmotic flow. However, mobility of nonionic surfactants due to the adsorption of ions onto the surface of the micelle has been reported by Swedberg (1990a). The observed increase in selectivity of separations involving nonionic surfactants depends heavily on the interaction of the analyte with the micelle. The analyte can interact with nonionic micelles utilizing the following mechanisms. The hydrophobic portion of the molecule can partition into the micelle by a mechanism that resembles the solvophobic interactions of reverse-phase HPLC, or the molecule can selectively adsorb onto the surface of the micelle (Hjertén *et al.,* 1989; Swedberg, 1990a; Terabe *et al.,* 1989b). Nevertheless, the utility of nonionic surfactants is limited, compared to that of their ionic counterparts.

C. Methods Development

The selectivity in MECC is dependent on the characteristics of the surfactant, the characteristics of the aqueous component of the electrophoretic buffer, the characteristics of the analytes, and the instrumental conditions under which the analyses are conducted. The characteristics of the surfactant added to the electrophoretic buffer that affect selectivity include concentration of the surfactant, charge of the polar head group, and the length of the hydrophobic tail. The characteristics of the aqueous component of the electrophoretic buffer that influences selectivity include pH, ionic strength, conductivity, buffer type, and the amount and type of organic modifier.

Except for the work of Foley (1990), no specific optimization routines have been developed for the separation of solutes by MECC. Several years of empirical experience suggest some general guidelines that should prove useful for developing many separations. For most separations, a good starting point for SDS concentration is 50 mM, with an electrophoretic buffer of a 20 mM phosphate–borate mixture adjusted to a pH of 7. After running the separation, the following guidelines apply:

1. If the separation results in long migration times with good resolution, perform one or more of the following: decrease the SDS concentration, increase the voltage, increase the pH, or shorten the capillary. The limitation therein is Joule heating. For 50-μm capillaries, the current should not exceed approximately 100 μA, or excessive band broadening may occur.

2. If the separation results in long separation times with poor resolution, use an organic modifier. The major role of the organic modifier is to alter the solute's micelle–water partition coefficient, particularly for hydrophobic species. The modifier permits the analyte to spend more time in the bulk aqueous phase, thereby increasing its migration speed. Too much modifier will result in disruption of the micelle, changing the mode of separation to CZE. Adding from 5 to 25% organic modifier is usually sufficient to enhance resolution.

3. If the separation results in short migration times with poor resolution, increase the SDS concentration. The practical limit of SDS concentration is 150 mM, after which Joule heating may become problematic. For certain separations, the resolution can be improved by adding an organic modifier.

4. If the separation results in short migration times with moderate resolution, increase the capillary length.

5. If the separation results in inadequate selectivity, even with the use of organic modifiers, change the surfactant type. For example, it may be useful to switch from SDS to a bile-salt surfactant.

6. If the separation results in short separation times with good resolution, no changes are necessary.

Whereas further fine tuning may be required after following these guidelines, these basic concepts have been useful for developing a wide variety of separations by MECC.

D. Cyclodextrins

Another form of electrokinetic chromatography utilizes cyclodextrins (CD). Cyclodextrins are macrocyclic oligosaccharides formed from the enzymatic digestion of starch by bacteria (Biyorak *et al.*, 1989). These compounds are formed with 6, 7, or 8 glucopyranose units and are referred to as α-, β-, or γ-CD, respectively. They are torus shaped and have a relatively hydrophobic internal cavity. The hydrophobic interior permits the formation of inclusion complexes with analytes that are capable of conforming to the interior size dimensions of the CD. If the analyte is too large, no complex is formed. If it is too small, the molecular contact with the CD may not be strong enough to affect the separation or the spectroscopy. Complexation can occur owing to hydrogen bonding, Van der Waals forces or hydrophobic interactions. No bonds are formed during complexation. Unlike micellar media, the CD complex can be stereoselective because of its optically active carbohydrate structure. Like micellar media, the CDs form a heterogeneous phase within the aqueous solution. The important chemical characteristics of CDs are shown in Table VII. For some applications (notably chiral recognition), the solubility of the CDs must be enhanced to 100 mM with the addition of 7 M urea. Neutral CDs such as α, β, and γ are only useful to separate solutes with a net charge. When complexed with the CD, the mobility can be sufficiently modified to effect a separation. Neutral compounds are separable only with a charged CD.

Terabe *et al.* (1985a) reported on electrokinetic chromatography, using 2-O-carboxymethyl-β-CD as a buffer additive. They reported capacity factors and distribution coefficients for about 40 substituted benzenes, showing good selectivity for structural isomers. Among the benefits of using CDs are high efficiency and improved fluorescence quantum yields. High efficiency may be the result of lowering the diffusion coefficients of the analytes by formation of the higher-molecular-weight complex. Another benefit of inclusion may be the lowering of capillary-wall adsorption. Elevation of the fluorescence quantum yields by a factor of 10 have been reported (Liu

Table VII   **Characteristics of Cyclodextrins**[a]

	Type of CD		
Parameter	α	β	γ
Molecular weight	972	1135	1297
Diameter of cavity (Å)	4.7–6	8	10
Volume of cavity (Å)	176	346	510
Solubility (g/100 ml, 25°C)	14.5	1.85	23.2
Molecules per unit cell	4	2	6

[a] Data from American Chemical Society (1989).

et al., 1990). The synthesis of tagging agents with the appropriate stearic and spectroscopic properties appears promising for the design of high-sensitivity assays.

IV. Isotachophoresis

Isotachophoresis (ITP) is an enigma in the United States. Automated instrumentation has been available since the mid-1970s, and the technique is highly regarded in Europe. Some important complementary features exist between ITP and CZE that are relevant to small-molecule separations. Following an introductory description, a potentially important application of ITP will be described.

Unlike CZE, the ITP buffer system is heterogeneous. The capillary is first filled with a *leading electrolyte,* the mobility of which is greater than that of any of the sample components to be determined. Following sample injection, the sample end of the capillary is placed in a *terminating electrolyte,* the ionic mobility of which is lower than that of any of the sample components. On application of the voltage, separation occurs in the boundary region between the leading and terminating electrolytes. Stable zone boundaries form between analytes, based on their individual mobilities.

A typical ITP run is shown in Fig. 5, employing both ultraviolet (UV) and conductivity detection. The stair-step nature of the conductivity trace is typical. The conductivity at the flow-cell electrodes declines as the leader and higher-mobility ions pass through the cell. The UV trace shows gaps between some of the bands. Note that peaks 1, 3, 5, and 6 have no or low UV absorbance. These peaks could serve as *spacers* in ITP/UV to help improve the separation.

Isotachophoresis has two characteristics, the combination of which is unique to electrophoretic methods: (1) all bands move at the same velocity, and (2) the bands are focused. The focusing attribute may be of particular importance. For example, highly mobile bands have high conductivity and, as a result, exhibit a lower voltage drop across the band. Since the mobility is the product of the conductivity and the voltage drop, and conductivity and voltage drop are inversely proportional, the individual band velocities are self-normalizing. Focusing is also a consequence of velocity normalization. For example, if a band diffuses into a neighboring zone, it will either speed up or slow down, based on the field strength encountered, and rejoin the original band.

The data presentation mode of ITP is unusual compared to LC or CE. The *y*-axis contains the qualitative information about the solute. The height of the step is proportional to the mobility of the solute when conductivity detection is employed. The *x*-axis relates to the concentration. The major advantage of ITP is that dilute solutes are focused into narrow zones.

Modern instrumentation can be employed for ITP with either coated or uncoated capillaries (Thormann, 1990). Because ITP is a focusing technique, it is possible to use large-diameter capillaries without the deterioration in resolution that is characteristic of CZE.

One approach to the general detection problem in CZE is to use ITP for trace enrichment (Dolnick *et al.,* 1990; Foret *et al.,* 1990). Both of these methods employ a

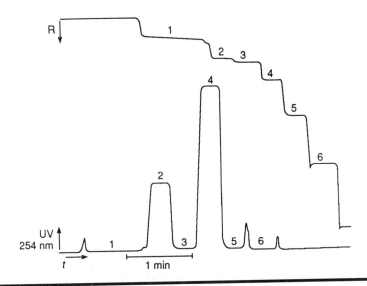

Figure 5 ITP with UV (254 nm) and conductivity detection. Conditions: capillary, PTFE, 30 cm × 550 μm; leader, 10 mM hydrochloric acid titrated to pH 9.1 with Ammediol; terminator, 10 mM β-alanine adjusted to pH 10 with saturated barium hydroxide; current, 210 μA reduced to 60 μA for detection; injection volume, 10 μl corresponding to (1) carbonate (from CO_2); (2) benzoate, 0.61 μg; (3) glu, 0.97 μg; (4) N-benzoyl-alanine, 0.73 μg; (5) gln, 0.73 μg; (6) ala, 0.45 μg. Reprinted with permission from Stehle *et al.* (1986), Elsevier Scientific Publishers.

preconcentration capillary, though the method of Dolnick *et al.* (1990) is more adaptable to commercial instrumentation. Trace enrichment of 200-fold may be possible with these techniques. Isotachophoresis can be viewed as a sophisticated form of stacking, and is covered in Chapter 3.

V. Chemically Modified Capillaries

A frequent problem associated with capillary electrophoresis in bare silica is adsorption of solutes on the capillary wall. Since the wall is anionic at most pH values, the problem is most severe for cationic solutes, although hydrophobic binding mechanisms can present problems for neutral and anionic species. In order to obtain efficient and reproducible separations, the interactions between these solutes and the capillary wall need to be suppressed or at least tightly controlled. Various approaches have been developed to reduce and or eliminate the ability of the solute to interact with the wall of the capillary. One approach is to adjust the pH of the buffer to a value of pH 2.0 or lower (McCormick, 1988), thereby suppressing the wall charge and reducing the interaction between the solute and the wall. Another approach is to adjust the pH of the electrophoretic buffer to a value above the pI of a protein (Lauer and McManigill,

1986). In this case, both the solute and the capillary wall are negatively charged, thereby reducing interactions due to coulombic repulsion. In addition to adjusting the pH of the electrophoretic buffer, a modifier can be added to the electrophoretic buffer, which can mask the effect of the charged silanol sites (Lauer and McManigill, 1986; Foret *et al.*, 1988). A properly chosen modifier will have a greater affinity for wall interaction than the solutes that are being separated. Surfactants, such as SDS, have been reported to help reduce the hydrophobic interactions between the solute and the capillary wall (Weinberger *et al.*, 1990). Finally, the solute–wall interactions can be reduced or eliminated by chemically modifying the surface of the capillary. Indeed, debate between chemists favoring additives and those favoring chemical treatment of the wall will prevail for some time, until data favoring one approach over the other are unequivocal.

Quartz capillaries have been modified with dextran (Herren *et al.*, 1987), diol (Herren *et al.*, 1987), methylcellulose (Hjertén, 1967, 1985; Herren *et al.*, 1987), octadecylsilane (Tsuda *et al.*, 1982), polyacrylamide (Hjertén, 1985) and various molecular-weight analogs of polyethylene glycol (Herren *et al.*, 1987).

Fused-silica capillaries have been modified with an aryl-pentafluoro group (Swedberg, 1990b), epoxy-diol (Bruin *et al.*, 1989b), glycero-glycidoxypropyl (McCormick, 1988), glycol (Jorgenson, 1984), maltose (Bruin *et al.*, 1989b), polyacrylamide (Dolnik *et al.*, 1989; Wainwright, 1990), polyethylene glycol (Terabe *et al.*, 1986; Bruin *et al.*, 1989a; Lux *et al.*, 1990), polyethyleneimine (Towns and Regnier, 1990), polymethylsiloxane (Terabe *et al.*, 1986; Lux *et al.*, 1990), polyvinylpyrrolidinone (McCormick, 1988), and trimethylchlorosilane (Jorgenson and Lukacs, 1981b, 1983; Balchunas and Sepaniak, 1987).

Small molecules such as derivatized amines and amino acids, nucleoside phosphates, aromatic acids, purine derivatives, nucleobases, phenolic compounds, and antiinflammatory drugs have been analyzed using chemically modified fused-silica capillaries (Table VIII). At Searle, a proprietary derivatized purine compound (DPC), currently under development, has been analyzed utilizing a chemically modified capillary. The capillary was kindly donated by Dr. John Stobaugh (Professor of Pharmaceutical Chemistry, University of Kansas). Unlike fused-silica capillaries, this novel capillary possesses a fixed charge on the wall. Consequently, the magnitude of the EOF was found to be relatively independent of pH (between pH 3 and 7.5). In addition to the fixed charge site, this new capillary contains a hydrophilic surface, which should minimize wall interactions.

Before the development and validation of an HPCE method for the determination of DPC, separations on bare silica were attempted. An MECC separation of DPC and its degradation products is shown in Fig. 6. The purine samples were degraded in separate pH 2 and pH 7 buffers for 1 week at 100°C. Samples aged for 1 week in these buffers gave peak area recoveries of 102 and 93% of the parent compound, respectively. The major degradation product eluted after the parent compound and was baseline resolved. The results of these degradation studies were confirmed by an independent HPLC method.

When the stability determination of DPC was repeated using a second non-

Table VIII Analysis of Small Molecules Utilizing Modified Fused-Silica Capillaries

Chemical modification	Electrophoretic buffer	Type of detection	Solutes analyzed	Reference
Polyacrylamide	Glutamic Acid/γ-aminobutyric acid	Ultraviolet	Twelve nucleoside phosphates	Dolnik et al. (1989)
Polyacrylamide	Borate-boric acid, pH 8.4 Phosphate-borate, pH 7.0 Mes-Tris, pH 6.1	Ultraviolet	Naproxen, ibuprofen, tolmetin	Wainright (1990)
Polyethylene glycol	SDS, sodium tetraborate, sodium dihydrogen phosphate, pH 7.0	Ultraviolet	Resorcinol, phenol, nitroaniline, nitrobenzene, toluene, 2-naphthol, Sudan III	Terabe et al. (1986)
Polyethylene glycol	SDS, phosphate, pH 7.0	Ultraviolet	Theobromine, theophylline, caffeine, uric acid	Lux et al. (1990)
Polyethylene glycol	SDS, phosphate, pH 7.0	Ultraviolet	Uridine, cytidine, guanosine, adenosine	Lux et al. (1990)
Polymethylsiloxane	SDS, sodium tetraborate, sodium dihydrogen phosphate, pH 7.0	Ultraviolet	Resorcinol, phenol, nitroaniline, nitrobenzene, toluene, 2-naphthol, Sudan III	Terabe et al. (1986)
Polymethylsiloxane	SDS, phosphate, pH 7.0	Ultraviolet	Theobromine, theophylline, caffeine, uric acid	Lux et al. (1990)
Polymethylsiloxane	SDS, phosphate, pH 7.0	Ultraviolet	Uridine, cytidine, guanosine, adenosine	Lux et al. (1990)
Trimethyl-chlorosilane	Phosphate, pH 7.0	Fluorescence	Dansylated derivatives of amino acids, asparagine, isoleucine, threonine, methionine, serine, alanine, glycine	Jorgenson and Lukacs (1981b, 1983)
Trimethyl-chlorosilane	SDS, sodium monohydrogen phosphate, with and without 2-propanol	Fluorescence	NBD-derivatized amines, ethylamine, propylamine, butyl-amine, cyclohexylamine, hexylamine, coumarin (540 Å)	Balchunas and Sepaniak (1987)

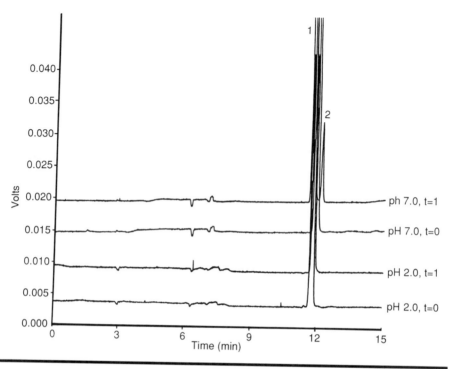

Figure 6 MECC of the derivatized purine compound and solution degradation products resulting from aging for 1 week at 100°C in pH 2 and pH 7 solutions. Conditions: capillary, unmodified fused silica, 50 cm (length to detector) × 50 μm i.d.; voltage, 30 kV; buffer, 25 mM phosphate, 50 mM SDS; detection, 200 nm; injection, vacuum, 5 sec. (1) DPC, (2) degradation product.

modified fused-silica capillary, the resolution between the parent compound and the degradation product was lost, and only a single peak was observed. The cause of this anomalous behavior is not totally known. Other separations by MECC have been shown to be robust (Weinberger and Albin, 1991; Fujiwara *et al.*, 1988; Nishi *et al.*, 1990b). This unsatisfactory behavior could be caused by, among other things, capillary nonuniformity or extreme sensitivit of the drug to buffer composition or temperature.

Nevertheless, in order to improve the reproducibility of the HPCE method for the separation of DPC, the nonmodified, fused-silica capillary was replaced with the chemically bonded capillary previously described. Separations on the treated capillary were satisfactory by CZE, so it was unnecessary to add SDS to the electrophoretic buffer.

A comparison of the separation of several of the related impurities from DPC and its major solution degradation product on the chemically modified capillary, and a bare silica capillary utilizing the same electrophoretic buffer, is shown in Fig. 7. Both capillaries gave sufficient resolution of parent compound and degradation products. The separations were complete in 5 and 7 min for the modified and bare silica capillaries, respectively. The peak asymmetry values were 1.8 and 2.5, respectively. Efficiencies were also superior on the treated capillary (Fig. 7).

Another problem associated with methods developed on bare silica capillaries involves migration time integrity. Often, during the course of multiple runs, the migration time of an analyte can change in a systematic manner. This can be the result of factors such as buffer depletion, buffer evaporation, or detrimental modification of the surface of the capillary. In this particular illustration, the latter case is likely, since the migration time of derivatized purine with the treated capillary varied by less than 0.1 min over a 10-hr run.

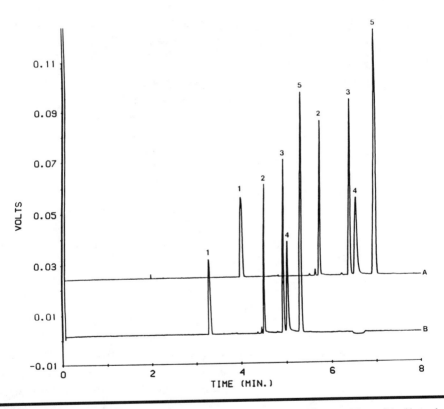

Figure 7 CZE of a derivatized purine compound (DPC) and its related impurities/degradation products on (A) untreated and (B) treated fused-silica capillaries. Conditions: capillaries, 50 cm (length to detector) × 50 μm i.d.; voltage, 25 kV; detection, 200 nm; injection, vacuum, 5 sec; buffer, 20 mM phosphate, pH 7.0, 20 mM tetrabutylammonium phosphate. (1,4,5) synthetic process impurities, (2) solution degradation product, (3) DPC.

VI. Method Validation

To ensure the performance of a method for its intended use, the method must be validated with respect to system suitability parameters, linearity, accuracy, precision, selectivity, sensitivity and, when appropriate, ruggedness and analyte stability. Derivatized purine compound was chosen as a model compound to determine whether an HPCE method could be validated for the determination of this compound in an aqueous solution. Since this compound is still in the early stages of development, the HPCE method was validated only with respect to the parent compound. The procedure for validating an HPCE method for the determination of the derivatized purine compound in an aqueous solution and its utilization for determining the stability of the test article will be described.

Capillary zone electrophoresis has been employed for quantitative analysis of quinine (Altria and Simpson, 1988b), *o*-phthalaldehyde-derivatized amines (Liu *et al.*, 1988), peptides (Cobb and Novotny, 1989), and riboflavin-5´-phosphate (Kenndler *et al.*, 1990). It has also been employed for quantitative analysis of solutes in various matrices: ferulic acid in dog plasma (Fujiwara and Honda, 1986), lithium in human serum (Huang *et al.*, 1988), methotrexate in human serum (Roach *et al.*, 1988), hippuric acid in uremic serum (Schoots *et al.*, 1990), benzylpenicillin in tablets and injectable preparations (Hoyt and Sepaniak, 1989), carboxylic acids in dairy products (Huang *et al.*, 1989), and active ingredients in over-the-counter pain, cold, and allergy medications (Wainright, 1990).

Micellar electrokinetic capillary chromatography has been utilized for the quantitation of neat chemicals, as well as for the determination of solutes in various matrices. These include the determination 2,5-dichlorophenol as the neat chemical (Otsuka *et al.*, 1987), quantitation of solutes in various matrices, including pyridoxic acid in human urine (Burton *et al.*, 1986), cefpiramide in human plasma (Nakagawa *et al.*, 1988, 1989), aspoxicillin in human plasma (Nishi *et al.*, 1990b), ingredients of antipyretic analgesic tablets (Fujiwara and Honda, 1987b), ingredients of a vitamin injection (Fujiwara *et al.*,1988), porphyrins in urine (Weinberger *et al.*, 1990), and ingredients in cold medicine (Nishi *et al.*, 1990c).

In order to verify the performance of a capillary electrophoresis system, a system-suitability test must be performed. This test contains parameters that are designed to measure the overall system performance and not the variation due to sample preparation. The three parameters included in the system-suitability test for DPC by MECC are migration time reproducibility, peak-response reproducibility, and resolution.

The system precision was determined by injecting a standard solution six times. For the six standard injections, the mean, standard deviation (SD), and the percentage relative standard deviation (%RSD) were calculated for the peak area, peak height, and the migration time. As in HPLC, the specifications for the variation of peak response using HPCE must be less than or equal to 2% for the analysis to continue. The results, presented in Table IX, are characteristic of a well-controlled method.

One approach to reduce instrumental and injection bias is to use an internal standard to correct for migration and peak-area variations (Liu *et al.*, 1988). Another ap-

Table IX   System-Suitability Results for the
Analysis of the Purine Compound

Injection number	Peak area (V · min)	Peak height (volts)	Migration time (min)
1	0.0016581	0.0616875	4.44
2	0.0016770	0.0619623	4.45
3	0.0016814	0.0624123	4.45
4	0.0016722	0.0629878	4.46
5	0.0016754	0.0628537	4.47
6	0.0016849	0.0628060	4.47
Mean	0.0016748	0.0624516	4.46
SD	0.00000934	0.00052907	0.012
% RSD	0.55	0.85	0.27

proach for correcting HPLC retention times in separations for the identification and tracking of impurities and degradation products in long-term stability studies is the use of relative retention time (impurity retention time/parent retention time). The extension of this procedure to HPCE has not been reported for stability studies but was employed for developing a relative migration scale of forensic drugs (Weinberger and Lurie, 1991). The peak area response variation also can be corrected, if necessary, with a relative peak-area response (Huang *et al.*, 1989).

The resolution of the system is normally determined for the two closest-eluting components. Referring to Fig. 7, this would be peaks numbered 3 and 4. Peak 4 results from a synthetic process impurity. Since this drug is so early in its development, the process impurity is not available in sufficient quantity to be utilized in the resolution standard. The solute represented by peak 2 is a solution degradation product and has not been isolated. Therefore, components represented by peaks 1, 3, and 5 were incorporated into the resolution standard. The resolution standard is injected at the beginning and at the end of each analysis. As the synthetic process impurity and the degradation product become available, these components will be included in the resolution standard.

In addition to the system-suitability test, the preparation of the standard that is used for the quantitation of the sample is verified. The standard preparation is verified by preparing in duplicate, standards at the same concentration. One of the standards is utilized as the analytical standard (external standard for sample quantitation), and the other standard is utilized to verify the preparation of the external standard. The average peak-area responses of the two standards are compared, and this result is converted into a percentage. The results of the percentage standard determination (drug recovery) for six different runs ranged from 99.1 to 101.1% (mean 99.6 ± 0.8). As in HPLC assays, the specification for the percentage standard determination by HPCE was established as ± 2% (i.e., range 98–102%).

A well-controlled and validated method will have a suitable linear dynamic range of response versus solute concentration. For CZE, the following detection modes have been satisfactory: ultraviolet absorbance (Fujiwara and Honda, 1986; Altria and Simpson, 1988b; Hoyt and Sepaniak, 1989; Schoots *et al.*, 1990), conductivity (Huang *et al.*, 1988, 1989), and laser-induced fluorescence (Roach *et al.*, 1988). For MECC, linearity has been established for ultraviolet absorbance (Fujiwara and Honda, 1987b, Fujiwara *et al.*, 1988; Nakagawa *et al.*, 1989), and laser-induced fluorescence (Burton *et al.*, 1986).

In evaluating system linearity, a standard curve is constructed using four or more solute concentrations to establish the limits of the usable range of the method. Linear regression is performed, and the bias corresponding to each standard is determined. The slope of the regression line, its variance, the y-intercept, its confidence limits, and the correlation coefficient are also reported.

For the linearity study of DPC, a five-level standard curve was prepared. The concentration levels ranged from 0.008 to 0.08 mg/ml. The results were as follows: the y-intercept of 0.000012 could not be distinguished from zero, the bias of any point about the line was less than 4%, and the correlation coefficient was 0.9991. With the linearity established, it was determined that a solute concentration of 0.04 mg/ml could be utilized as a single-point external standard.

The precision and the accuracy of the method was determined utilizing a 2-day, single-analyst protocol. Spiked samples at two different dosing concentrations were prepared in triplicate and analyzed in duplicate. A standard was placed before the first sample and after every third sample. All results were calculated using either a bracketing standard or a nearest mode of quantitation. The value used to characterize the precision of the analysis is the % RSD. The accuracy is characterized by the average analytical recovery. A summary of the precision and the accuracy results are presented in Table X. The within-run precision of approximately 1% and the between-run precision of approximately 3% was adequate for the intended use of the method. In all cases, the amount of the DPC recovered was at least 98.5%.

Once the CZE method was developed and validated, the stability of DPC was determined at dosing concentrations of 100 mg/ml and 0.1 mg/ml. Duplicate samples

Table X   **Summary of Validation Data**

Level	Within-run RSD	Between-run RSD	Total RSD	Accuracy, mean recovery
		Precision		
100 mg/ml	0.98	2.61	2.79	99.8%
0.1 mg/ml	1.14	2.96	3.18	100.1%

were prepared on days 0, 1, 2, 5, and 7. The stability results are given in Table XI. For the 100 mg/ml dosing concentration, the results of linear regression analysis for the recovery of DPC versus time gave a lower t-value of the slope (-0.65) than the standard statistical value for significance (-1.860), indicating that the observed downward trend has no practical significance. It can be concluded that aqueous solutions of the purine solute at a dosing concentration of 100 mg/ml are stable for at least 7 days when stored at ambient conditions.

For the 0.1 mg/ml dosing concentration, linear regression of the recovery of DPC versus time yielded a positive t-test ($t = -2.15$), indicating a statistically significant downward trend. After 7 days, the predicted recovery of the DPC was calculated to be 98.4%, with a corresponding 95% lower confidence level (LCL) of 96.9%.

Because aqueous solutions of DPC are stable for at least 7 days at dosing concentrations of 0.1 mg/ml and 100 mg/ml, when stored at ambient conditions, it can be concluded that all dosing concentrations between 0.1 and 100 mg/ml are also stable for at least 7 days when stored under comparable conditions.

Table XI   **Aqueous Stability of the Purine Compound**

		Percentage recovery	
Day	Sample	100 mg/ml	0.1 mg/ml
0	1	101.0	100.5
	2	104.1	100.3
1	1	101.4	99.1
	2	102.1	101.3
2	1	102.9	102.4
	2	99.9	101.4
5	1	102.3	100.1
	2	101.0	97.1
7	1	100.8	97.6
	2	102.2	99.3
Intercept		102.1	101.0
Slope		−0.0985	−0.368
t-Ratio (slope)		−0.65	−2.15
Degrees of freedom		8	8
t (0.05, df)		−1.860	−1.860
Predicted recovery (day 7)		101.4	98.4
95% Lower confidence limit (LCL) (day 7)		100.7	96.9

VII. Applications

In this section, a number of applications are highlighted, including pharmaceutical, clinical, amino acid, and chiral-recognition separations.

A. Pharmaceuticals

Pharmaceutical analysis encompasses a broad range of measurement processes, including (1) bulk drug purity determination; (2) drug impurity measurements; (3) optical purity determination; (4) active ingredients determination in various dosage forms; (5) drug stability studies; (6) dissolution efficacy; (7) bioavailability; (8) therapeutic drug levels; (9) pharmacokinetics; (10) metabolic pathways; (11) illicit drug identification; and (12) class separations (e.g., penicillins, barbiturates). At the present state of the art of HPE, not all of these facets of drug analysis can be accomplished all of the time.

The purity of riboflavin-5´-phosphate (vitamin B_2 phosphate) was determined by CZE (Kenndler *et al.*, 1990). The basis for selecting CZE was interferences in HPLC and spectrophotometric techniques. With fluorescence detection, the amount of riboflavin-5´-phosphate found to be present in all the samples was between 72 and 75%. The impurities of riboflavin, other riboflavin monophosphates, and riboflavin diphosphates were present in the samples at levels of about 7, 16, and 4%, respectively.

The determination of impurities in bulk drugs generally requires sensitivity to the 0.1% impurity level. Whether or not this can be accomplished depends on the limits of detection (LOD) of the drug and its impurities. Much above 1 mg/ml, the peak-area linear range of HPCE is exceeded because of field perturbation (Thormann, 1983; Mikkers *et al.*, 1979; Weinberger and Albin, 1991). If a 1 mg/ml solution of parent drug is used, then the LOD must be less than 1 µg/ml for the impurities. Not all drugs will meet this criterion without some form of trace enrichment. An example of a drug substance that meets this criterion is shown in Fig. 8. The LOD of the parent drug, naproxen, is about 60 ng/ml.

Oral and intravenous pharmaceutical formulations have been separated by CZE. The dosage forms for the oral formulations studied include tablets (Fujiwara and Honda, 1987b; Hoyt and Sepaniak, 1989; Wainright, 1990; Weinberger and Albin, 1991) and granules (Nishi *et al.*, 1990c). The active ingredients that have been quantitated in these oral formulations include acetaminophen, acetylsalicylic acid, benzylpenicillin, caffeine, chlorpheniramine, ethenzamide, *p*-acetamidophenol, phenylpropanolamine, pseudoephiramine, salicylamide, and tipepidine. In the analysis of various cold medicine tablets (Wainright, 1990), it was found that the filling material that was used in the tablet preparation did not interfere with the separation of the active components. For the intravenous formulations, various vitamins (Fujiwara *et al.*, 1988) and benzylpenicillin (Hoyt and Sepaniak, 1989) have been reported. There is substantial activity in this area, and many examples of dosage-form analysis are forthcoming.

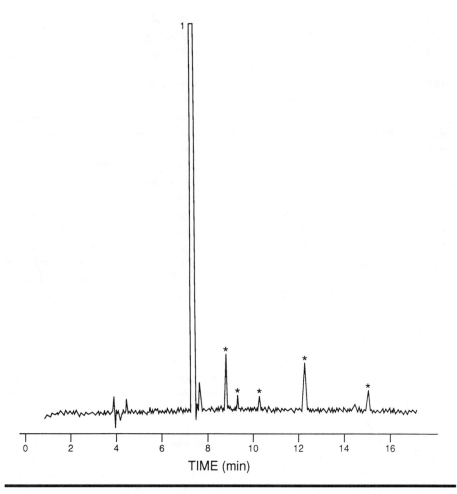

Figure 8   MECC separation of naproxen (1 mg/ml) and impurities (1 µg/ml). Conditions: capillary, 50 cm (length to detector) × 50 µm; buffer, 25 mM sodium borate, 25 mM SDS, 10% methanol, pH 9.5; field strength, 417 V/cm; temperature, 40°C; detection, 230 nm; injection, 2 sec vacuum (7 nl).

The analysis of drugs in biological fluids (i.e., serum, plasma, and urine) is difficult for the following reasons: (1) the drug is present at low concentrations (ng/ml); (2) the drug must be separated from its metabolites; and (3) the drug may bind to the proteins that are present in the sample matrix. High-performance CE has been utilized for the analysis of drugs in human serum (Roach *et al.,* 1988; Schoots *et al.,* 1990), plasma (Krivankova and Bocek, 1990; Nakagawa *et al.,* 1988, 1989; Nishi *et al.,* 1990b) and urine (Burton *et al.,* 1986). In general, unless derivatization, trace enrichment, or laser fluorescence is employed, the sensitivity of CE methods may be inadequate for the analysis of drugs in biological fluids. Several drugs have been analyzed utilizing MECC in which the plasma samples were directly injected. The drugs that have been analyzed using the technique are cefpiramide (Nakagawa *et al.*, 1988, 1989) and aspoxicillin (Nishi *et al.*, 1990b). The SDS in the electrophoretic buffer appears to solubilize the plasma proteins. As a result of this solubilization process, the SDS will displace the drug from the plasma proteins. In addition, the micelle–protein complex exhibits a high negative charge density; thus, the electrophoretic mobility of this complex is reduced, and the drugs of interest elute before the plasma proteins. It appears that another function of the SDS is to prevent the adsorption of plasma proteins to the capillary wall. One note of caution with regard to direct injection procedures is that unless treated capillaries are used, the risk of protein binding to the capillary wall is substantial. Validation studies with a variety of clinical serum or plasma specimens suffering various pathologies (hyperlipidemia, etc.) have yet to be studied.

The power of MECC for illicit drug screening is shown in Fig. 9. Whereas the separation is excellent, confirmation by mass spectroscopy is required. Although HPCE has been interfaced to the mass spectrometer, there have been no reports of using MECC as the separation mode.

B. Amino Acids

With few exceptions, it is necessary to derivatize amino acids (AA) to gain sufficient sensitivity, even by HPLC. Good separations have been reported for pre-capillary derivatized AAs (Otsuka *et al.*, 1985; Liu *et al.*, 1988) and post-capillary derivatized AAs (Tsuda *et al.*, 1988; Rose and Jorgenson, 1988). Kuhr and Yeung (1988) employed indirect fluorescence detection of native AAs. The sensitivity of some of these methods appears sufficient for protein hydrolysate analysis. The sensitivity at this time is insufficient to perform phenylthiohydantoin (PTH)-AA separations off a protein sequencer. At HPCE '91, Nickerson and Jorgenson (1991) substituted fluorescein isothiocyanate (FITC) as the Edman reagent to improve sensitivity.

C. Clinical Analyses

The sensitivity issue plays a significant role when measuring endogenous substances. In the μM range, most applications pose no substantial problem. Measurements in the nM range and below face similar problems, as in some forms of drug analysis.

A few methods for measuring endogenous components in biological fluids have recently appeared. Jellum *et al.* (1991) presented a paper at HPCE '91 describing the use of HPCE for diagnosing various metabolic disorders, for example: homocystinuria, cystinuria, glutathione (GSH) synthetase deficiency, and adenyl-succinase deficiency. The HPCE of the derivatized sulfur-containing amino acids was a quick and simple alternative to classical amino acid analysis (AAA). The major advantage of HPCE compared to HPLC or AAA was speed. The high-speed HPCE assay permits

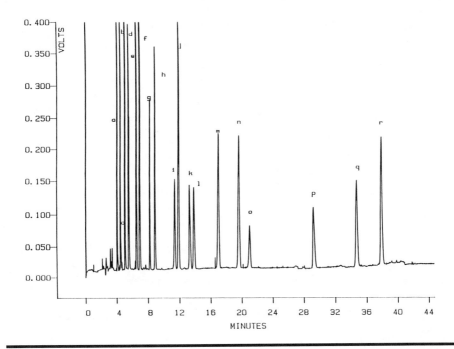

Figure 9 MECC forensic drug screen. Conditions: capillary, 25 cm × 50 μm i.d.; voltage, 20 kV; temperature, 40°C; buffer, 85 mM SDS, 8.5 mM phosphate, 8.5 mM borate, 15% acetonitrile, pH 8.5; detector wavelength, 210 nm; sample concentration, 250 μg/ml of each drug. (a) psilocybin, (b) morphine, (c) phenobarbital, (d) psilocin, (e) codeine, (f) methaqualone, (g) LSD, (h) heroin, (i) amphetamine, (j) Librium, (k) cocaine, (l) methamphetamine, (m) lorazepam, (n) diazapam, (o) fentanyl, (p) PCP, (q) cannabidiol, (r) Δ⁹-THC. Reprinted with permission from Weinberger and Lurie (1991), American Chemical Society.

the routine screening, in a cost-effective fashion, of patient samples for many treatable ailments. Figure 10 illustrates the CZE separation of GSH in red blood cells from a healthy patient and a patient suffering from GSH-synthetase deficiency.

Weinberger *et al.* (1990) illustrated the separations of urinary porphyrins from a patient suffering from porphyria cutanea tarda. The separation times were two to three times faster than the traditional method using gradient-elution HPLC. A separation of a standard and patient sample is shown in Figure 11.

D. Chiral Recognition

Chiral recognition of racemic mixtures continues to be an active area of research in gas chromatography, liquid chromatography, and of late, capillary electrophoresis.

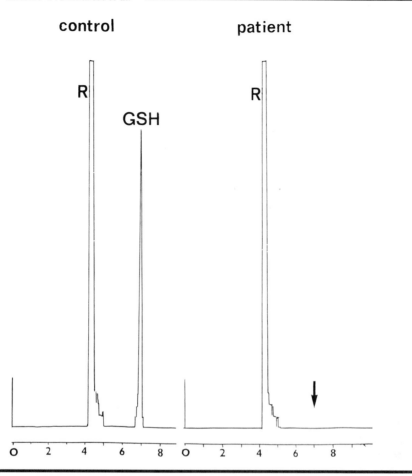

Figure 10   CZE of monobromobimane-labeled glutathione in red blood cells from a healthy patient (left) and a patient suffering from GSH-synthetase deficiency (right). Conditions: capillary, 40 cm (length to detector) × 100 μm; buffer, 50 m*M* phosphate, pH 7.5; voltage, 24 kV; detection, fluorescence, 375 nm excitation, 480 nm emission; injection, electrokinetic. R, reagent peak; GSH, glutathione. Reprinted with permission from Jellum *et al,* (1991). Elsevier Scientific Publishers.

Regardless of the separation technique employed, chiral recognition is obtained in one of three ways:

1. formation of diastereomers by additives to the mobile-phase or carrier electrolyte;
2. formation of diastereomers through interaction with a stationary phase or heterogeneous carrier electrolyte; or

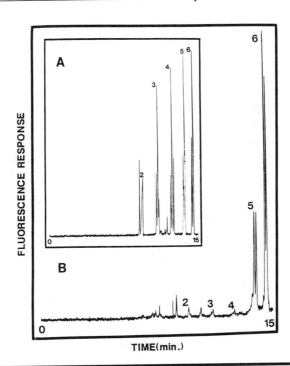

Figure 11   MECC of urine from a patient suffering from porphyria cutanea tarda. Conditions: capillary, 50 cm (length to detector) $\times$ 50 μm; buffer: 100 mM SDS, 20 mM (cyclohexylamino)propanesulfonic acid (CAPS), pH 11; voltage, 20 kV; temperature, 45°C; injection, 2 sec vacuum; detection: fluorescence, 400 nm excitation (xenon arc), emission wavelengths >550 nm. (A) photodegraded standard, 5 nmol/ml; (B) patient sample. (2) coproporphyrin; (3) pentacarboxyl porphyrin; (4) hexacarboxyl porphyrin isomers; (5) heptacarboxyl porphyrin; (6) uroporphyrin. Reprinted with permission from Weinberger *et al,* (1990). Elsevier Scientific Publishers.

   3. pre-column (capillary) derivatization with an optically pure derivatizing
      reagent.

   In the latter two cases, diastereomer formation is dynamic and occurs by electrostatic and/or hydrophobic mechanisms. In pre-capillary derivatization, the separation is usually performed by MECC, and the separation occurs based on differences in hydrophobicity between the diastereomers.
   Resolution of amino acids is important from the standpoint of the control of peptide synthesis and geological dating. One of the best examples of this work involves the copper(II)-aspartame support electrolyte (Gozel *et al.,* 1987). Chiral recognition is based on the formation of a ternary complex between Cu(II), L-aspartame, and amino

acids. Chelation of metal ions probably occurs through formation of a six-membered ring consisting of Cu(II), the α-amino, and the β-carboxy groups of the aspartyl residue. When the amino acid is added to the electrolyte, it can replace one aspartame ligand. Chiral selectivity occurs, since the stability constants are slightly different for each optical form of the AA. At neutral pH, the Cu complex is positively charged and will move faster toward the negative electrode than will the neutral free amino acids. Then the Cu(II) stability constant will determine the electrokinetic migration times of the AA enantiomers.

Capillary preparation is more severe than that with most other forms of CZE. Metal ions are stripped with 0.1 M phosphoric acid for several hours. The surface is reactivated with 10 mM potassium hydroxide and equilibrated with buffer solution for 10 hr. Neutral amino acids will not be resolved by this technique. The addition of surfactant (20 mM sodium tetradecyl sulfate) to the run buffer permits the resolution of neutral amino acids and gives more retention for arginine.

Micellar electrokinetic capillary chromatography seems a more versatile means of separating enantiomers, since the hydrophobic component of the separation mechanism provides for compound recognition as well as optical recognition (Dobashi *et al.*, 1989; Nishi *et al.*, 1989d; Terabe *et al.*, 1989; Otsuka and Terabe, 1990). Chiral surfactants naturally occur, as represented by bile salts. Both pH and structure of the bile salt affect chiral resolution. It appears that compounds with a relatively rigid structure can be optically resolved using this method, though it is still difficult to predict which substances can be resolved.

The optical purity of trimetoquinol hydrochloride was determined utilizing MECC (Nishi *et al.*, 1990d). Chiral recognition was obtained with the optically pure surfactant sodium taurodeoxycholate. The R-isomer of the drug could be detected at a concentration level of 1.0% with respect to the S-isomer. After analyzing five batches of the drug, these researchers were able to determine that the S-isomer of the drug has a potency value of greater than 99%.

Cyclodextrins are useful as buffer and gel additives for chiral recognition. It is quite reasonable to expect that compounds resolvable on cyclobond (ASTEC, Whippany, NJ) HPLC columns can also be resolved by capillary electrophoresis with chiral additives but with greater resolution.

Sydor and Mularz (1991) presented optimization studies for chiral recognition of antihistamines and adrenergic drug substances at the HPCE '91 meeting. Separations of pheniramines are shown in Fig. 12. The L-isomers always elute first, presumably because the D-isomer forms a more stable inclusion complex.

The last approach for the determination of enantiomers involves pre-capillary derivatization. The advantages of this approach are more predictable chiral recognition and additional sensitivity, owing to the tagging agent. Much like conventional derivatization, if primary and secondary amines are involved, good tagging agents are commercially available. For example, the optically pure ethyl analog of FMOC (fluorenylmethyl chloroformate, known as FLEC, reacts rapidly and gives derivatives separable by MECC.

This particular proprietary compound has three chiral centers, so eight peaks

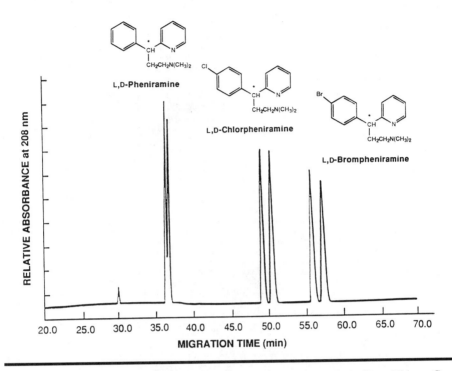

Figure 12 Chiral recognition of pheniramines with cyclodextrin buffer additives. Conditions: capillary, 50 cm (length to detector) × 50 μm; buffer, 150 mM sodium phosphate, pH 2.5, 5 M urea, 50 mM β-cyclodextrin; field strength, 208 V/cm; injection, 7.5 sec, vacuum; detection, 208 nm. Reprinted with permission from Sydor and Mularz (1991).

should be resolved. The electropherogram is shown in Fig. 13. The first peak at 34.57 min is excess FLEC, so only five peaks are separable (LC can separate six peaks, derivatizing a different portion of the molecule). This particular separation used 75 mM SDS, 15 mM borate buffer, pH 8.8, with 25% DMF as modifier. Other modifiers did not give as good resolution.

Whereas many aspects of chiral recognition by capillary electrophoresis are still under investigation, the utility of capillary electrophoresis for optical isomer assay has been demonstrated and reduced to practice in many labs.

VIII. Conclusion

In a historical sense, perhaps HPCE is in the stage of development equivalent to vintage 1973 HPLC. Yet a rich variety of separation mechanisms and detection schemes have already been developed. Intriguing methodologies for improving sensitivity and reducing wall interactions are presently under development. Until a few years ago,

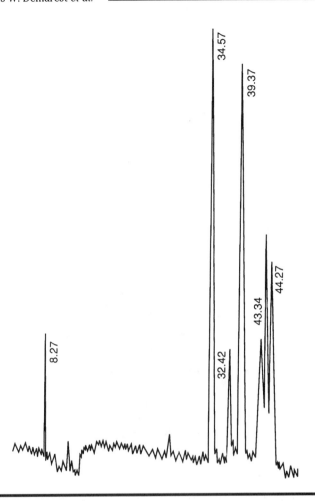

Figure 13   MECC separation of a FLEC-derivatized drug substance containing 3 chiral centers. Conditions: capillary, 50 cm (length to detector) × 25 μm; buffer, 75 m*M* SDS, 15 m*M* borate, pH 8.8, 25% dimethylformamide; detection, 260 nm.

electrophoresis was recognized as a technique exclusive to large biomolecules. It is quite clear that recent work has shown the impact and utility of HPCE for the separation of small molecules. It is also clear that HPCE will play a formidable role in small-molecule separations in the years to come.

References

Ahnoff, M., and Einarsson, S. (1990). *In* "Chiral Liquid Chromatography" (W. J. Lough, ed.), pp. 40–80. Blackie and Sons, London.

Altria, K. D., and Simpson, C. F. (1987). *Chromatographia* **24**, 527–532.

Altria, K. D., and Simpson, C. F. (1988a). *Anal. Proc.* **25**, 85.

Altria, K. D., and Simpson, C. F. (1988b). *J. Pharm. Biomed. Anal.* **6**, 801–807.

American Chemical Society (1989). "Luminescence Applications in Biological, Chemical, and Hydrological Sciences." ACS Symposium Series 383.

Armstrong, D., and Nome, F. (1981). *Anal. Chem.* **53**, 1662–1666.

Atamna, I. Z., Metral, C. J., Muschik, G. M., and Issaq, H. J. (1990a). *J. Liq. Chromatogr.* **13**(13), 2517–2528.

Atamna, I. Z., Metral, C. J., Muschik, G. M., and Issaq, H. J. (1990b). *J. Liq. Chromatogr.* **13**(16), 3201–3210.

Balchunas, A. T., and Sepaniak, M. J. (1987). *Anal. Chem.* **59**, 1466–1470.

Balchunas, A. T., Swaile, D. F., Powell, A. C., and Sepaniak, M. J. (1988). *Sep. Sci. Technol.* **23**, pp. 1891–1904.

Biyorak, L. A., Patonay, G., and Warner, I. M. (1989). *In* "Luminescence Applications in Biological Chemical, Environmental, and Hydrological Sciences" (M. C. Goldberg, ed.) pp. 167–169, American Chemical Society, Washington, DC.

Bruin, G. J. M., Chang, J. P., Kuhlman, R. H., Zegers, K., Kraak, J. C., and Poppe, H. (1989a). *J. Chromatogr.* **471**, 429–436.

Bruin, G. J. M., Huisden, R., Kraak, J. C., and Poppe, H. (1989b). *J. Chromatogr.* **480**, 339–349.

Burton, D. E., Sepaniak, M. J., and Maskarinec, M. P. (1986). *J. Chromatogr. Sci.* **24**, 347–351.

Burton, D. E., Sepaniak, M. J., and Maskarinec, M. P. (1987). *J. Chromatogr. Sci.* **25**, 514–518.

Bushey, M. M., and Jorgenson, J. W. (1989a). *J. Microcol. Sep.* **1**(3), 125–130.

Chervet, J. P., Van Soest, R. E. J., and Ursem, M. (1991). Presented at HPCE '91, San Diego, California.

Cobb, K. A., and Novotny, M. (1989). *Anal. Chem.* **61**, 2226–2231.

Cohen, A. S., Terabe, S., Smith, J. A., and Karger, B. L. (1987). *Anal. Chem.* **59**, 1021–1027.

CRC (1965). "Handbook of Physics and Chemistry," 46th Ed. CRC, Cleveland, Ohio.

Dobashi, A., Ono, T., Hara, S., and Yamaguchi, J. (1989). *Anal. Chem.* **61**, 1986–1988.

Dolnick, V., Liu, J., Banks, J. F., Jr., Novotny, M., and Bocek, P. (1989). *J. Chromatogr.* **480**, 321–330.

Dolnick, V., Cobb, K., and Novotny, M. (1990). *J. Microcol. Sep.* **2**, 127–131.

Foley, J. P. (1990). *Anal. Chem.* **62**, 1302–1308.

Foret, F., Deml, M., and Bocek, P. (1988). *J. Chromatogr.* **452**, 601–613.

Foret, F., Sustacek, V., and Bocek, P. (1990). *J. Microcol. Sep.* **2**, 229–233.

Fuchs, M., Timmoney, P., and Merion, M. (1991). Presented at HPCE '91, San Diego, California.

Fujiwara, S., and Honda, S. (1986). *Anal. Chem.* **58**, 1811–1814.

Fujiwara, S., and Honda, S. (1987a). *Anal. Chem.* **59**, 487–490.

Fujiwara, S., and Honda, S. (1987b.) *Anal. Chem.* **59**, 2773–2776.

Fujiwara, S., Iwase, S., and Honda, S. (1988). *J. Chromatogr.* **447**, 133–140.

Garner, T. W., and Yeung, E. (1990). *J. Chromatogr.* **515**, 639–644.

Gorse, J., Balchunas, A. T., Swaile, D. F., Sepaniak, M. J. (1988). *J. High Res. Chromatogr.* **11**, 554–559.

Gozel, P., Gassman, E., Michaelson, H., and Zare, R. N. (1987). *Anal. Chem.* **59**, 44–49.

Griest, W. H., Maskarinec, M. P., and Row, K. H. (1988). *Sep. Sci. Technol.* **23**, 1905–1914.

Gross, L., and Yeung, F. G. (1990). *Anal. Chem.* **62**, 427–431.

Guttman, A., Paulus, A., Cohen, A. S., Grinberg, N., and Karger, B. L. (1988). *J. Chromatogr.* **488**, 41–53.

Herren, B. J., Shafer, S. G., Alstine, J. V., Harris, J. M., and Snyder, R. S. (1987). *J. Colloid Interface Sci.* **115**, 46–55.

Hjertén, S. (1967). *J. Chromatogr. Rev.* **9**, 122–219.

Hjertén, S., (1985). *J. Chromatogr.* **347**, 195–198.

Hjertén, S., Valtcheva, L., Elenbring, K., and Eaker, D. (1989). *J. Liq. Chromatogr.* **12**(13), 2471–2499.

Hoyt, A. M., Jr., and Sepaniak, M. J. (1989). *Anal. Lett.* **22**, 861–877.

Huang, X., Gorden, M. J., and Zare, R. N. (1988). *J. Chromatogr.* **425**, 385–390.

Huang, X., Luckey, J. A., Gorden, M. J., and Zare, R. N. (1989). *Anal. Chem.* **61**, 766–770.

Jellum, E., Thorsrud, A. K., and Time, E. (1991). *J. Chromatogr.* **559**, 455–465.

Jorgenson, J. W. (1984). *Trends Anal. Chem.* **3**, 51–54.

Jorgenson, J. W., and Lukacs, K. D. (1981a). *J. Chromatogr.* **218**, 209–216.
Jorgenson, J. W., and Lukacs, K. D. (1981b). *Anal. Chem.* **53**, 1298–1302.
Jorgenson, J. W., and Lukacs, K. D. (1983). *Science* **222**, 266–272.
Karger, B. L., Snyder, L. R., and Horvath, C. (1973). "An Introduction to Separation Science." Wiley, New York.
Kenndler, E., Schwer, C., and Kaniansky, D. (1990). *J. Chromatogr.* **508**, 203–207.
Krivankova, L., and Bocek, P. (1990). *J. Microcol. Sep.* **2**, 80–83.
Kuhr, W. G., and Yeung, E. S. (1988). *Anal. Chem.* **60**, 1832–1834.
Lauer, H. H., and McManigill, D. (1986). *Anal. Chem.* **58**, 166–170.
Lehninger, A. L. (1970). "Biochemistry." Worth Publishers, New York.
Liu, J., Cobb, K. A., and Novotny, M. (1988). *J. Chromatogr.* **468**, 55–65.
Liu, J., Banks, J. F., Jr., and Novotny, M. (1989). *J. Microcol. Sep.* **1**(3), 136–141.
Lux, J. A., Yin, H., and Schomburg, G. (1990). *J. High Res. Chromatogr.* **13**, 145–147.
McCormick, R. M. (1988). *Anal. Chem.* **60**, 2322–2328.
Merion, M., Aebersold, R. H., and Fuchs, M. (1991). Presented at HPCE '91, San Diego, California.
Mikkers, F. E. P., Everaerts, F. M., and Verheggen, Th. P. E. M. (1979). *J. Chromatogr.* **169**, 1–20.
Nakagawa, T., Oda, Y., Shibukawa, A., and Tanaka, H. (1988). *Chem. Pharm. Bull.* **36**, 1622–1625.
Nakagawa, T., Oda, Y., Shibukawa, A., Fukuda, H., and Tanaka, H. (1989). *Chem. Pharm. Bull.* **37**, 707–711.
Nickerson, B., and Jorgenson, J. W. (1991). Presented at HPCE '91, San Diego, California.
Nishi, H., Fukuyama, T., Matsuo, M., and Terabe, S. (1989a). *J. Microcol. Sep.* **1**(5), 234–241.
Nishi, H., Tsumagari, N., Kakumoto, T., and Terabe, S. (1989b). *J. Chromatogr.* **465**, 331–343.
Nishi, H., Tsumagari, N., Kakumoto, T., and Terabe, S. (1989c). *J. Chromatogr.* **477**, 259–270.
Nishi, H., Tsumagari, N., and Terabe, S. (1989d). *Anal. Chem.* **61**, 2434–2439.
Nishi, H., Fukuyama, T., and Matsuo, M. (1990a). *J. Microcol. Sep.* **2**, 234–240.
Nishi, H., Fukuyama, T., and Matsuo, M. (1990b). *J. Chromatogr.* **515**, 245–255.
Nishi, H., Fukuyama, T., Matsuo, M., and Terabe, S. (1990c). *J. Chromatogr.* **498**, 313–323.
Nishi, H., Fukuyama, T., Matsuo, M., and Terabe, S. (1990d). *J. Chromatogr.* **515**, 233–243.
Otsuka, K., and Terabe, S. (1989). *J. Microcol. Sep.* **1**(3), 150–154.
Otsuka, K., and Terabe, S. (1990). *J. Chromatogr.* **515**, 221–226.
Otsuka, K., Terabe, S., and Ando, T. (1985). *J. Chromatogr.* **332**, 219–226.
Otsuka, K., Terabe, S., and Ando, T. (1987). *J. Chromatogr.* **396**, 350–354.
Rasmussen, H. T., and McNair, H. M. (1989). *J. High Resol. Chromatogr.* **12**, 635–636.
Rasmussen, H. T., Goebel, L. K., and McNair, H. M. (1990). *J. Chromatogr.* **517**, 549–555.
Roach, M. C., Gozel, P., and Zare, R. N. (1988). *J. Chromatogr.* **426**, 129–140.
Rose, D. J., and Jorgenson, J. W. (1988). *J. Chromatogr.* **447**, 117–131.
Row, K. H., Griest, W. H., Maskarinec, M. P. (1987). *J. Chromatogr.* **409**, 193–203.
Schoots, A. C., Verheegen, T. P. E. M., De Vries, P. M. J. M., and Everaerts, F. M. (1990). *Clin. Chem.* **36**, 435–440.
Scrilec, M., and Cline Love, L. J. (1980). *Anal. Chem.* **52**, 1559–1564.
Snopek, J., Jelinak, T., and Smolkova-Keulemansoka, E. (1988). *J. Chromatogr.* **452**, 571–590.
Stehle, P., Bahsitta, H.-P., and Furst, P. (1986). *J. Chromatogr.* **370**, 131–138.
Sydor, W., and Mularz, E. (1991). Presented at HPCE '91, San Diego, California.
Swedberg, S. A. (1990a). *J. Chromatogr.* **503**, 449–452.
Swedberg, S. A. (1990b). *Anal. Biochem.* **185**, 55–56.
Terabe, S., and Isemura, T. (1990a). *J. Chromatogr.* **515**, 667–676.
Terabe, S., and Isemura, T. (1990b). *Anal. Chem.* **62**, 650–652.
Terabe, S., Otsuka, K., Ichikawa, K., Tsuchiya, A., and Ando, T. (1984). *Anal. Chem.* **56**, 113–116.
Terabe, S., Ozaki, H., Otsuka, K., and Ando, T. (1985a). *J. Chromatogr.* **332**, 211–217.
Terabe, S., Otsuka, K., and Ando, T. (1985b). *Anal. Chem.* **57**, 834–841.
Terabe, S., Utsumi, H., Otsuka, K., Ando, T., Inomata, T., Kuze, S., and Hanaoka, Y. (1986). *J. High Res. Chromatogr.* **9**, 666–670.
Terabe, S., Otsuka, K., and Ando, T. (1989a). *Anal. Chem.* **61**, 251–260.

Terabe, S., Shibuta, M., and Miyashita, Y. (1989b). *J. Chromatogr.* **480**, 403–411.

Thormann, W. (1983). *Electrophoresis* **4**, 383–390.

Thormann, W. (1990). *J. Chromatogr.* **516**, 211–217.

Towns, J. K., and Regnier, F. E. (1990). *J. Chromatogr.* **516**, 69–78.

Tsuda, T., Nomura, K., and Nakagawa, G. (1982). *J. Chromatogr.* **248**, 241–247.

Tsuda, T. (1987). *J. High Res. Chromatogr.* **10**, 622–624.

Tsuda, T., Korayashi, Y., Hori, A., Matsumoto, T., and Suzuki, O. (1988). *J. Chromatogr.* **456**, 375–381.

van Soest, R. E. J., Chervet, J. P., and Salzmann, J. P. (1991). Presented at HPCE '91, February, 1991, paper number PM 23.

Wainright, A. (1990). *J. Microcol. Sep.* **2**(4), 166–175.

Wallingford, R. A., and Ewing, A. G. (1988). *J. Chromatogr.* **441**, 299–309.

Wallingford, R. A., and Ewing, A. G. (1989). *Anal. Chem.* **61**, 98–100.

Wallingford, R. A., Curry, P. D., Jr., and Ewing, A. G. (1989). *J. Microcol. Sep.* **1**(1), 23–27.

Weinberger, R., and Albin, M. (1991). *J. Liq. Chromatogr.* **14**, 953–972.

Weinberger, R., and Lurie, I. S. (1991). *Anal. Chem.* **63**, 823–827.

Weinberger, R., Sapp, E., and Moring, S. (1990). *J. Chromatogr.* **516**, 271–285.

Troubleshooting Guide to Capillary Electrophoresis Operations

Joel C. Colburn and Paul D. Grossman

This appendix is intended to help the reader identify and rectify a variety of problems that can interfere with proper operation of CE instrumentation. This appendix is not meant to replace an operator's manual. (The table below is an adaptation of a similar table contained in the user's manual for the 270A Capillary Electrophoresis System, Applied Biosystems, Inc., and is adapted with the permission of Applied Biosystems, Inc.)

Symptom	Possible causes	Action
Poor mobility reproducibility	A. Temperature fluctuation	1. Measure capillary oven temperature
		2. Check temperature control
	B. Capillary not washed	Check hydrodynamic flow
	C. Ion depletion of buffers	Replace buffers
	D. Sample – wall interaction	Wash capillary thoroughly
Poor efficiency	A. Wall sticking	Wash capillary thoroughly
	B. Too much Joule heating	1. Reduce voltage
		2. Decrease buffer concentration
	C. High-conductivity sample buffer	1. Dilute sample in water
		2. Remove salt from sample
		3. Increase running buffer concentration
	D. Buffer siphoning	Equalize buffer reservoir liquid levels
	E. Too large an injection volume	Reduce injection time

(Continued)

Symptom	Possible causes	Action
Poor sensitivity	A. Incorrect injection volume	Longer injection time
	B. Aging detector lamp	Replace lamp
	C. Incorrect reporting device settings	Check settings
	D. Incorrect detector range	Check detector setting
	E. Incorrect detector wavelength	1. Confirm wavelength setting
		2. Confirm wavelength accuracy
Poor quantitative reproducibility	A. Incorrect reporting device	Check settings
	B. Insufficient signal-to-noise ratio	1. Set detector to more-sensitive range
		2. Longer injection time
	C. Sample evaporation (particularly sample volumes < 20 μl)	1. Larger sample volumes
		2. Keep samples covered or cold
Gradual change in current	A. Different cathode/anode buffers	1. Confirm buffer identity
		2. May be normal if buffers are different
	B. Stacking using large injection volumes	Normal condition
	C. Temperature change	Confirm capillary temperature
	D. Voltage change	Confirm capillary voltage using a buffer of known conductivity
Current fluctuations	A. Bubbles	1. Visual check for bubbles at outlet
		2. Flush capillary with media
	B. Voltage fluctuations	Confirm capillary voltage using a buffer of known conductivity
	C. Temperature fluctuations	Monitor capillary temperature
No current	A. Plug in capillary	1. Confirm by pumping through a marker (a bubble)
		2. Replace capillary
		3. Trim capillary ends
	B. Empty capillary	Check buffer reservoirs
	C. Broken capillary	1. Visual check for bubbles at outlet
		2. Replace capillary
	D. Safety interlock not closed	Close interlock
	E. Wrong buffer	Confirm buffer composition
Noisy baseline	A. Wrong detector wavelength	1. Confirm wavelength setting
		2. Confirm wavelength accuracy
	B. Capillary not washed (vacuum failure?)	1. Confirm by pumping through a marker (a bubble)
		2. Wash reservoir empty
	C. No sample injection	1. Confirm sample in vial
		2. Confirm capillary extends into sample
		3. Confirm injection time
	D. Particles (sharp spikes)	1. Filter or replace buffers
		2. Wash capillary
		3. Filter samples
	E. Air bubbles (pinholes or broken capillary)	1. Confirm by pumping through a marker (a bubble)
		2. Replace capillary
	F. Incorrect detector range setting	Confirm range

Symptom	Possible causes	Action
Noisy baseline	G. Aging detector lamp	Replace lamp
	H. Data system problem	Check recorder connections, settings, and cable shielding
Flat baseline	A. Capillary not aligned in detector	Align capillary
	B. No voltage	1. Confirm voltage setting
		2. Ensure capillary ends are immersed in buffer
	C. No sample injection	1. Confirm sample in vial
		2. Confirm capillary extends into sample
		3. Confirm injection time
	D. Incorrect detector wavelength	1. Confirm wavelength setting
		2. Confirm wavelength accuracy
	E. Dead detector lamp	Replace lamp
	F. Plug in capillary	1. Confirm by pumping through a marker (a bubble)
		2. Replace capillary
		3. Trim capillary ends
	G. Data system problem	Check recorder connections, settings, and cable shielding

Index

347